ENGINEERING FUNDAMENTALS FOR PROFESSIONAL ENGINEERS' EXAMINATIONS

ENGINEERING FUNDAMENTALS FOR PROFESSIONAL ENGINEERS' EXAMINATIONS

Third Edition

Lloyd M. Polentz, P.E.

McGraw-Hill Book Company

New York St. Louis San Francisco Auckland Bogotá
Hamburg Johannesburg London Madrid Mexico
Milan Montreal New Delhi Panama
Paris São Paulo Singapore
Sydney Tokyo Toronto

Library of Congress Cataloging-in-Publication Data

Polentz, Lloyd M.
Engineering fundamentals for professional
engineers' examinations.

Rev. ed. of: Engineering fundamentals for
professional engineers' examinations. 2nd ed.
c1979.
Includes index.
1. Engineering—Examinations, questions, etc.
I. Polentz, Lloyd M. Engineering fundamentals for
professional engineers' examinations. II. Title.
TA159.P6 1987   620   86-15376
ISBN 0-07-050393-1

1234567890   BKP/BKP   893210987

ISBN 0-07-050393-1

The editors for this book were Betty Sun and Laura Givner,
the designer was Naomi Auerbach, and the production
supervisor was Annette Mayeski. It was set in Century Schoolbook
by Techna Type, Inc.

Printed and bound by The Book Press.

To my parents
Gustave E. Polentz
Ethel M. Polentz

CONTENTS

PREFACE

P-1 PURPOSE OF THIS BOOK

The primary purpose of this third edition of *Engineering Fundamentals for Professional Engineers' Examinations* is to help engineers review the fundamental principles of engineering and the methods of applying those principles to the solution of problems and to help them prepare for the Fundamentals of Engineering, or Engineer-in-Training (EIT), examination. A secondary purpose is to help engineers review their engineering fundamentals for a better grasp of the principles of engineering to aid them in their professional activities as practicing engineers or to prepare for graduate study in engineering. Many engineers in all three categories have satisfactorily achieved their desired goals through the use of the first two editions of this book. The expanded coverage of this third edition will provide an even more satisfactory method of achieving those goals.

It might also be noted that a number of engineers have found the information contained in this book of considerable help in preparing for the professional branch, or Principles and Practice of Engineering, examinations.

P-2 EXAMINATION COVERAGE AND FORM

The Fundamentals of Engineering, or Engineer-in-Training (EIT), examination is an 8-hour examination which is given in two parts—4 hours in the morning and 4 hours in the afternoon. It is a multiple-choice examination, and all questions and problems are followed by five possible answers, only one of which is correct.

The morning session consists of 140 multiple-choice problems and questions. All 140 questions must be answered. An examinee who does not know the correct answer is encouraged to guess, and no points are subtracted for wrong answers. The subjects which may be included in the morning session include the following, in alphabetical order: chemistry, computer programming, dynamics, electrical circuits, engineering economics, fluid mechanics, materials science, mathematical modeling of engineering systems, mathematics, mechanics of materials, statics, structure of matter, and thermodynamics and heat transfer.

The afternoon session is composed of groups of problems covering four required subjects and two optional subjects. There are 15 questions in engineering mechanics (statics, dynamics, and mechanics of materials), 15 questions in mathematics, 10 questions in electrical circuits, and 10 questions in engineering economics, for a total of 50 points. These are the four required subjects. In addition, the examinee must select two additional subjects from the five following categories with 10 questions in each: computer programming, electronics and electrical machinery, fluid mechanics, mechanics of materials, and thermodynamics and heat transfer. The 20 points in the two elective categories added to the 50 points in the required categories give a total of 70 points for the afternoon session.

The subdivisions of the main topics—mathematics, statics, dynamics, mechanics of materials, fluid mechanics, chemistry, thermodynamics and heat transfer, electrical circuits, and engineering economics—are all described in detail in the body of this book. The subdivisions of the five minor or specialty subjects are as follows: computer programming—number systems, algorithms, computer organization, and problem solving with FORTRAN and BASIC; materials science—physical properties, chemical properties, electrical properties, atomic bonding, crystallography, phase diagrams, processing tests and properties, diffusion, and corrosion; mathematical modeling of engineering systems—models, feedback, block diagrams, system functions, Laplace transforms, steady-state and transient performance, frequency response, step and impulse response, stability criteria, and state variable descriptions; structure of matter—nuclei, atoms, and molecules; and electronics and electrical machinery—solid-state devices and circuits and electrical machinery.

A word might be in order here regarding metric units. Problems are given in the examination in metric units. However, the same problem is also given in U.S. customary units, and the examinee may work either version. There has not yet been any requirement for the examinee to be skilled in the use of metric units. The subject of metric units is discussed in Chap. 2, Section 2-10, "Dimensional Units," and examples of calculations using metric units are given there and at a few other places throughout the text.

More detailed information regarding the fundamentals examination can be obtained from the National Council of Engineering Examiners (NCEE) at P.O. Box 1686 (State Road 210), Clemson, South Carolina 29633-1686. The NCEE prepares the fundamentals examinations which are given by all states and territories except for the state of Illinois. All the different states which use the NCEE examinations give those examinations during the same 3-day span in April and October-November of each year, although some states, or areas within a state, may give the fundamentals examination only once a year. For information regarding the specific examination and registration requirements for your state, contact your state board office. The address is available from the NCEE. Remember that your state board is the only official source of information regarding the fundamentals examination. Information from friends or acquaintances may or may not be correct and could get you into trouble. You should contact your state board early, as the required lead time for applying to take the examination may be quite long.

P-3 HOW TO TAKE THE FUNDAMENTALS EXAMINATION

First, of course, you should get a good night's sleep before the exam; do your celebrating later. You should get to the examination room 15 to 30 minutes before the specified starting time so that you can find your way around, locate the rest room, and get settled comfortably into position before the examination papers are handed out. Be sure that you have brought all necessary paraphernalia with you. Plan this in advance, and assemble the items the day before the examination. You should include the following in your kit as well as anything else that you may think that you will need: a pencil with extra leads, an eraser, a calculator with spare batteries, a watch or a desk clock, a straightedge, triangles, a scale, a compass, a protractor, and reference books. This is an open-book examination, and the examinee may use textbooks, handbooks, bound reference materials, silent calculators, or slide rules. However, unbound tables or notes are not allowed, nor are writing tablets. Sufficient scratch paper is provided in the test booklets.

In regard to reference books, you should take only a few—the few that you are familiar with and which you think that you might need. As a suggestion, you might take this book plus, perhaps, a few of your college texts or reference books with which you are familiar, and a good general type of handbook such as Marks or Eschbach. You should be able to carry everything you need in a briefcase or attaché case. As an aside, you will probably see one or more "hopefuls" come to the examination room with toy wagons loaded with books. These people will usually arrive quite early. These examinees will select their seats and carefully arrange their libraries in front of them on tables somewhere in a special order. You will also note that each book will have numerous tabs glued to the pages. On receiving their examination papers, these individuals will look at the problems, select one, and then quickly reach out and grasp a book from the carefully organized library. They will frantically open the book at tab after tab, put the book back in its place, yank out another one, and repeat the process. When, and if, they finally find the exact solution to a particular problem, you will see a look of satisfaction or relief in their faces, and they will rapidly copy down the solution on the examination paper. These individuals will spend the entire examination period paging through books looking for answers and will never try to actually work a problem. Sometimes they will be successful in finding enough worked-out examples to pass an examination. Usually they will still be frantically turning pages and searching for answers when the final bell rings. This provides a beautiful example of how *not* to take or prepare for an examination. If these hopefuls had spent the same amount of time studying and reviewing engineering principles that they spent inspecting and tabulating their libraries of books, they would be much better prepared for the examination.

As noted above, the examination is divided into two parts. The morning session contains 140 questions, all of which are to be answered. There are available 240 minutes in the morning. From this can be subtracted some 30 minutes for reading the instructions—always read the instructions first, carefully—and personal needs, leaving 210 minutes to do 140 problems, or $1\frac{1}{2}$ minutes per problem. Don't exceed that amount of time. Those questions to which you know the answer, you will find that you can answer in less than the allotted time. The others you will probably not answer correctly even if you spend 5 or more minutes on each. The law of diminishing returns applies here, and every minute that you waste on a problem which you cannot answer will cost you valuable points because time will run out before you get to those last few questions which you can answer. Glance at your watch or clock occasionally to make sure that you are not spending too much time on an individual problem, and hurry over those you can answer, leaving the others blank. Then, about 15 minutes before the end of the period, rush over the blanks and make your best guess for those. Do not leave any of the 140 questions unanswered; after all, you have a one-in-five chance of hitting the correct answer on a random basis, and there is no penalty for guessing.

The afternoon session contains 70 questions and problems. This gives a little over 3 minutes per problem. That means that you can afford to spend 45 minutes on the required section on engineering mechanics.

Watch yourself and don't exceed that amount of time. Similarly, spend no more than 45 minutes on mathematics, 30 minutes on electrical circuits, 30 minutes on engineering economics, and 30 minutes on each of the two elective subject categories. You will find that you will finish well within the allotted time for those subject categories for which you are well prepared. Then, when you have completed the five categories, go back and do the best that you can with the gaps that you have left, remembering to guess at the answers to those questions to which you do not know the answers.

As a suggestion, whenever you can, draw a figure. This will apply for many problems in mathematics, statics, dynamics, mechanics of materials, electrical circuits (particularly phasor diagrams), fluid mechanics, and thermodynamics. In some cases, particularly in some of the categories, you may not be able to draw a meaningful figure, but in those cases where you can, you will often find that the figure will help to point out a method of analysis and will also provide a rough check of the validity of the answer through scaling of the figure. One word of caution: Be sure to draw the figure to scale. An out-of-scale figure can sometimes be a detriment rather than a help. This can be graphically illustrated with the aid of the ancient mathematical fallacy that any triangle is isosceles.

Refer to Fig. P-1. It is given that triangle ABC is scalene. Construct the bisector of angle B and the perpendicular bisector of base AC. The two lines will intersect at a point D. From point D construct line DE perpendicular to side BC and line DF perpendicular to side AB. This gives two right triangles, I and II. Angle $\alpha = \beta$ and $\rho = \mu$, since $180 - 90 - \alpha = \rho$ and $180 - 90 - \beta = \mu$. The hypotenuses are equal: $BD = BD$. Therefore, the two right triangles are congruent, which gives $BF = BE$ and $FD = DE$.

Next consider triangles III and IV. Line AD equals line DC—lines

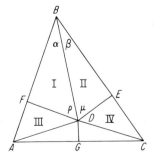

Figure P-1

drawn from the same point on a perpendicular bisector to the ends of the line bisected are equal. This gives two right triangles in which the hypotenuse and one leg of one triangle are equal to the hypotenuse and one leg of the other right triangle; therefore, the two right triangles are congruent. Since triangles III and IV are congruent, side AF = side EC. It follows, then, that $BF + FA = BE + EC$ and the triangle is isosceles. Thus all triangles are isosceles! This can't be true, so what is wrong with the proof? There is nothing wrong with the proof; all the statements are valid. The error lies in the sketch. The two bisectors will not intersect inside the triangle unless it is isosceles, and then they will be collinear. If the sketch is distorted only slightly, a proposterous conclusion can be drawn by a perfectly legitimate analysis. So be careful to draw your sketches to scale!

P-4 TEXT COMPOSITION AND COVERAGE

The composition and contents of this book are based on the review of the contents of a large number of fundamentals examinations, more than 20 years of teaching a course in engineering fundamentals to help engineers review the fundamental principles of engineering, and communications with a number of people who have taken the fundamentals examination.

The emphasis in past fundamentals examinations has been on the application of the fundamental engineering principles, and this application has been stressed throughout the preparation of this book. Memorization of formulas has served to get many a college student through a given course with a passing grade, and some opportunistic individuals have occasionally been able to squeak through a fundamentals examination by copying solutions from published texts and review books. Such practices, however, are not practicable for a comprehensive type of examination such as the fundamentals examination. This is also true of the solution of many of the problems which confront engineers in the practice of their professions. There are too many different cases and too many different types of problems to permit memorization of a formula for each case, or discovery in the literature of exactly the proper formula that fits the case at hand. It is actually easier, and certainly much more satisfactory, to learn how to apply the fundamental principles of engineering and to use those principles to develop a relationship to fit the case at hand. This has been the guiding philosophy employed throughout the preparation of this book.

In working with engineers, and helping them review their engineering fundamentals through the past three decades, the author has found that the most effective method of teaching consists of essentially three steps as follows:

1. Explain the engineering principle.
2. Illustrate the application of the principle by means of an appropriate problem or problems.
3. Supply problems for individual practice and review.

This three-step sequence—explain, illustrate, and practice—is the pattern followed in this book. The principles are first discussed, then the application of those principles is illustrated through the solution of illustrative problems taken from past examinations. A group of sample problems is given at the end of each chapter for the reader to solve for practice. In addition, solutions to those sample problems, with explanations, are also provided. These sample problems have been selected from previous fundamentals examinations or modeled after problems given in those examinations. Almost all the problems have been presented in the present-day, multiple-choice format. The majority of the problems are of the longer, more involved type given in the afternoon session of the examination. Many of the problems will be found to be more difficult than those given in present-day examinations, and some will probably take longer than 3 minutes to solve. However, they are good practice and illustrate important principles. In some cases, if you find these problems to be more difficult than those you will be confronted with in the examination room, that is all to the good, since you will then be better prepared to handle the examination problems.

In addition to the large number of afternoon-session-type problems, there are 10 of the shorter, morning-session-type problems given at the end of each chapter.

No effort has been made to cover all the subjects that the NCEE lists in its guide for preparation for the fundamentals examination. Some are so general that it would take a book larger than this to cover just one category. See, for example, the NCEE listings of the sub-subjects given above for materials science, mathematical modeling of engineering systems, and structure of matter. No attempt has been made to cover the special field of computer programming, including discussions of algorithms, number systems, and explanations of programming in FORTRAN and BASIC. Nor has any attempt been made to describe methods of designing solid-state circuitry, while the discussion of electrical machinery has been kept within a reasonable space limit. In the case of the broad general subjects, the space required to give any meaningful discussion of those subjects could not be justified in comparison with the relatively small point value attributable to those subjects. Moreover, they are so broad in nature that the odds would be very slim of touching on any point covered in the examination. In the cases of computer programming and solid-state circuit design, those are specialized subjects. If you know them, fine, you may pick up a few easy points, but if you don't know them, no short discussion is going to make you sufficiently conversant with either of them to enable you to answer any more than very superficial questions regarding either of the two subjects.

This book has been written to help engineers review subjects with which they are already familiar. It has not been written to serve as an introduction. If you find that the text proceeds too rapidly, or if you are very rusty, it is recommended that you avail yourself of a book devoted exclusively to the subject causing you difficulty. In most cases your college texts will be adequate as references. If you don't have any reference texts yourself, it may be necessary to borrow or buy them. Be sure that you page a text before you buy it to make sure that it is understandable and covers the subject matter that you are interested in. The latest text is not necessarily the best. The fundamental principles of engineering have not changed in many years, and an older book may be more clearly written. Engineering is not a great deal freer from fadism than are other branches of education.

This book is devoted to helping engineers prepare for the technical portion of the fundamentals examination. No information has been included on registration procedures or requirements. This type of information is well covered in the book *How to Become a Professional Engineer,* third edition, by John Constance, McGraw-Hill Book Company, Inc., New York.

P-5 HOW TO USE THIS BOOK

As previously noted, the contents of this book follow this useful sequence: explain, illustrate, and practice and review. It is suggested that you follow the same sequence in your review. First study the chapter, starting with Chap. 1, and study the illustrative examples until you understand the principles. Then work the sample problems. Do the best that you can with each problem, reviewing the text material as necessary, before you refer to the solution. It is recommended that you work all the sample problems. The one most frequent recommendation by those who have taken the fundamentals examination, especially those who have been out of school for a few years, is to work as many problems as possible to prepare for the examination. Many engineers who have been out of school for a few years have not worked any problems for some while and thus have forgotten how to solve a problem in an expeditious manner. There have been cases where experienced engineers, particularly those who have switched into management, and have not adequately prepared for the examination, have been completely stymied when confronted by the need to work out solutions to the problems given on the examination. They have forgotten how to proceed, and, as a result, lose valuable time organizing their thoughts preparatory to working the problem that confronts them. The only way to forestall such a dilemma is to prepare yourself in advance by working a large number of sample problems during your period of preparation.

P-6 CONCLUSIONS AND ACKNOWLEDGMENTS

To recap, contact your state board early, preferably about 6 months before you plan to take the fundamentals examination, and when you review in preparation for the examination, work as many problems as possible.

Many people have been helpful in the preparation of this revision. Foremost among those is my wife Rose, whose moral support and understanding of the time demands of a project of this magnitude are greatly appreciated. I should also like to express my appreciation to those who took the time to point out the occasional errors which crept into the second edition, especially those who did so in a helpful and constructive manner. I should also like to thank those instructors and students who offered suggestions as to how coverage of the material might best be expanded and improved.

All the material in this book has been checked and rechecked, especially the solutions to the sample problems. There is still a chance, however, that some errors may have remained undetected. If you find any such errors, I should appreciate your bringing them to my attention. I am grateful to all those who have written me in the past and hope that future users of this book will also send me their comments.

Lloyd M. Polentz

ENGINEERING FUNDAMENTALS FOR PROFESSIONAL ENGINEERS' EXAMINATIONS

1
MATHEMATICS

1-1 EXAMINATION COVERAGE

In the Fundamentals of Engineering Examination, mathematics is treated as a major subject in the morning session and is also one of the required subjects in the afternoon session.

The subdivisions of the major subject which may be covered, and which should be reviewed, include the following:

Trigonometry

Analytic geometry

Differential calculus

Integral calculus

Differential equations

Probability and statistics

Linear algebra

1-2 ALGEBRA

The basic algebra requirement includes factoring, solution of quadratic equations, and solution of simultaneous equations. Also included are laws of exponents and their application (e.g., logarithms).

The second-order binomial expansion should be remembered: $(a + b)^2 = a^2 + 2ab + b^2$. Another basic relationship which is often encountered is $(a + b)(a - b) = a^2 - b^2$. More sophisticated factoring can be accomplished by division.

Quadratic equations can sometimes be solved by factoring, but often they cannot. For these cases, the general solution of a quadratic equation can be used to determine the answer. The general form of the quadratic equation can be written as $ax^2 + bx + c = 0$. A solution to this general form can be obtained by completing the square and equals:

$$x = \frac{-b \pm \sqrt{b^2 - 4ac}}{2a}$$

To illustrate the application of this relationship, consider the following problem.

- What are the roots of the equation $17x^2 + 41x - 74 = 19$?
  (a) 3.84 and 1.42   (b) -1.21 and 2.63   (c) -3.84 and 1.42   (d) -1.42 and 2.63   (e) 2.63 and 3.84

First restate the problem in the proper form, and then solve:

$$17x^2 + 41x - 93 = 0$$

Then

$$x = \frac{-41 \pm \sqrt{41^2 - 4 \times 17 \times (-93)}}{2 \times 17}$$
$$= -1.21 \pm 2.63$$
$$= -3.84 \quad \text{or} \quad 1.42$$

Note that both roots found in this case are real, but if the term $b^2 - 4ac$ is less than zero, both roots will be imaginary. The correct answer is (c).

1-3 LOGARITHMS AND EXPONENTS

There are two systems of logarithms in use: the common, or briggsian, logarithms, which have the base 10, and the natural, or napierian, logarithms, which have the base e. The same principles of operation apply to both systems, since both are merely applications of the theory of exponents. A logarithm to the base 10 will be abbreviated "log," and a natural logarithm will be abbreviated "ln" throughout this work. We have then, by definition, that if $a = \log B$, $10^a = B$; and if $m = \ln P$, $e^m = P$.

- What is the logarithm of 243 to the base 3?
  (a) 1   (b) 2   (c) 3   (d) 4   (e) 5

Restate the problem: $\log_3 243 = A$; then $3^A = 243$. Next take the logs of both sides: $A \log 3 = \log 243$ and $A = (\log 243)/(\log 3) = 5$. The correct answer is (e).

■ Given $\log_5 M = \log_{25} 4$, find M.

   (a) 1 (b) 2 (c) 3 (d) 4 (e) 5

Restatement gives $\log M$ to the base 5 $= \log 4$ to the base 25. Let $\log_5 M = A$, then $5^A = M$ and $A = \log_{25} 4$, which gives $4 = 25^A$. Taking logs of both sides gives $A = (\log 4)/(\log 25) = 0.4307$ and $M = 5^{0.4307} = 2.00$. The correct answer is (b).

1-4 TRIGONOMETRY

Some of the more basic principles with which the examinee should be familiar are the definitions of the various functions and their relationships to one another, the law of sines, the law of cosines, double-angle relationships, and the basic right triangles.

The various functions can be explained by means of the unit circle. In Fig. 1-1, OP is the radius and is equal to 1:

$$\sin \theta = PM = \text{opposite side divided by the hypotenuse}$$

$$\cos \theta = OM = \text{adjacent side divided by the hypotenuse}$$

$$\tan \theta = \frac{\sin \theta}{\cos \theta} = \frac{PM}{OM}$$

By proportion, $PM/OM = TA/OA = TA$; so $\tan \theta = TA$.

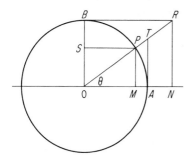

Figure 1-1

The law of sines, which holds for any triangle (Fig. 1-2), is easily remembered and need only be stated:

$$\frac{\sin A}{a} = \frac{\sin B}{b} = \frac{\sin C}{c}$$

The law of cosines is another important relationship.

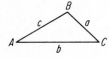

Figure 1-2

- Given a triangle with sides $a = 4$, $b = 5$, and $c = 8$ with opposite angles of A, B, and C, respectively, find angles A and C.

  (a) 24.15 and 54.9°   (b) 30.75 and 125.10°   (c) 24.15 and 125.1°   (d) 30.75 and 54.9°   (e) 54.9 and 125.10°

First solve for angle A, using the law of cosines:

$$16 = 25 + 64 - 80 \cos A \qquad A = \cos^{-1} 0.9125 = 24.15°$$

Then, using the law of sines, $(\sin A)/a = 0.10227$, $\sin B = 0.5113$, $\sin C = 0.8182$, and arcsin $0.8182 = 54.90°$. However, the sketch in Fig. 1-3 shows that 54.90° is not the correct value for angle C. The arcsine of 0.8182 also equals $180 - 54.90 = 125.10°$. The correct answer is (c).

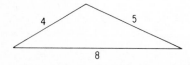

Figure 1-3

The trigonometry requirement also includes manipulation of the various functions algebraically with their various equivalents. No general set of rules can be laid down to cover all cases; the reader is urged to become familiar with the more important equalities and different methods of substitution.

- What is $\sec \theta \sin^3 \theta \cot \theta$ equal to?

  (a) $\sin \theta$ (b) $\cos \theta$ (c) $\cos^2 \theta$ (d) $\sin^2 \theta$ (e) $\tan \theta$

Substitution gives $1/(\cos\theta) \times \sin^3\theta \times \cos\theta/\sin\theta = \sin^2\theta$, which also equals $1 - \cos^2\theta$. The correct answer is (d).

1-5 ANALYTIC GEOMETRY

A few of the fundamentals of analytic geometry should be reviewed, primarily those pertaining to straight lines and circles.

The equation of a straight line through the origin (Fig. 1-4) can be written as

$$\frac{y}{x} = \text{constant} = \tan\theta \qquad \text{or} \qquad y = mx$$

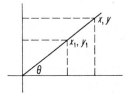

Figure 1-4

Alternatively, using points x and y, we find that $y = (y_1/x_1)x$. The more general case is one in which the line does not pass through the origin. The line may then be specified by one point and the angle or by two points (Fig. 1-5).

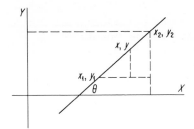

Figure 1-5

If two points are given, the slope equals $(y_2 - y_1)/(x_2 - x_1) = m$, or $m = (y - y_1)/(x - x_1)$, so the equation of the line is

$$\frac{y_2 - y_1}{x_2 - x_1} = \frac{y - y_1}{x - x_1}$$

or $y = y_1 + [(y_2 - y_1)/(x_2 - x_1)](x - x_1)$, which is of the form $y = mx + c$.

If one point and the slope of the line are given, the equation of the line is $y - y_1 = m(x - x_1)$. The special cases where one intercept or more is given will be seen to fit easily one of the cases discussed.

The slope of a perpendicular to a line can be seen to equal the negative reciprocal of the slope of the line (Fig. 1-6).

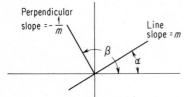

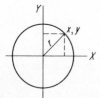

Figure 1-6

We know that $\tan \beta = -\tan(180 - \beta)$. From Fig. 1-6 we see that $180 - \beta = 90 - \alpha$ and the tangent of $(90 - \alpha) = 1/\tan \alpha$. Combining these gives us $\tan \beta = -(1/\tan \alpha)$, or the slope of a perpendicular to a line equals the negative reciprocal of the slope of the line.

■ The equation of the straight line through the point $(4, 7)$ at an angle of $45°$ with the horizontal is

(a) $y = 3x + 4$ (b) $y = 4x + 3$ (c) $y = 2x + 2$
(d) $y = x + 3$ (e) $x = y + 3$

The equation of a straight line is $y = mx + c$, where m is the slope of the line or the tangent of the angle that the line makes with the horizontal axis. For this case, $m = \tan 45° = 1.00$, so $y = x + c$. Substitution of the coordinates of the given point gives $7 = 4 + c$, or $c = 3$. The required equation is $y = x + 3$, and the correct answer is (d).

The equations of circles are as easily derived as those of a straight line. Given a circle with its center at the origin, the equation will be $x^2 + y^2 = r^2$, which can readily be seen from Fig. 1-7. If the center is not at the origin, the equation changes to $(x - a)^2 + (y - b)^2 = r^2$.

Figure 1-7

This relationship follows readily from Fig. 1-8.

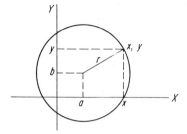

Figure 1-8

■ Find the points of intersection of the following two curves:

$$x - 7y + 25 = 0 \quad \text{and} \quad x^2 + y^2 = 25$$

(a) (3, 4) and (−4, 3)   (b) (5, 2) and (4, 3)   (c) (4, 3) and (−3, 4)   (d) (10, 5) and (−3, 4)   (e) (−3, −4) and (−4, 3)

The points of intersection can be determined by solving the two equations simultaneously. The correct answer is (d).

■ Which of the following is the equation of the tangent to a circle whose equation is $x^2 + y^2 = 25$ if the slope of the tangent is $-\frac{3}{4}$:

(a) $3y + 4x = 25$ (b) $3x + 4y = 25$ (c) $y = x + 6$
(d) $x = y + 15$ (e) $4y - 3x = -25$

The given circle has its center at the origin, and the required tangent will be perpendicular to the diameter whose slope is the negative reciprocal of the slope of the tangent. The slope of the diameter would then be $\frac{4}{3}$, and the equation of the diameter would be $y = \frac{4}{3}x$, since the diameter line would go through the origin.

The diameter line would intersect the circle at two points:

$$y^2 = 25 - x^2 \quad y^2 = (\tfrac{16}{9})x^2$$
$$(\tfrac{16}{9})x^2 = 25 - x^2 \quad x = \pm 3$$

The two points of intersection would be $x = 3$, $y = 4$; and $x = -3$, $y = -4$. The tangent line selected would then go through point (3, 4) or point (−3, −4) and have a slope of $-\frac{3}{4}$. Since the slope of a line also equals $(y - y_1)/(x - x_1)$, the equation of the required tangent is $(y - 4)/(x - 3) = -\frac{3}{4}$, or $(y + 4)/(y + 3) = -\frac{3}{4}$, which reduce to $4y + 3x = 25$ and $4y + 3x = -25$. The correct answer is (b).

The general equation of a parabola is $y = Ax^2 + Bx + C$, which can also be written in the form $y = A(x + D)^2 + E$.

This is a quadratic curve, and there will be two values of x for each value of y. The curve will be concave downward if A is positive and concave upward if A is negative. The curve will be symmetrical about its vertical axis. If the equation of the parabola is of the form $x = Ay^2 + By + C$, it will be symmetrical about its horizontal axis.

■ A support cable in the shape of a parabola is attached to anchor posts 150 ft apart. The two attachment points are at the same elevation. The center point of the cable is midway between the two support points and is 30 ft below them. Which of the following is the closest to the distance that the cable would be below the elevation of the attachment points at a distance of 25 ft from the attachment points measured on the horizontal?

(a) 20.0 ft   (b) 15.3 ft   (c) 18.6 ft   (d) 16.7 ft
(e) 21.1 ft

First sketch the parabola, as shown in Fig. 1-9. Draw the X and Y axes through the center point of the cable. The coordinates of the three points will then be

Point M: $x = -75, y = 30$

Point N: $x = 75,\quad y = 30$

Point O: $x = 0,\quad y = 0$

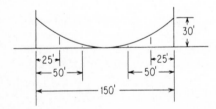

Figure 1-9

From the general form of the equation for a parabola $y = Ax^2 + Bx + C$, and the three given points, we can write the following three equations:

$$30 = 5625A - 75B + C$$
$$30 = 5625A + 75B + C$$
$$0 = C$$

The third equation says that $C = 0$, and the first two equations give us that $B = 0$ and that $A = 0.00533$. The required equation for the parabola is then

$$y = 0.00533x^2$$

At a distance of 25 ft from an attachment point $x = 50$ or -50 and $y = 13.33$ ft. The distance of the two points below the attachment points would then equal $30.00 - 13.33 = 16.67$ ft. The correct answer is (d).

There are three types of conic sections, as shown in Fig. 1-10: the

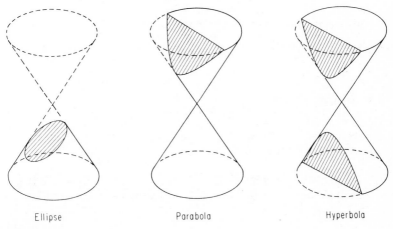

Ellipse   Parabola   Hyperbola

Figure 1-10

ellipse, the parabola, and the hyperbola. The general form of the equation of an ellipse is

$$\frac{(x - a)^2}{h^2} + \frac{(y - b)^2}{k^2} = 1$$

where the center is at (a, b) and the major axis is parallel with the X axis. The major axis is $2h$ long, and the length of the minor axis is $2k$. If the center of the ellipse is at the origin, this equation reduces to the more common form of

$$\frac{x^2}{h^2} + \frac{y^2}{k^2} = 1$$

The general form of the equation can be reduced to

$$(x - a)^2 + c(y - b)^2 = d$$

This is the same form as the equation of a circle except for the numerical modifier c of the $(y - b)^2$ term.

The general equation of a hyperbola is

$$\frac{(x - a)^2}{h^2} - \frac{(y - b)^2}{k^2} = 1$$

where the center is at (a, b) and the transverse axis is parallel to the X axis.

This equation can be modified to the form

$$(x - a)^2 - c(y - b)^2 = d$$

1-6 PROBABILITY

The probability of an event occurring plus the probability of it not occurring must equal 1 (it either occurs or it doesn't). An illustration of this is afforded by the following problem:

- If a sharpshooter can hit a flying bird one out of five times and shoots at five different flying birds, what is the probability of a hit?

  (a) 100%   (b) 89%   (c) 74%   (d) 67%   (e) 59%

The sharpshooter has four chances out of five of missing with each shot; thus the odds of missing five times consecutively equal $(\frac{4}{5})^5 = 0.32768$. The probability of a hit (i.e., probability of not missing) thus equals $1 - 0.32768 = 0.67232$. The correct answer is (d).

- If a person tosses eight coins into the air, what is the probability that four will land heads up and four will land tails up?

  (a) 1.00   (b) 0.500   (c) 0.617   (d) 0.125   (e) 0.273

This problem can be worked with the aid of the following binomial formula. The probability that an event will occur exactly k times out of n tries equals

$$\frac{n!}{k! \, (n - k)!} \, p^k (1 - p)^{n-k}$$

where
k = exact number of times that the event will occur
n = the total number of trials
p = the probability that the event will occur on any one trial

The case being considered is the same as if one coin were flipped eight times. The probability that a head (or tail) would come up on any one try is $\frac{1}{2}$. Substitution in the formula then gives

$$\frac{8!}{4! \times 4!} \left(\frac{1}{2}\right)^4 \left(\frac{1}{2}\right)^4 = 0.273$$

The correct answer is (e).

1-7 CALCULUS

The calculus requirement can include differentiation, integration, and calculation of areas, area moments, and maxima and minima.
A few of the more common derivatives should be remembered:

$d\ ax = a\ dx \qquad a$ is a constant

$d\ (u + v) = du + dv \qquad u$ and v are variable

$d\ (uv) = u\ dv + v\ du \qquad u$ and v are variable

$d\ x^n = nx^{(n-1)}\ dx$

$d\ e^x = e^x\ dx$

$d\ \sin x = \cos x\ dx$

$d\ \cos x = -\sin x\ dx$

$d\ \ln x = dx/x$

Integration is the opposite of differentiation; thus we have

$\int a\ dx = ax + c \qquad$ where c = the constant of integration

$\int x^n\ dx = (1/n + 1)x^{(n+1)} + c \qquad$ (except for $n = -1$)

$\int dx/x = \ln x + c$

$\int e^x\ dx = e^x + c$

$\int \sin x\ dx = -\cos x + c$

$\int \cos x\ dx = \sin x + c$

One more important integral which should be remembered is $\int u\ dv = uv - \int v\ du$, where u and v are both functions of the same variable.

■ Find $\int x \cos x \, dx$.

(a) $\sin x + \cos x$ (b) $\sin x + x \cos x$ (c) $x \sin x + \cos x$
(d) $x \sin x + x \cos x$ (e) $\sin^2 x + \cos x$

Let $u = x$; then $du = dx$. Let $dv = \cos x \, dx$; then $v = \sin x$; $uv - \int v \, du = x \sin x - \int \sin x \, dx = x \sin x + \cos x + c$. To check, $d(x \sin x + \cos x) = \sin x \, dx + x \cos x \, dx - \sin x \, dx$, or $d(x \sin x + \cos x) = x \cos x \, dx$. The correct answer is (c).

1-8 MAXIMUM AND MINIMUM

Whenever an equation gives y as a function of x [abbreviated $y = f(x)$], it can be represented by a curve [we shall assume that the function $y = f(x)$ is continuous and differentiable] when a series of points is plotted on a graph (Fig. 1-11). The derivative dy/dx is the limit of the fraction $\Delta y / \Delta x$ as Δx approaches zero.

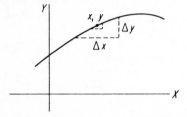

Figure 1-11

As we can see from Fig. 1-12, whenever the curve becomes a maximum or a minimum, the slope of the curve at the particular point of maximum or minimum is horizontal. In other words, the tangent to the curve at the particular point where the curve is either a maximum or a minimum will be parallel with the x axis and will, therefore, have a slope of zero,

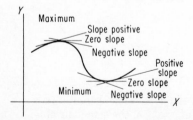

Figure 1-12

since the tangent of 0° (and of 180°) is zero. This means that at any point of maximum or minimum of the function $y = f(x)$ the derivative at that point, dy/dx, will equal zero.

To determine whether the point at which $dy/dx = 0$ is a maximum or a minimum, take the second derivative of the function at the particular point; if d^2y/dx^2 is positive, the point is a minimum; if d^2y/dx^2 is negative, the point is a maximum.

- It is desired to make a closed cylindrical container to hold a volume of 50 gal. The material to be used for the ends costs three times as much as the material to be used for the body of the container. What dimensions will give the most economical container?

  (a) $1.50d \times 3.78h$ (b) $1.62d \times 3.24h$ (c) $1.35d \times 4.67h$ (d) $1.42d \times 4.25h$ (e) $1.38d \times 4.47h$

The required volume is $50/7.48 = 6.68$ ft^3, a constant:

$$\text{Volume} = 6.68 = \pi r^2 h$$
$$\text{Surface area} = 2\pi r^2 + 2\pi rh$$
$$\text{Cost} = 2\pi r^2 \times 3M + 2\pi rhM = 6\pi r^2 M + 2\pi rhM$$

We need an equation for cost in terms of either r or h. We have a relationship between r and h in the equation for volume, since the volume is a constant. We can solve for h in terms of r and then substitute in the relationship for cost:

$$h = \frac{V}{\pi r^2}$$

$$C = 6\pi r^2 M + 2\pi r \frac{V}{\pi r^2} M = 6\pi r^2 M + 2\frac{V}{r} M$$

Now differentiate with respect to the independent variable r and set equal to zero:

$$\frac{dC}{dr} = 12\pi rM - 2VMr^{-2} = 0$$

$$r^3 = \frac{V}{6\pi} = 0.35439 \qquad r = 0.7077 \text{ ft}$$

The diameter of the container would be 1.415 ft, and the height would be 4.245 ft. The material cost would equal $28.36M$. The correct answer is (d).

1-9   RATE PROBLEMS

Previous examinations have included a number of problems on the rate of radioactive decay. There have been a number of rate problems in past examinations both of the radioactive-decay type and of the salt problem type.

Radioactive-decay problems can be handled either by using the compound-interest law or by setting up the differential equation and solving it. As an example, take the following problem.

■ A radioactive substance has a half-life of 25 days. How long will it take to lose 90 percent of its radioactivity?

(a) 45 days   (b) 68 days   (c) 76 days   (d) 83 days
(e) 91 days

Solve first by using the compound-interest equation (see Chap. 9 for an explanation of the equation). The amount of radioactivity after n periods $= A_0 (1 - i)^n$. In this case the amount of radioactivity is decreasing so the "interest rate" will be negative. Fifty percent remains after 25 days, so

$$0.50 = (1 - i)^{25}$$
$$\ln 0.50 = 25 \ln (1 - i) \qquad 1 - i = 0.9726$$

so the time required for only 10% to remain would be given by the relationship

$$0.10 = 0.9726^n \qquad n = \frac{\ln 0.10}{\ln 0.9726} = 83.0 \text{ days}$$

This problem can also be solved by means of a differential equation:

$$\frac{dA}{dt} = -kA \qquad \text{The rate of decay is proportional to the amount of radioactive material}$$

$$\frac{dA}{A} = -k \, dt \qquad \ln A = -kt + C \qquad \text{or} \qquad A = A_0 e^{-kt}$$

Note that it is necessary to solve for k. When $A/A_0 = 0.50$, $t = 25$; so $0.50 = e^{-25k}$ and $k = 0.0277$, giving the equation for decay

$$A = A_0 e^{-0.0277t}$$

For $A/A_0 = 0.10$, $t = \ln 0.10/(-0.0277) = 83.0$ days. The correct answer is (d).

Sample Problems

1-1 A bombardier is sighting on a target on the ground directly ahead. If the bomber is flying 2 mi above the ground at 240 mi/h, how fast must the sighting instrument be turning when the angle between the path of the bomber and the line of sight is 30°?
(a) 30°/min   (b) 45°/min   (c) 32.5°/min   (d) 31°/min
(e) 29°/min

1-2 A transformer core is to be built up of sheet-steel laminations, using two different strip widths, x and y, so that the resultant symmetrical cross section will fit within a circle of diameter D. Compute the values of x and y in terms of D so that the cross section of the core will have maximum value (see Fig. 1P-2).

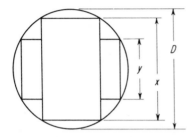

Figure 1P-2

(a) $y = 0.75D$ $x = 0.53D$ (b) $y = 0.85D$ $x = 0.53D$
(c) $y = 0.85D$ $x = 0.67D$ (d) $y = 0.58D$ $x = 0.76D$
(e) $y = 0.53D$ $x = 0.85D$

1-3 It is desired to measure the width of a river. A baseline 1000 ft long is laid out along one bank. An object on the other bank opposite the baseline is sighted on from one end of the baseline and found to make a 60° angle with it. The same object is sighted on from the other end of the baseline and found to make a 30° angle with it. Calculate the width of the river.
(a) 433 ft   (b) 348 ft   (c) 462 ft   (d) 324 ft   (e) 409 ft

1-4 Find the equation of the parabola in the form $y = f(x)$ which passes through the points (1, 2), (3, 20), and (4, 35).
(a) $y = x^2 - x + 1$ (b) $y = 2x^2 + x - 1$ (c) $y = 2x^2 + x - 1$
(d) $y = x^2 + 2x - 1$ (e) $x = y^2 - 3y + 1$

1-5 The base and altitude of a rectangle are 5 and 4 in, respectively. At a certain instant they are increasing continuously at the rate of 2 and 1 in/s, respectively. At what rate is the area of the rectangle increasing at that instant? (Solve by calculus.)
(a) 13 in²/s   (b) 15 in²/s   (c) 16 in²/s   (d) 11 in²/s
(e) 9.5 in²/s

1-6 Write the equation of the straight line that passes through the point (0, 6) and the point (−3, −8).
(a) $y = 7x - 9$ (b) $x = 4y + 3$ (c) $3y = 14x + 18$
(d) $6x = 4y - 6$ (e) $4y = 12x + 6$

1-7 A roll of tape is unrolled on a plane surface at the rate of 6 ft/s of length. Assuming a true circular roll, if the tape is ⅛ in thick, at what rate is the radius of the roll decreasing when the radius is 1 ft? (Solve by calculus.)
(a) $dr/dt = 0.13$ (b) $dr/dt = -0.13$ (c) $dr/dt = -0.12$
(d) $dr/dt = -0.14$ (e) $dr/dt = -0.15$

1-8 A tract of land 400 by 300 ft is to be crossed diagonally (see Fig. 1P-8) by a road 100 ft wide. What area comprises the right of way?

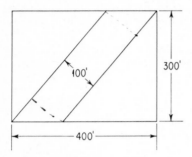

Figure 1P-8

(a) 30,200 ft²   (b) 50,300 ft²   (c) 38,700 ft²   (d) 43,800 ft²
(e) 40,100 ft²

1-9 Secant⁻¹ 1.40 = ?
(a) 37.2°   (b) 44.4°   (c) 46.2°   (d) 36.8°   (e) 34.9°

1-10 A gasoline tank consists of a horizontal cylinder 1 ft in diameter and 5 ft long. Determine the number of gallons in the tank when a gauge rod in the plane of the vertical diameter shows a depth of 4 in in the tank (1 gal = 231 in³).
(a) 8.57 gal   (b) 9.31 gal   (c) 10.12 gal   (d) 11.68 gal
(e) 12.03 gal

1-11 $\displaystyle\int_{\pi/2}^{\pi} \cos x\, dx = ?$
(a) 0   (b) 1.0   (c) −1.0   (d) 0.707   (e) −0.707

1-12 Ship A is sailing due south at 15 mi/h, and a second ship B 32 mi south of A is sailing due east at 12 mi/h. At what rate are they approaching or separating at the end of 1 h?
(a) 5.34 mi/h   (b) 6.27 mi/h   (c) −6.27 mi/h
(d) −5.34 mi/h   (e) −4.37 mi/h

1-13 Given: $y = wlx/2 - wx^2/2$. Determine the area between this curve and the x axis for values of x between 0 and x = 1.
(a) $wl/4 - w/6$ (b) $wl/8 - w/6$ (c) $wl/6 - w/8$
(d) $wl/6 - w/4$ (e) $wl/8 - w/4$

1-14 Integrate: $\displaystyle\int_0^L \frac{dx}{\sqrt{7 + 6x - x^2}}$

(a) $\sin^{-1}(0.25l - 0.75)$ (b) $(7 + 6x - x^2)^{1/2}$ (c) $2(7 + 6x - x^2)^{1/2}$

(d) $\sin^{-1}(0.25l + 0.75) + 3.99$

(e) $\sin^{-1}(0.25l - 0.75) - 3.99$ (angles in radians)

1-15 Determine which of the following is closest to the angle between the curve $(x + 2)^2 + (y - 3)^2 = 95$ and the line $2x + 3y = 12$.

(a) $78.5°$ (b) $70.8°$ (c) $64.23°$ (d) $84.7°$ (e) $81.7°$

1-16 From one corner of a pasture, one fence bears due east and the other bears S62°E. The owner wishes to build a new fence diagonally across the corner starting at a point 800 ft due east of it. If the owner has only 1200 ft of fencing, where will the new fence intersect the one that runs southeast?

(a) 1567 ft   (b) 1687 ft   (c) 1784 ft   (d) 1846 ft   (e) 1934 ft

1-17 The equation of a parabola is $y^2 = 16x$. Find the equation of a chord through the points on the curve whose abscissas are 1 and 4 and whose ordinates are positive.

(a) $y = 6x + 4$ (b) $3y = 4x + 8$ (c) $4y = 2x - 6$

(d) $2y = 5x + 7$ (e) $6y = 3x - 6$

1-18 The equation of a curve is $4a^2y = (x^3/3) - 2ax^2 + 5a^2$. Determine the points where the curve is parallel to the x axis.

(a) $(4.0a, -2.3a)$ (b) $(4.8a, -2.3a + 1.25)$ (c) $(4a, -2.7a + 1.25)$

(d) $(0, 1.3a + 1.5)$ (e) $(4.5a, 1.75)$

1-19 Determine the volume (in cubic units) generated by the rotation about the x axis of the area bounded by the segment of the parabola $y^2 = 8x$ between the origin of the x axis and the ordinate $x = 6$.

(a) 523   (b) 487   (c) 453   (d) 447   (e) 432

1-20 A rectangular plot of ground has two adjacent sides along Highways 20 and 32. In the plot is a small lake, one end of which is 256 ft west from Highway 20 and 108 ft north from Highway 32. Find the length of the *shortest* straight path which cuts across the plot from one highway to the other and passes by, i.e., is just tangent to or touches, the end of the lake.

(a) 300 ft   (b) 400 ft   (c) 500 ft   (d) 600 ft   (e) 487 ft

1-21 A person is buying two square lots of unequal size which lie adjacent to each other. Together the lots have a total area of 12,500 ft². To embrace the two lots in a single enclosure would require 500 ft of fence. Determine the dimensions of each lot.

(a) $100 \times 100, 50 \times 50$ (b) $60 \times 60, 80 \times 80$ (c) $83 \times 83, 83 \times 83$

(d) $90 \times 90, 40 \times 40$ (e) $110 \times 110, 30 \times 30$

1-22 Prove that $\sec^2 \theta = 1 + \tan^2 \theta$.

1-23 Evaluate the following definite integral:

$$\int_0^{\pi/2} \frac{\cos \theta \, d\theta}{1 + \sin^2 \theta}$$

(a) $\tfrac{1}{4}$ (b) π (c) $\pi/2$ (d) $\pi/3$ (e) $\pi/4$

1-24 Determine the length of the catenary from $x = 0$ to $x = x_1$, using the exponential equation $y = \frac{1}{2}a(e^{x/a} + e^{-x/a})$.

(a) $(a/2)(e^{x/a} - e^{-x/a})$ (b) $2a(e^{x/a} - e^{-x/a})$ (c) $a(e^{-x/a} - e^{x/a})$
(d) $(a/2)(e^{-x/a} - e^{x/a})$ (e) $a(e^{-x/a} + e^{x/a})$

1-25 Along the bank of a stream of water two points A and B are laid out 100 yd apart. A tree is visible from points A and B across the stream and very near to the opposite bank of the stream. The angle at A subtended by the distance from B to the tree is found to be 30°, and the angle at B subtended by the distance from A to the tree is 105°. With these data calculate the width of the stream.

(a) 70.7 yd   (b) 62.9 yd   (c) 84.3 yd   (d) 72.1 yd   (e) 68.3 yd

1-26 Water is flowing into a conical reservoir 20 ft deep and 10 ft across the top, at the rate of 15 (ft³/min.). Find how fast the surface is rising when the water is 8 ft deep.

(a) 1.4 ft/min   (b) 2.7 ft/min   (c) 1.2 ft/min   (d) 3.1 ft/min
(e) 0.91 ft/min

1-27 In the expression $R = re^{\tan x}$, where r is a constant and e is the natural logarithm base, what value of x will give $R = 0$?

(a) $\pi/2$ (b) $-\pi$ (c) $-\pi/2$ (d) $-\pi/4$ (e) π

1-28 Two ships leave the same port at the same time, one sailing due northeast at a rate of 6 mi/h and the other sailing due north at a rate of 10 mi/h. Find the distance between the two ships after 3 h of sailing.

(a) 18.6 mi   (b) 19.2 mi   (c) 21.5 mi   (d) 22.1 mi   (e) 22.8 mi

1-29 Solve for x in the expression $\log_4 3 + \log_4 (x - \frac{5}{3}) = 2$.

(a) 4   (b) 3   (c) 7   (d) 6   (e) $\frac{8}{3}$

1-30 Two sides of a triangle are 84 and 48 ft. The angle opposite the 84-ft side is 35°20′. Find the length of the third side.

(a) 97.1 ft   (b) 118.4 ft   (c) 127.3 ft   (d) 112.6 ft   (e) 99.7 ft

1-31 A trapezoidal trough has a bottom twice as wide as the slant length of a side. What angle does the side make with the bottom for maximum capacity of the trough?

(a) 12°   (b) 111.5°   (c) 112.4°   (d) 118°   (e) 97.3°

1-32 The half-life of a radioactive element is the time required for half a given quantity of that element to disintegrate into a new element. For example, it takes 1600 years for half a given quantity of radium to change into radon. In another 1600 years, half the remainder will have disintegrated, leaving one-quarter of the original amount. The half-life of radium is therefore said to be 1600 years. If the rate of disintegration of a radioactive material is 35 parts per 100 every hour, what is the half-life of this material?

(a) 2.8 h   (b) 1.9 h   (c) 1.6 h   (d) 1.4 h   (e) 1.2 h

1-33 Find the coordinates of the maximum ordinate on the curve

$$y + 10 = 3x^2 - 2x$$

(a) $-\frac{1}{3}, -\frac{31}{3}$ (b) 1, -9 (c) no max   (d) 2, -2 (e) $-1, -5$

1-34 Given the curve $y = x^3/100 - 5x/4$. The area between the tangent to the curve at the point $x = 5$, the X axis, and the line $x = 5$ is most nearly equal to which of the following (in square units)?
(a) 25.1   (b) 32.0   (c) 18.8   (d) 19.2   (e) 28.6

1-35 A cylindrical tank having a flat top and bottom is to have a capacity of 10,000 ft³. What should be its diameter and length if the surface area is to be a minimum?
(a) 18.2×28.3 (b) 23.4×23.4 (c) 20.1×25.7 (d) 28.3×18.2
(e) 25.2×21.8

1-36 A surveyor runs a line $AC = 380.00$ ft. With 0 as the midpoint on the line AC, the surveyor lays out a 20° curve CB. What is the distance between A and B? (See Fig. 1P-36.)

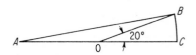

Figure 1P-36

(a) 385 ft   (b) 391 ft   (c) 362 ft   (d) 374 ft   (e) 379 ft

1-37 If x approaches zero in the expression $y = (5x^3 - \cos x + e^x)/(x^2 + 2x)$, then what does y approach as a limit?
(a) 0   (b) 1   (c) $\frac{1}{2}$ (d) $\frac{2}{3}$ (e) $\frac{3}{4}$

1-38 Find the value of $\int_0^{\pi/2} \sin \theta \cos \theta \, d\theta$.

(a) 0   (b) 1   (c) $\frac{1}{2}$ (d) $\frac{2}{3}$ (e) $\frac{3}{4}$

Multiple-Choice Sample Quiz

You can expect to find problems of this type in the morning session of the fundamentals examination. There will be 140 of these to complete in 4 hours, so speed is essential. Try to complete the 10 problems below in 12 minutes or less for practice. For each question, select the correct answer from the five given possibilities.

1M-1 The $\log_{10}$ of 2 is 0.301030. The $\log_{10}$ of $\frac{1}{2}$ is:
(a) $9.301030 - 10$ (d) $9.69897 - 10$
(b) $0.301030/2$ (e) $1/0.301030$
(c) $1 - 0.301030$

1M-2 Which of the following is incorrect?
(a) $\cos^2 A = \tan A \cot A$
(b) $\sin A = \cos A \tan A$
(c) $\cos 2A = \cos^2 A - \sin^2 A$
(d) $2 \sin^2 A = 1 - \cos 2A$
(e) $\sin (A + B) = \sin A \cos B + \cos A \sin B$

1M-3 The logarithm of the number -0.342 is:
   (a) positive   (d) a complex number
   (b) negative   (e) not a real number
   (c) zero

1M-4 The value of tan $(A + B)$, where tan $A = \frac{1}{3}$ and tan $B = \frac{1}{4}$ is (A, and B are acute angles):
   (a) $\frac{7}{12}$ (d) $\frac{7}{13}$
   (b) $\frac{1}{11}$ (e) none of the above
   (c) $\frac{7}{11}$

1M-5 If $x^{3/4} = 8$, x equals:
   (a) 6   (d) 16
   (b) 9   (e) 20
   (c) -9

1M-6 If $\log_A 10 = 0.250$, *then* $\log_{10} A$ equals:
   (a) 4   (d) 0.25
   (b) 0.50   (e) 10
   (c) 2

1M-7 The csc of $960°$ is equal to:
   (a) $-2\sqrt{3}/3$ (d) -2
   (b) 1   (e) $-\frac{3}{2}$
   (c) $\frac{1}{2}$

1M-8 If $i = \sqrt{-1}$, then i^{27} equals:
   (a) 0   (d) 1
   (b) i (e) -1
   (c) $-i$

1M-9 If $x = \log_B N$, then:
   (a) $x = N^B$ (d) $N = B^x$
   (b) $B = x^N$ (e) $N = x^B$
   (c) $x = B^N$

1M-10 If x equals $+6$ and -4, the equation satisfying both these values would be:
   (a) $2x^2 + 3x - 24 = 0$ (d) $x^2 - 4x - 32 = 0$
   (b) $x^2 + 10x - 24 = 0$ (e) $x^2 - 3x - 18 = 0$
   (c) $x^2 - 2x - 24 = 0$

2
STATICS

2-1 EXAMINATION COVERAGE

Statics is covered in the morning session and is one of the required subjects in the afternoon session.

The items covered in the section on statics can include the following:

Vector algebra

Two-dimensional equilibrium

Three-dimensional equilibrium

Concurrent force systems and resultants

Centroid of area

Moment of inertia

Analysis of internal forces

Friction effects

Equilibria of rigid bodies

Couples and moments

The subject of statics can be summarized by the statement of Newton's first law of motion: *When a body is at rest or moving with constant speed in a straight line, the resultant of all of the forces exerted on the body is zero.*

$$\Sigma F_x = 0 \qquad \Sigma F_z = 0$$
$$\Sigma F_y = 0 \qquad \Sigma M = 0$$

or the summation of all the forces in the X, Y, and Z planes must equal zero and the summation of all moments must also equal zero. These conditions are both necessary and sufficient.

2-2   VECTORS

Forces are vectors, and both the magnitude of the force and its direction of application must be given before the force is completely defined. A vector quantity is one which has both magnitude and direction. If only the magnitude is given, the quantity is not a vector but a scalar quantity. A vector quantity can be represented graphically by an arrow of the correct length, pointed in the proper direction. Vector quantities can be added vectorially by moving the vector to be added parallel to itself until the tail or starting point of the vector to be added (the second vector) just touches the head or ending point of the vector to which it is to be added (the first vector). The vector sum will then be a vector starting from the tail of the first vector and ending at the head of the second vector. This process can be repeated for any number of vectors. The final sum of all the vectors will be the vector starting at the tail (or beginning) of the first vector and ending at the head (or ending) of the last vector. This is shown graphically in Fig. 2-1.

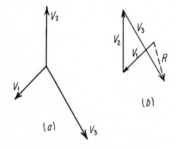

Figure 2-1

Any vector can be divided into two or more components. For mathematical calculations, it is usually convenient to divide the vector into its components in the X and Y directions for two-dimensional problems and into its X, Y, and Z components for three-dimensional problems. The two-dimensional case is illustrated in Fig. 2-2. The resultant component forces obtained are the X and Y components of the resultant of these three forces and are easily calculated with $R = \sqrt{R_x^2 + R_y^2}$ and the angle equal to $\tan^{-1}(R_y/R_x)$.

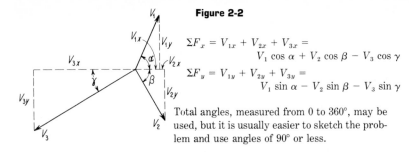

Figure 2-2

$\Sigma F_x = V_{1x} + V_{2x} + V_{3x} =$
$\qquad V_1 \cos \alpha + V_2 \cos \beta - V_3 \cos \gamma$

$\Sigma F_y = V_{1y} + V_{2y} + V_{3y} =$
$\qquad V_1 \sin \alpha - V_2 \sin \beta - V_3 \sin \gamma$

Total angles, measured from 0 to 360°, may be used, but it is usually easier to sketch the problem and use angles of 90° or less.

2-3   COPLANAR FORCES

All determinate statics problems can be solved by the application of the two requirements that the summation of all forces and the summation of all moments must equal zero. A considerable saving in time and effort can often be made, however, by the application of the principle that if three forces are in equilibrium the forces must be coplanar and must also be either parallel or concurrent.

- Find reactions R_1 and R_2 for the system shown in Fig. 2-3 ($W = 100$ lb).

  (a) $R_1 = 73.2$ lb, $R_2 = 51.8$ lb   (b) $R_1 = 51.8$ lb, $R_2 = 73.2$ lb   (c) $R_1 = 62.5$ lb, $R_2 = 57.7$ lb   (d) $R_1 = 81.3$ lb, $R_2 = 47.1$ lb   (e) $R_1 = 78.3$ lb, $R_2 = 50.0$ lb

This problem can be solved by constructing the force triangle as shown in Fig. 2-4:

$$\frac{W}{\sin \alpha} = \frac{R_1}{\sin 45°} = \frac{R_2}{\sin 30°} \qquad \alpha = 180 - (30 + 45) = 105°$$

$$R_1 = \frac{\sin 45°}{\sin 105°} \times W = \frac{0.707}{0.966} \times 100 = 73.2 \text{ lb}$$

$$R_2 = \frac{\sin 30°}{\sin 105°} \times W = \frac{0.500}{0.966} \times 100 = 51.8 \text{ lb}$$

The correct answer is (a).

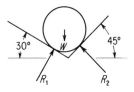

Figure 2-3

Figure 2-4

2-4 FREE-BODY DIAGRAMS

When working a problem concerning the equilibrium of a body or a juncture of forces, it is necessary to consider *all* the forces acting on the particular body or point under investigation. It is just as important to omit any force which does not act on the body being investigated, since one force too many or one force too few would destroy the conditions of equilibrium. The first step in attacking a problem in statics should then be the determination of just which forces are acting upon the body or juncture under investigation. The diagram of all these forces is called a "free-body diagram."

A "free body" is any system in static equilibrium. When a system is in static equilibrium, all internal forces are in balance, so no effect of any internal force can penetrate the boundary of a free body. The internal forces can thus all be ignored. Only the external forces must be considered. It should be noted that weight is always an external force since weight is a force caused by the gravitational attraction of the earth.

2-5 METHODS OF JOINTS AND SECTIONS

- Determine the force [in pounds of compression (lb C) or pounds of tension (lb T)] in member *CD* in Fig. 2-5.

  (*a*) 3000 lb C   (*b*) 4000 lb T   (*c*) 2318 lb T   (*d*) 4000 lb C
  (*e*) 3000 lb T

The three external forces acting on the framework are the reactions R_A and R_G and the applied force of 4000 lb. The vertical force at G, R_G, may be found by taking moments about point A:

$$\Sigma M_A = 0 = -4000 \times 10 \cos 30° + 30 R_G$$

$$R_G = \frac{34{,}641}{30} = 1154.7 \text{ lb}$$

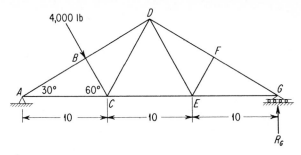

Figure 2-5

From $\Sigma F_y = 0$,

$$A_y + 1154.7 - 4000 \cos 30° = 0$$

$$A_y = 2309.4 \text{ lb}$$

From $\Sigma F_x = 0$,

$$A_x = -4000 \sin 30° = -2000 \text{ lb}$$

Construct a free-body diagram about point A (Fig. 2-6). Here we have two unknowns, the forces in AB and AC. These can be determined by satisfying the two equilibrium relationships, $\Sigma F_x = 0$ and $\Sigma F_y = 0$:

$$\Sigma F_x = 0 = -2000 + AC + AB \cos 30°$$

$$\Sigma F_y = 0 = 2309.4 + AB \sin 30°$$

Here, $AB = -2309.4/\sin 30° = -4618.8$ lb, so the assumed direction of action of the force in member AB was incorrect, and AB is in compression. Here also, $AC = 2000 - AB \cos 30° = 6000$ lb tension. We now have two unknown forces meeting at point B and three unknown forces meeting at point C. Since we have only two equations at our disposal ($\Sigma F_x = 0$ and $\Sigma F_y = 0$), we next solve for the forces acting at point B (Fig. 2-7).

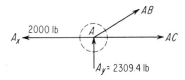

Figure 2-6

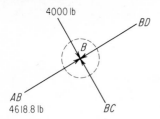

4000 lb

Figure 2-7

Point B is taken as a pin joint, so it can withstand no torque. Therefore, the applied force of 4000 lb must be resisted entirely by member BC. The force in BC will then equal 4000 lb compression. Similarly, the force in BD will equal 4618.8 lb compression.

There are now three unknown forces at point D and two at point C, so the forces acting at point C can now be determined (Fig. 2-8):

$$\Sigma F_x = 0 = -6000 + 4000 \cos 60° + CD \cos 60° + CE$$

$$\Sigma F_y = 0 - 4000 \sin 60° + CD \sin 60°$$

which gives $CD = 4000$ lb tension. The correct answer is (b).

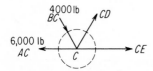

Figure 2-8

Note that the force in member EF can be determined by inspection by examining the joint at F. No torsion can be resisted at a pin joint, so no force vertical to member DG can be tolerated. Thus the force in member EF must equal 0. Then, since there is no vertical force at pin E, the force in member ED must also equal 0.

The force in member CD could also have been determined by the "method of sections." Draw a free-body diagram about the section of the truss containing member CD. There are three unknown forces acting on the free body, the forces in members BD, CD, and CE.

$$\Sigma F_y = 0 = 2309 - 0.866 \times 4000 + 0.866CD - 0.500BD$$

$$\Sigma F_x = 0 = -2000 + 0.500 \times 4000 - 0.866BD + 0.500CD + CE$$

$$\Sigma M_A = 0 = -4000 \times 0.866 + CD \times 0.866 \qquad CD = 4000 \text{ lb T}$$

The third equation was all that was necessary to solve for CD; the three equations enable us to solve for the forces in *BD* (4618 lb compression) and *CE* (2000 lb tension) also.

2-6 CABLE SUPPORTS

Refer to Fig. 2-9. It is required to find the following:

- 1. What is the tension in the cable?
    (a) 250 lb   (b) 289 lb   (c) 314 lb   (d) 400 lb   (e) 417 lb
- 2. What is the distance *S*?
    (a) 10.0 ft   (b) 12.1 ft   (c) 11.9 ft   (d) 11.4 ft
    (e) 10.9 ft

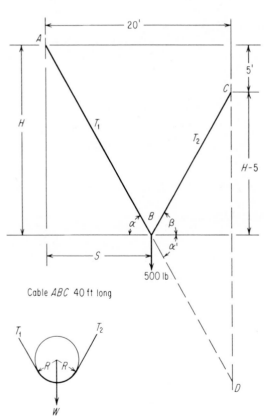

Cable *ABC* 40 ft long

Figure 2-9

Take the pulley as a free body in static equilibrium. The summation of moments about the axis of the pulley must then equal zero, or $T_2R - T_1R = 0$. Thus $T_1 = T_2$, and the tension is the same over the length of the cable. Similarly, since the summation of forces in the X direction must equal zero,

$$T \cos \beta - T \cos \alpha = 0 \quad \text{and} \quad \alpha = \beta$$

Vertical angles are equal so $\alpha = \alpha'$. Then triangles BCE and BDE are congruent and the distance AD equals 40 ft. Then $\alpha = \cos^{-1}(\frac{20}{40}) = 60°$:

$$T_1 \sin \alpha + T_2 \sin \beta = 500$$

but $\alpha = \beta$ and $T_1 = T_2$, so $2T \sin \alpha = 500$ and $T = 250/\sin 60° = 289$ lb.

From the geometry shown in Fig. 2-9, $40 \sin \alpha = 2H - 5$, which gives $H = 19.82$ ft. Since $S = H/\tan \alpha$, $S = 11.44$ ft. The length of cable section T_1 equals 22.89 ft and the length of section T_2 equals 17.11 ft. The correct answers are: part 1—(b), part 2—(d).

2-7   CENTER OF GRAVITY AND CENTROID OF AREA

The center of gravity equals the centroid of the forces of gravitation, which equals the center of mass.

▪ Referring to Fig. 2-10, what is the distance S from the center of gravity of W_1 to the center of gravity of the system?

(a) 6.2 ft   (b) 7.3 ft   (c) 8.4 ft   (d) 8.7 ft   (e) 7.0 ft

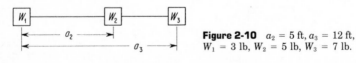

Figure 2-10 $a_2 = 5$ ft, $a_3 = 12$ ft, $W_1 = 3$ lb, $W_2 = 5$ lb, $W_3 = 7$ lb.

We can calculate this distance as follows:

$$S = \frac{\Sigma a\, \Delta W}{\Sigma \Delta W} = \frac{a_2 W_2 + a_3 W_3}{W_1 + W_2 + W_3}$$

Taking moments about the center of gravity of W_1, the distance from the center of gravity of W_1 to the center of gravity of the system is

$$S = \frac{5 \times 5 + 12 \times 7}{3 + 5 + 70} = 7.27 \text{ ft}$$

The correct answer is (b).

For a plane surface, which has no weight, the term "center of gravity" is replaced by the "centroid of the area," which gives the mathematical expression $\Sigma r\, \Delta A = 0$.

An illustration is afforded by the determination of the centroid of a triangle (see Fig. 2-11):

$$S = \frac{\displaystyle\int_0^h x\, dA}{A} \qquad dA = b'\, dx$$

$$b' = \frac{x}{h}\, b \qquad \text{giving } dA = \frac{b}{h}\, x\, dx$$

$$A = \tfrac{1}{2}bh$$

$$= \frac{\displaystyle\int_0^h x\, b/h\, x\, dx}{\tfrac{1}{2}bh} = \frac{b/h \displaystyle\int_0^h x^2\, dx}{\tfrac{1}{2}bh} = \frac{\tfrac{1}{3}h^3}{\tfrac{1}{2}h^2} = \tfrac{2}{3}h$$

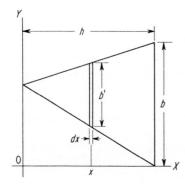

Figure 2-11

or the centroidal axis of a triangle is two-thirds the distance from the apex to the base.

▪ What is the location of the horizontal centroidal axis of the built-up beam shown in Fig. 2-12, measured from the bottom end of the beam?

(a) 4.5 in (b) 5.1 in (c) 3.0 in (d) 3.9 in (e) 4.2 in

$$\bar{y} = \frac{\Sigma s\, \Delta A}{A} = \frac{\tfrac{1}{2}(10 \times 1) + (1 + 3)(6 \times 3) + (1 + 6 + 1)(2 \times 4)}{10 \times 1 + 6 \times 3 + 4 \times 2}$$

$$= 3.91 \text{ in}$$

The correct answer is (d).

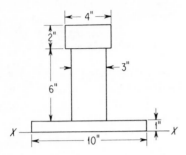

Figure 2-12

To summarize, the resultant of the weight of a body or system will pass through one point in the body, or the body extended, for all orientations of the body; this point is defined as the center of gravity of the body or system. The centroid of an area corresponds to the center of gravity of a thin plate and may be determined as the center of gravity is determined, by assuming the area to be a thin plate.

2-8 FRICTION AND FRICTION ANGLE

A frictional force will always be in such a direction as to resist motion. It is a passive force and can never exceed whatever force may be applied to an object. It can never cause motion; it may vary from zero to a maximum. The maximum frictional force which can be exerted on an object is equal to the coefficient of friction times the force *normal* to the surface along which the frictional force is exerted. This is easily demonstrated in its simplest form by the forces acting on a block resting on an inclined plane (Fig. 2-13). The frictional force is equal to $W \cos \varphi \times$

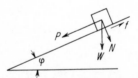

Figure 2-13 μ = coefficient of friction between block and plane.

μ, or the normal force times the coefficient of friction between the block and the plane. The force parallel to the face of the plane tending to cause the block to slide down the plane is equal to $W \sin \varphi$. If the angle of the plane is adjusted so that slipping impends, the value of the angle φ is termed the "friction angle." In this case the frictional force resisting motion will be just equal to the force component $W \sin \varphi$. This gives the relationship $W \sin \varphi = \mu W \cos \varphi$, which reduces to $\mu = \tan \varphi$, or the coefficient of friction between the block and the plane is equal to the tangent of the friction angle. This concept of a friction angle is extremely useful and considerably simplifies the working of the more complicated friction problems. It is possible to add the friction angle to whatever mechanical angle is being considered (or subtract the friction angle from the mechanical angle) to obtain an equivalent angle of a frictionless system, and the answer can be readily calculated. The method is illustrated as follows:

- Referring to Fig. 2-14, how large a horizontal force F must be applied to a block of weight W resting on an incline of angle α to cause motion to impend up the slope? The coefficient of friction between the block and the slope is μ.

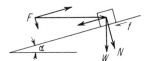

Figure 2-14

The force acting up the plane parallel to it equals $F \cos \alpha$. The forces resisting motion are the force down the slope due to the weight of the block, equal to $W \sin \alpha$, and the frictional resistance equal to the force exerted by and on the block normal to the plane—N times the coefficient of friction μ. N is made up of two components, $W \cos \alpha$ and $F \sin \alpha$.

Adding the forces acting, we have

$$F \cos \alpha = W \sin \alpha + (W \cos \alpha + F \sin \alpha)\mu$$

and if we substitute $\tan \varphi$ for its equivalent, μ, we get

$$F \cos \alpha = W \sin \alpha + (W \cos \alpha + F \sin \alpha) \frac{\sin \varphi}{\cos \varphi}$$

$$F \cos \alpha \cos \varphi = W \sin \alpha \cos \varphi + W \cos \alpha \sin \varphi + F \sin \alpha \sin \varphi$$

which gives

$$F(\cos \alpha \cos \varphi - \sin \alpha \sin \varphi) = W(\sin \alpha \cos \varphi + \cos \alpha \sin \varphi)$$

which reduces to

$$F \cos (\alpha + \varphi) = W \sin (\alpha + \varphi)$$

$$F = W \tan (\alpha + \varphi)$$

which is the same result we should have obtained had we considered the slope angle to be α + the friction angle, or $(\alpha + \varphi)$, and considered the surfaces to be frictionless.

Utilizing the concept of friction angle, resketch the problem as shown in Fig. 2-15. In this case we get

$$F \cos (\alpha + \varphi) = W \sin (\alpha + \varphi)$$
$$F = W \tan (\alpha + \varphi)$$

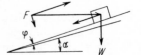

Figure 2-15

Similarly, it can be shown that to find the force to just hold the block in place we can consider the friction angle as being subtracted from the slope angle and again treat the system as if it were frictionless (see Fig. 2-16):

$$F \cos (\alpha - \varphi) = W \sin (\alpha - \varphi)$$
$$F = W \tan (\alpha - \varphi)$$

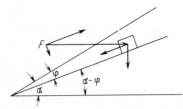

Figure 2-16

By adding or subtracting the friction angle, it is possible to incorporate the effects of friction without treating the frictional components separately.

One implication of the friction-angle concept is immediately discernible; i.e., if the friction angle is greater than the angle of the slope, the system will be statically stable.

Figure 2-17 shows the condition of a statically stable wedge where

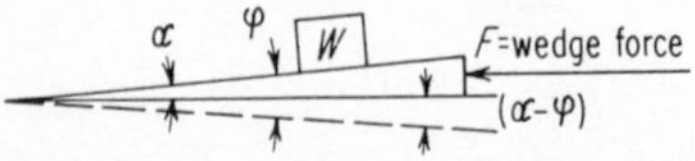

Figure 2-17

$\varphi > \alpha$ and the weight will not slip. On the other hand, Fig. 2-18 shows a wedge which is not statically stable. The weight would slide as if on a frictionless plane with a slope of $\alpha - \varphi$; or the wedge would shoot out if the force F were removed. Also, if the sum of the slope angle plus the friction angle equals or exceeds 90°, it would be impossible to push the block up the slope with a force which is applied horizontally.

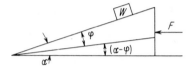

Figure 2-18

A word of caution should be given about the use of the friction angle: *Do not change the direction of the application of the force.*

A more involved type of wedge and friction problem is illustrated in Fig. 2-19.

Figure 2-19

- For the wedge system shown, find the force F required to raise the weight W ($W = 1000$ lb, $\alpha = 5°$). The coefficient of friction for all surfaces is equal to $\mu = 0.3$.

  (a) 584 lb   (b) 638 lb   (c) 708 lb   (d) 793 lb   (e) 813 lb

Applying the principle of friction angle to the problem, we can construct a new figure which will incorporate the effects of the friction angles and give us an equivalent frictionless system (Fig. 2-20). Here we can de-

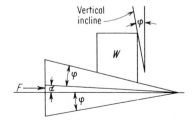

Figure 2-20

termine the relationship between F and W by means of the conservation of energy (or equivalence of work); i.e.,

$$FS = Wh_{\text{eff}}$$

where h_{eff} represents the effective elevation. We further obtain

$$h_{\text{eff}} = S[\tan(\alpha + \varphi) + \tan\varphi] + h'$$

where h' is the additional elevation due to the movement to the left which results from the movement of the upper right corner of the weight in the figure along the back angle of the vertical incline. These quantities are shown in Fig. 2-21. Then

$$\frac{M}{\sin\varphi} = \frac{h}{\sin\gamma} \qquad h' = M\sin(\alpha + \varphi)$$

$$\gamma = 90 - (\alpha + 2\varphi)$$

$$\sin\gamma = \cos(\alpha + 2\varphi)$$

$$h' = \frac{\sin\varphi}{\sin\gamma} h \sin(\alpha + \varphi) = \frac{\sin\varphi \times \sin(\alpha + \varphi)}{\cos(\alpha + 2\varphi)} h$$

Then

$$h_{\text{eff}} = S[\tan(\alpha + \varphi) + \tan\varphi]\left[1 + \frac{\sin\varphi}{\cos(\alpha + 2\varphi)}\sin(\alpha + \varphi)\right]$$

$$F = W[\tan(\alpha + \varphi) + \tan\varphi]\left[1 + \frac{\sin\varphi}{\cos(\alpha + 2\varphi)}\sin(\alpha + \varphi)\right]$$

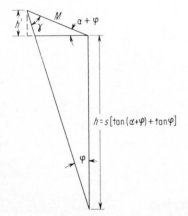

$$h = s[\tan(\alpha + \varphi) + \tan\varphi]$$

Figure 2-21

If $W = 1000$ lb, $\mu = 0.3$, and $\alpha = 5°$, we obtain

$$F = W(\tan 21.7° + \tan 16.7°)\left(1 + \frac{\sin 16.7°}{\cos 38.4°} \sin 21.7°\right)$$

$$= W(0.398 + 0.300)\left(1 + \frac{0.288}{0.784} \times 0.370\right) = 793 \text{ lb}$$

The correct answer is (d).

2-9 JACKSCREW

One other example is the derivation of the mechanical efficiency of a jackscrew (Fig. 2-22). First assume a square thread, with a pitch P:

$$\text{Pitch} = \frac{1}{\text{No. of threads/in}}$$

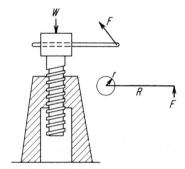

Figure 2-22

Also consider the screw to be single-threaded (lead equal to pitch). The coefficient of friction is equal to μ, and the mean thread diameter is D_m. From Fig. 2-23 we can see that if we unwind the thread, the lead angle λ is easily determined; it is equal to $\tan^{-1}(\text{lead}/\pi D_m)$. The friction angle φ is equal to $\tan^{-1}\mu$. The equivalent angles of frictionless screws are $\lambda + \varphi$ if the load is being raised and $\lambda - \varphi$ if the load is being lowered.

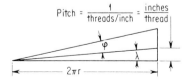

Figure 2-23

A torque Tq, in pound-inches (lb·in), applied to the jackscrew will supply a force F equal to $2Tq/D_m$ at the thread:

$$Tq = FR \qquad \text{lb·in}$$

The equivalent force at a distance of 1 in from the axis is then equal to the torque, and the work done per revolution of the screw equals $2\pi \times$ Tq. This also equals the work done in raising the weight a distance equal to the pitch P plus the work done against friction, or

$$2\pi\, Tq = \pi 2r \tan(\lambda + \varphi)W$$

$$Tq = \frac{D_m \tan(\lambda + \varphi)}{2}\,W = r\tan(\lambda + \varphi)W = FR$$

If $\varphi > \lambda$, the jackscrew will be statically stable and will require no force on the handle to maintain the load in its position. On the other hand, if $\lambda > \varphi$, the weight will lower by itself if all force is removed from the handle. This case is similar to that of the wedge, and the effective angle is $\lambda - \varphi$ (Fig. 2-24).

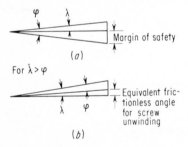

For $\lambda > \varphi$

(a) Margin of safety

(b) Equivalent frictionless angle for screw unwinding

Figure 2-24 Torque required to lower weight; $FR = Wr \tan(\varphi - \lambda)$.

2-10 DIMENSIONAL UNITS

There is an effort under way to promote a standardized system of units throughout the world through the International Bureau of Weights and Measures (BIPM). The system of units proposed by that international bureau is termed the "International System of Units" and is abbreviated as the "SI" system. Problems in the EIT examination are sometimes given in both the English and the SI systems of units. The examinee may work either version of the problem.

The term "lb/ft³" is used in the U.S. system and is defined as specific weight, symbolized by w. There is no corresponding unit in the SI system. Density is usually designated by the symbol ρ (Greek letter rho) and

Quantity	SI unit	U.S. customary unit
force	newton (N)	pound (lb)
mass	kilogram (kg)	slug
length	meter (m)	foot (ft)
pressure	pascal (Pa)	pounds per square inch (lb/in^2)
density	kilograms per cubic meter (kg/m^3)	slugs per cubic foot ($slugs/ft^3$)
acceleration	meters per second per second (m/s^2)	feet per second per second (ft/s^2)
moment	newton-meter ($N \cdot m$)	pound-foot or pound-inch ($lb \cdot ft$ or $lb \cdot in$)

equals w/g, where g is the acceleration of gravity, at 32.17 ft/s^2 or 980.665 cm/s^2.

Some of the principal conversion factors between the SI and the U.S. Customary systems are listed here:

1 in = 2.54 cm or 25.4 mm (this is an exact conversion)

1 N = 0.2248 lb, or 1 lb = 4.44822 N

1 lb of weight corresponds to 0.45359 kg of mass, or 0.0311 slugs

1 slug = 14.594 kg

1 lb/in^2 = 6894.757 Pa, or 1 megapascal (M Pa) = 145.04 lb/in^2

1 bar = 10^5 Pa = 14.504 lb/in^2

■ As an example of a problem in SI units, refer to Fig. 2-3. Find reactions R_1 and R_2 for the system shown, assuming that the mass of W is 45.4 kg.

(a) R_1 = 325.9 N, R_2 = 230.5 N   (b) R_1 = 230.4 N, R_2 = 325.6 N   (c) R_1 = 278.0 N, R_2 = 250.6 N   (d) R_1 = 361.6 N, R_2 = 209.5 N   (e) R_1 = 348.3 N, R_2 = 222.4 N

The force exerted downward by the 45.4-kg mass equals, from Newton's second law, $F = ma$:

$$45.4 \text{ kg} \times 9.805 \text{ m/s}^2 = 445.2 \text{ N}$$

$$R_1 = \frac{\sin 45°}{\sin 105°} \times 445.2 = 325.9 \text{ N}$$

$$R_2 = \frac{\sin 30°}{\sin 105°} \times 445.2 = 230.5 \text{ N}$$

38   Statics

So the correct answer is (a). The problem could also have been worked by converting the mass to slugs, which would give a mass of 45.4/14.594 = 3.111 slugs or a weight of 100 lb. The forces would then equal 73.2 and 51.8 lb, corresponding to 325.6 and 230.4 N, which are the same as the values calculated directly in the SI system, considering rounding off.

The problem can be worked directly in the given set of units, or the units can be converted from the SI system to the U.S. system or vice versa. The problem can then be solved in the more familiar system and the answer converted to whichever system of units is desired.

Sample Problems

2-1 What is the maximum weight of the box that can be held in place by the force as shown in Fig. 2P-1?

(a) 200 lb   (b) 300 lb   (c) 400 lb   (d) 500 lb   (e) 600 lb

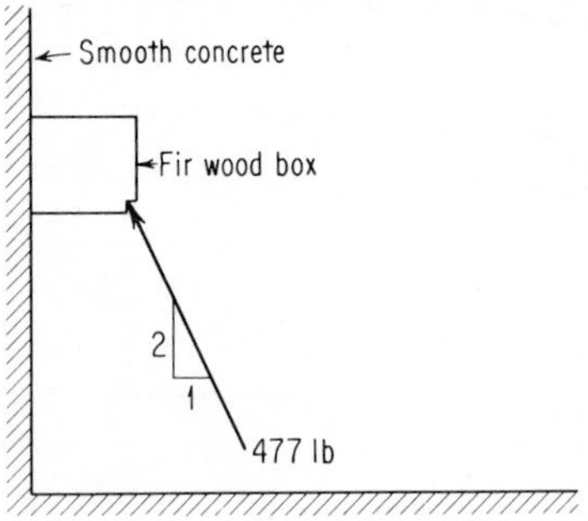

Smooth concrete

Fir wood box

2

1

477 lb

Figure 2P-1

2-2 A cable 25 ft long is connected between two points 20 ft apart horizontally and 5 ft apart vertically. If a 500-lb load is suspended on the cable by use of a frictionless pulley, how far from the upper support will the pulley come to rest?

(a) 8.66 ft   (b) 12.00 ft   (c) 13.33 ft   (d) 15.2 ft   (e) 16.1 ft

2-3 How far along the centerline from side A of the area shown in Fig. 2P-3 is the centroid?

(a) 4.3 in   (b) 5.0 in   (c) 6.2 in   (d) 3.7 in   (e) 4.7 in

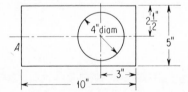

4"diam

$2\frac{1}{2}$"

5"

A

3"

10"

Figure 2P-3

2-4 A jackscrew having eight threads to the inch, i.e., $\frac{1}{8}$-in threads, is operated manually by a handle. The point at which the handle is gripped moves through a 12-ft circumference. If friction is disregarded, what force must be applied at the grip on the handle to just raise a load of 2300 lb?
(*a*) 4.7 lb   (*b*) 1.7 lb   (*c*) 2.0 lb   (*d*) 3.1 lb   (*e*) 2.7 lb

2-5 For the truss loaded as shown in Fig. 2P-5, find the force in member U_4L_4.
(*a*) 15,000 lb T   (*b*) 0   (*c*) 14,140 lb T   (*d*) 13,330 lb T
(*e*) 12,010 lb T

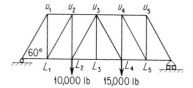

10,000 lb   15,000 lb   **Figure 2P-5**

2-6 The bracket shown in Fig. 2P-6 is fastened to a vertical wall at points B, C, and D. Point D is 3 ft higher than C, with member AB horizontal. Find the force in member AB.
(*a*) 608 lb   (*b*) 815 lb   (*c*) 425 lb   (*d*) 912 lb   (*e*) 878 lb

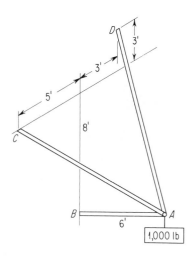

1,000 lb   **Figure 2P-6**

2-7 The coefficient of friction between the 10-lb block A and the 20-lb block B, shown in Fig. 2P-7, is 0.40. The coefficient of friction between B and the horizontal plane is 0.15. Determine the force P required to cause motion of B to impend to the right.
(*a*) 4.5 lb   (*b*) 9.8 lb   (*c*) 12.2 lb   (*d*) 10.1 lb   (*e*) 8.6 lb

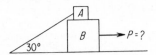

Figure 2P-7

2-8 A solid sphere weighing 1000 lb is held at rest on a 45° inclined plane by means of cables a, b, and c, of which a is attached to the center of the sphere. Solve for the tension in cable c. (See Fig. 2P-8.)

(a) 250 lb   (b) 898 lb   (c) 688 lb   (d) 732 lb   (e) 417 lb

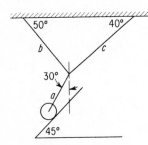

Figure 2P-8

2-9 Locate the centroid of the shaded area shown in Fig. 2P-9. Find the distance x from lower left corner.

(a) 3.5 in   (b) 3.2 in   (c) 3.8 in   (d) 4.0 in   (e) 3.0 in

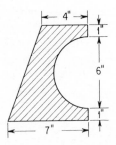

Figure 2P-9

2-10 A steel block is 8 in square and 12 in high. A cylindrical hole 6 in deep is to be drilled from the upper end. Compute the necessary diameter of the hole required so that the center of gravity of the block will be 5 in from the bottom of the block.

(a) 5.2 in   (b) 6.4 in   (c) 6.6 in   (d) 5.8 in   (e) 4.3 in

2-11 Find the force in member CH of the truss shown in Fig. 2P-11.

(a) 1020 lb C   (b) 980 lb C   (c) 1020 lb T   (d) 980 lb T
(e) 1117 lb T

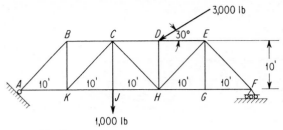

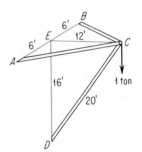

Figure 2P-11

2-12 Solve for the magnitude and direction of the force in member AC of the frame shown in Fig. 2P-12. Distances between lettered points are as indicated. Plane ABC is horizontal, and plane CDE is vertical.
(a) 680 lb T   (b) 760 lb T   (c) 810 lb T   (d) 840 lb T
(e) 860 lb T

Figure 2P-12

2-13 Figure 2P-13 shows how a weight can be raised by an ideal frictionless wedge. For what angle can a weight W be raised that is 10 times as large as the applied force F?
(a) 3.1°   (b) 4.2°   (c) 4.8°   (d) 5.2°   (e) 5.7°

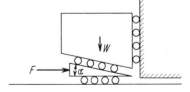

Figure 2P-13

2-14 Calculate the force in member $L_1 U_2$ in the truss shown in Fig. 2P-14.
(a) 1789 lb C   (b) 2236 lb C   (c) 2381 lb C   (d) 3211 lb C
(e) 2691 lb C

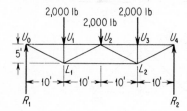

Figure 2P-14

2-15 A weight of 4000 lb is supported at the midpoint of cable *HNK*, which is 18 ft long (Fig. 2P-15). Assuming that the safe tension in the cable is 2600 lb, determine angle θ.

(*a*) 45°   (*b*) 36.9°   (*c*) 60°   (*d*) 50.3°   (*e*) 49.2°

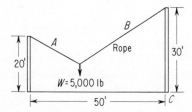

Figure 2P-15

2-16 Total length of ropes $A + B = 60$ ft (Fig. 2P-16). The weight on the sheave is free to move on the rope. Disregard the weight of the rope. Find the tensions in ropes A and B.

(*a*) 4520 lb   (*b*) 4780 lb   (*c*) 3920 lb   (*d*) 5130 lb   (*e*) 5690 lb

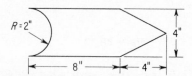

Figure 2P-16

2-17 What is the moment of inertia about the centroidal *Y-Y* axis of the figure shown in Fig. 2P-17?

(*a*) 225 in⁴   (*b*) 317 in⁴   (*c*) 437 in⁴   (*d*) 189 in⁴   (*e*) 292 in⁴

Figure 2P-17

2-18 A wooden beam is made up of full-dimension timbers, as shown in Fig. 2P-18. What is the moment of inertia about axis *X-X*?

(a) 1728 in⁴   (b) 1692 in⁴   (c) 1541 in⁴   (d) 1492 in⁴
(e) 1430 in⁴

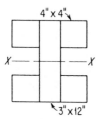

Figure 2P-18

2-19 Refer to Fig. 2P-19, which shows a Fink truss, and determine the force in
the member labeled *a*. Express your final answer in pounds, and state whether
the force is *tension* or *compression*.
(a) 18,400 lb T   (b) 10,000 lb T   (c) 14,700 lb T   (d) 16,500 lb T
(e) 17,250 lb T

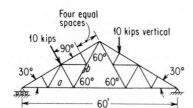

Figure 2P-19

2-20 A cantilever beam is hinged to the wall as shown in Fig. 2P-20. What are
the amount and direction (compression or tension) of force in member *A*?
(a) 10,000 lb T   (b) 15,000 lb T   (c) 20,000 lb T   (d) 25,000 lb T
(e) 30,000 lb T

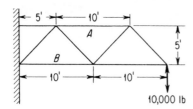

10,000 lb   Figure 2P-20

2-21 A tractor, as shown in Fig. 2P-21, is on level ground. What pull *X* will
remove all weight from the front wheels if the tractor is going up a slope
20° from horizontal?
(a) *W*   (b) 2*W*   (c) 3*W*   (d) 1.5*W*   (e) 2.7*W*

44   Statics

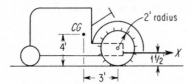

Figure 2P-21

2-22 Given a truss as shown in Fig. 2P-22, assume that the horizontal components of the reactions at A and B are equal. Determine the force in BD.
  (*a*) 0   (*b*) 0.35*P* C   (*c*) 0.35*P* T   (*d*) 0.71*P* T   (*e*) 0.71*P* C

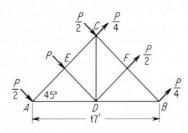

Figure 2P-22

2-23 Find the least force P just sufficient to move the wheel shown in Fig. 2P-23, radius r, which weighs W, over the curb of height h.
  (*a*) 100 lb   (*b*) 82 lb   (*c*) 67 lb   (*d*) 75 lb   (*e*) 60 lb

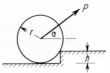

Figure 2P-23 $r = 12$ in, $h = 4$ in, $W = 100$ lb.

2-24 Find the moment of inertia about axis X-X through centroid of welded plate and angle girder shown in Fig. 2P-24.
  (*a*) 571 in⁴   (*b*) 428 in⁴   (*c*) 379 in⁴   (*d*) 482 in⁴   (*e*) 503 in⁴

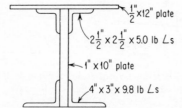

Figure 2P-24

2-25 The ends of a cable 21 ft long are fastened to two rigid overhead points 15 ft apart and in the same horizontal plane. A weight of 200 lb is attached to the cable 9 ft from one end. Determine the maximum force in the cable.
(*a*) 160 lb (*b*) 140 lb (*c*) 180 lb (*d*) 150 lb (*e*) 175 lb

2-26 Given the truss as shown in Fig. 2P-26, determine the force [in kilopounds (kips)] in member U_2L_4.
(*a*) 4.0 kips C (*b*) 5.7 kips C (*c*) 6.9 kips C (*d*) 6.1 kips C
(*e*) 4.6 kips C

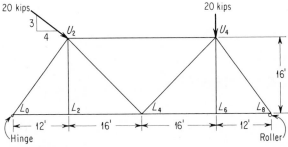

Figure 2P-26

2-27 The weights *A* and *B* (Fig. 2P-27) are supported by a continuous rope which is attached at points *C* and *D* and passes around frictionless pulleys as shown. Disregard the weight of the rope and pulleys. Find the position of the pulley *P* relative to point 0 for equilibrium.
(*a*) 3.0 ft (*b*) 4.0 ft (*c*) 3.9 ft (*d*) 3.5 ft (*e*) 4.2 ft

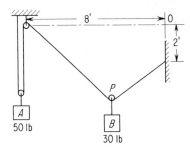

Figure 2P-27

2-28 A homogeneous rod weighing 50 lb is hinged at *A* and rests on a block at *B* (Fig. 2P-28). The block, weighing 3 lb, rests on a horizontal floor. If the coefficient of friction for all surfaces is one-third, what is the horizontal force *P* necessary to cause the block to slide to the left? Consider the hinge at *A* to be frictionless.
(*a*) 12 lb (*b*) 14 lb (*c*) 16 lb (*d*) 18 lb (*e*) 20 lb

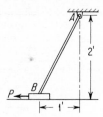

Figure 2P-28

2-29 What is the moment of inertia about its horizontal neutral axis of a section assembled from four 3- by 3- by 0.5-in angles and an 11- by 0.5-in plate as shown in Fig. 2P-29?

(a) 462 in⁴   (b) 413 in⁴   (c) 398 in⁴   (d) 376 in⁴   (e) 347 in⁴

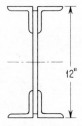

Figure 2P-29

2-30 Two smooth cylinders rest as shown in Fig. 2P-30 on an inclined plane and a vertical wall. Compute the value of the force F exerted by the vertical wall.

(a) 140 lb   (b) 180 lb   (c) 173.2 lb   (d) 78 lb   (e) 120 lb

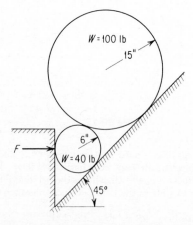

Figure 2P-30

2-31 For the slider arm in Fig. 2P-31, $a = 2$ in, $d = 4$ in, and the coefficient of friction is 0.20. What is the minimum dimension of X so that a force F will result in no movement of the slider arm?

    (a) 14 in   (b) 16 in   (c) 18 in   (d) 20 in   (e) 22 in

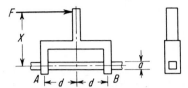

Figure 2P-31

2-32 Three springs, having spring constants of 5, 10, and 15 lb/in, respectively, are connected in series as shown in Fig. 2P-32. What is the magnitude of the combined spring constant?

    (a) 30 lb/in   (b) 15 lb/in   (c) 6.9 lb/in   (d) 4.2 lb/in
    (e) 2.7 lb/in

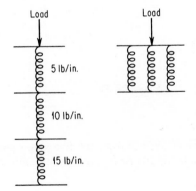

Figure 2P-32

2-33 Three people are carrying a heavy 12-ft log of circular cross section 8 in in diameter at the small end and 12 in in diameter at the large end. One person takes the small end of the log, and the other two place a bar under the log at a point so that each of the three carries an equal load. How far from the small end was the bar placed?

    (a) 8.0 ft   (b) 9.2 ft   (c) 10.2 ft   (d) 11.3 ft   (e) 9.9 ft

2-34 The weightless beam shown in Fig. 2P-34 is suspended at each end by cables at points 1 and 2. Find the tension F_1 in the cable supporting the beam at point 1.

    (a) 45.9 lb   (b) 44.8 lb   (c) 46.2 lb   (d) 45.1 lb   (e) 51.0 lb

48   Statics

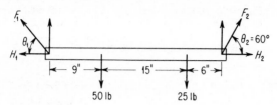

Figure 2P-34

2-35   The hoisting rig in Fig. 2P-35 is composed of two symmetrical column members bd and cd with pin joints at b and c and a steel cable ad. A vertical load L of 8 tons is supported at d. The distance ae is 21 ft and ef is 10.5 ft. The angle daf is 30°, def is 60°, and dbc is 60°. The distance $af = 31.5$ ft. Determine the tension or compression in the leg bd, F_{bd}.

(a) 14,000 lb C   (b) 16,000 lb C   (c) 18,000 lb C   (d) 20,000 lb C
(e) 12,000 lb C

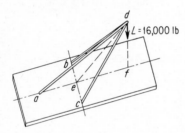

Figure 2P-35

Multiple-Choice Sample Quiz

For each question select the correct answer from the five given possibilities. For practice, try to complete the 10 problems below in 12 minutes or less.

2M-1   Which of the following is true with respect to the vector diagram in Fig. 2PM-1?

(a) $\vec{D} = \vec{A} + \vec{B} - \vec{C}$
(b) $2\vec{B} = \vec{A} + 2\vec{C}$
(c) $2\vec{A} = \vec{B} + \vec{D}$
(d) $\vec{C} = \vec{D} - \vec{A} - \vec{B}$
(e) $\vec{A} - \vec{B} = \vec{D} - \vec{C}$

Figure 2PM-1

2M-2 The reaction at point A in Fig. 2PM-2 is:
   (a) zero
   (b) 40 lb ↑
   (c) 40 lb ↓
   (d) 40 lb ↑ plus 400 ft·lb
   (e) 40 lb ↓ plus 400 ft·lb

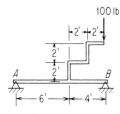

Figure 2PM-2

2M-3 One newton equals:
   (a) 10 dynes   (d) 10^2 dynes
   (b) 10^3 dynes   (e) 10^4 dynes
   (c) 10^5 dynes

2M-4 For a system to be in equilibrium, the sum of the external forces acting on the system must be:
   (a) equal to unity   (d) a maximum
   (b) indeterminant   (e) zero
   (c) infinite

2M-5 The vector which represents the sum of a group of force vectors is called the:
   (a) magnitude   (d) resultant
   (b) sum   (e) phase angle
   (c) force polygon

2M-6 Two weights are suspended on an inextensible weightless cord from frictionless pulleys as shown in Fig. 2PM-6. Which of the following most nearly equals the angle θ at equilibrium?
   (a) 60°   (d) 53°
   (b) 37°   (e) 82°
   (c) 75°

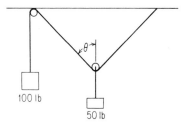

Figure 2PM-6

2M-7   A 50-lb pulley, supported as shown in Fig. 2PM-7, carries an inextensible cable supporting an additional 50-lb load. The force the beam exerts on the pulley shaft is:

(a) 50 lb up   (d) 150 lb up
(b) 100 lb up   (e) 150 lb down
(c) 100 lb down

50 lb   **Figure 2PM-7**

2M-8   A 100-lb weight hangs by a string from the ceiling. It is pulled by a horizontal force until the string makes an angle of 30° with the vertical. The tension in the string most nearly equals:

(a) 100 lb   (d) 110 lb
(b) 125 lb   (e) 120 lb
(c) 115 lb

2M-9   An 80-lb body lies on a 30° slope. The frictional force most nearly equals:

(a) 35 lb   (d) 40 lb
(b) 50 lb   (e) 60 lb
(c) 45 lb

2M-10 An object weighing 100 lb is held by two strings 4 ft long, the fixed ends of which are fastened to a ceiling 4 ft apart. What is the tension in each string?

(a) 58 lb   (d) 50 lb
(b) 62 lb   (e) 48 lb
(c) 60 lb

3

DYNAMICS

3-1 EXAMINATION COVERAGE

There are required problems in dynamics in both the morning and afternoon sessions.

The items which may be covered, and which should be reviewed, include the following:

Kinematics, including relative motion and curvilinear motion

Force, inertia, mass, and acceleration

Impulse and momentum

Work and energy

Particles and solid bodies

Linear motion and rotary motion

Dynamics is based on Newton's second law plus effects of Newton's other two laws. The three laws may be stated as follows:

1. Every body persists in its state of rest or of uniform motion in a straight line unless it is compelled by some force to change that state.
2. The rate of change of the momentum of a body is proportional to the force acting on it and it is in the direction of the force ($F = ma$).
3. Action and reaction are equal and opposite.

3-2 EQUATIONS OF LINEAR MOTION

Linear motion, or motion along a straight line, is called "velocity," where velocity is a vector quantity (having both magnitude and direction) and is equal to the rate of change of distance with time. The average velocity is equal to displacement/elapsed time, and if we take the distance covered as $s - s_0$ in the time interval $t - t_0$, the average velocity can be expressed as

$$v_{\text{avg}} = \frac{s - s_0}{t - t_0}$$

This can also be written as $v = \Delta s/\Delta t$, which, as Δt approaches zero, has the limiting value $v = ds/dt$. Similarly, the rate of change of velocity with time is termed "acceleration" and the limiting value of the acceleration, $a = dv/dt = d^2s/dt^2$. If we start with a constant acceleration, $a = dv/dt$, and integrate, we get $v = at + c_1$, where c_1 is the constant of integration. Replacing v by its equivalent, we have $ds/dt = at + c_1$, which, on integration, gives us $s = \frac{1}{2}at^2 + c_1t + c_2$, where c_2 is the constant resulting from the second integration. If $s = s_0$ and $v = v_0$ at time $t = 0$, these give us the general relationship for distance: $s = s_0 + v_0t + \frac{1}{2}at^2$.

If $s = 0$ at $t = 0$, this give us the more familiar relationship $s = v_0t + \frac{1}{2}at^2$. If the equation $v = v_0 + at$ is solved for t and the value substituted in the equation for s, we get $v^2 = v_0^2 + 2as$. Thus we can easily derive the four basic equations of linear motion:

$$v = v_0 + at$$

$$v^2 = v_0^2 + 2as$$

$$v_{\text{avg}} = \frac{v + v_0}{2}$$

$$s = v_0t + \tfrac{1}{2}at^2 \quad \text{or} \quad s = v_{\text{avg}}t$$

■ The speed of an automobile starting from rest increased in 18 s to 66 ft/s. Which of the following most nearly equals the rate of uniform acceleration?

(a) 2.8 ft/s²   (b) 3.1 ft/s²   (c) 3.3 ft/s²   (d) 3.5 ft/s²
(e) 3.7 ft/s²

From $v = v_0 + at$, $v = 66$ ft/s, $v_0 = 0$, $t = 18$ s, and

$$a = \tfrac{66}{18} = 3.67 \text{ ft/s}^2$$

the final speed in miles per hour equals 66 ft/s × 3600 s/h × $\frac{1}{5280}$ mi/ft = 45 mi/h.

$$s = \tfrac{1}{2}at^2 = \tfrac{1}{2} \times 3.67 \times 18^2 = 594 \text{ ft}$$

The correct answer is (e).

3-3   RELATIVE VELOCITY

Velocity is defined as the rate of change of distance (from some arbitrarily selected point) with time, or $v = ds/dt$. Thus, a velocity possessed by an object must be in reference to some other object or reference system. That is, *all velocities are relative.*

▪ A runaway railroad car is moving along a (frictionless) horizontal stretch of track at a constant velocity of 30 mi/h. A locomotive starts out in pursuit and reaches a speed of 60 mi/h when it is exactly 1 mi behind the runaway car. The engineer (locomotive engineer, that is) starts to decelerate the engine at a constant rate so that when it touches the car to couple with it, the locomotive will be going at the same rate of speed as the car. Which of the following most nearly equals the required rate of acceleration?

(*a*) -400 m/h²   (*b*) -425 m/h²   (*c*) -450 m/h²
(*d*) -475 m/h²   (*e*) -495 m/h²

This problem can be worked by relating all velocities to the surface of the earth, but it is more straightforward and simple to use relative velocities. The initial relative velocity equals $60 - 30 = 30$ mi/h $= u$. (See Fig. 3-1.) The final relative velocity equals zero:

$$u^2 = u_0^2 + 2aS$$

$$0 = 30^2 + 2a \times 1 \qquad \text{giving } a = -450 \text{ mi/h}^2$$

$$S = u_0t + \tfrac{1}{2}at^2 \quad \text{or} \quad S = u_{\text{avg}}t$$

$$u_{\text{avg}} = \tfrac{30}{2} = 15 \text{ mi/h} \qquad S = 1 \text{ mi} \qquad t = \tfrac{1}{15} \text{ h}$$

so the runaway car would travel an additional 2.0 mi before being caught. The correct answer is (*c*).

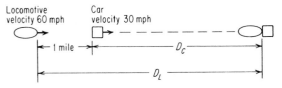

Locomotive
velocity 60 mph

Car
velocity 30 mph

Figure 3-1

3-4 FLIGHT OF A PROJECTILE

- If a bullet leaves a gun muzzle at a velocity of 2700 ft/s at an angle of 30° with the horizontal, which of the following most nearly equals the maximum height to which the bullet will rise?

(a) 28,000 ft (b) 26,000 ft (c) 24,000 ft (d) 22,000 ft
(e) 20,000 ft

First sketch the figure and determine the velocity components (Fig. 3-2). The vertical component of the velocity $v_y = v \sin \alpha$, and the component

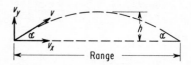

Figure 3-2

$v_x = v \cos \alpha$. In the vertical direction the bullet will be subjected to the acceleration of gravity, and if air resistance is disregarded, the acceleration will equal g. The time required for the bullet to reach its apogee can be determined from the fact that at the top the vertical velocity will equal zero. Then, from $v = v_0 + at$, we have $0 = v_y - gt$, or

$$\text{Time to apogee} = \frac{v_y}{g} = \frac{v \sin \alpha}{g}$$

The time for the bullet to fall back to the earth will equal the time to rise to the apogee, or the total time of flight will equal $2 \times [(v \sin \alpha)/g]$. The maximum height can be calculated from the relationship $v_{y,\text{avg}} \times t = h$, where $v_{y,\text{avg}} = (v_y + 0)/2$. The time t to rise to the height h has already been determined, and

$$h = \frac{v_y}{2} \frac{v_y}{g} = \frac{(v_y)^2}{2g}$$

The height h can also be determined from the relationship

$$v^2 = v_0^2 + 2as$$

which gives $(v_y)^2 = 2gh$, or $h = v_y^2/2g = (v^2 \sin^2 \alpha)/2g$. The range R will equal the distance traveled horizontally during the total time of flight at the constant horizontal velocity $v_x = v \cos \alpha$. Thus $R = v \cos \alpha \times (2v \sin \alpha)/g = 2[(v^2 \cos \alpha \sin \alpha)/g]$.

$$h = \frac{v^2 \sin^2 30°}{2g} = \frac{2700^2 \times 0.500^2}{2 \times 32.2} = 28,300 \text{ ft}$$

$$R = 2\frac{v^2 \cos \alpha \sin \alpha}{g} = 2\frac{2700^2 \times 0.866 \times 0.500}{32.2} = 196,000 \text{ ft}$$

The correct answer is (a).

3-5 MOMENTUM AND IMPULSE

The momentum of a body is equal to the product of its mass times its velocity, or $M = mv$. If no external force acts, the momentum of a system will remain constant. Since momentum is equal to $m \times v$, or is the product of a vector and a scalar, it is a vector quantity and thus must possess both magnitude and direction.

From $F = ma$ and $a = dv/dt$, we obtain $F\,dt = m\,dv$, from which we see that $F\,dt = mv_2 - mv_1$. The integral of $F\,dt$ is termed the "impulse," and the quantity $m(v_2 - v_1)$ is the change in momentum of mass m. This gives the relationship that impulse equals change in momentum, where impulse is also a vector quantity.

Take as an example a golf ball which is struck by a club. High-speed photography shows that the club is in contact with the ball for approximately one-half-thousandth of a second and the velocity of the ball is 250 ft/s immediately after being struck. What force is exerted by the club if the ball weighs $1\frac{2}{3}$ oz? Assume that the force is constant during the period of contact, so the impulse will equal $0.0005F$. The change in momentum of the golf ball will equal mv, or $[1.67/(16 \times 32.2)]250 = 0.810$ lb·s or 0.810 slug·ft/s. Since the change in momentum is equal to the impulse, $F = 0.810/0.0005 = 1620$ lb.

This problem could also have been worked using the relationships $F = ma$ and $a = \Delta v/\Delta t$, giving

$$F = \frac{1.67}{16 \times 32.2}\frac{250}{0.005} = 1620 \text{ lb}$$

- A plastic body weighing 100 lb is moving with a velocity of 20 ft/s and overtakes another plastic body, weighing 150 lb, moving in the same direction with a velocity of 15 ft/s. Which of the following most nearly equals the final velocity?

(a) 15 ft/s (b) 17 ft/s (c) 19 ft/s (d) 21 ft/s
(e) 23 ft/s

From the principle of the conservation of momentum, we have

$$m_1 v_1 + m_2 v_2 = (m_1 + m_2) v_r$$

The resulting velocity is

$$v_r = \frac{m_1 v_1 + m_2 v_2}{m_1 + m_2} = \frac{100 \times 20 + 150 \times 15}{100 + 150} = 17 \text{ ft/s}$$

The correct answer is (b).

The loss of kinetic energy is calculated as follows:

$$
\begin{aligned}
\text{KE initial} &= \tfrac{1}{2} m_1 v_1{}^2 + \tfrac{1}{2} m_2 v_2{}^2 \\
&= \frac{100}{2g} \times 20^2 + \frac{150}{2g} \times 15^2 \\
&= 1144 \text{ ft·lb} \\
\text{KE final} &= \tfrac{1}{2}(m_1 + m_2) v_r{}^2 \\
&= \frac{100 + 150}{2g} \times 17^2 \\
&= 1120 \text{ ft·lb}
\end{aligned}
$$

The loss in kinetic energy = 24 ft·lb.

Energy is not a vector quantity and direction has no bearing on the amount of kinetic energy possessed by a moving body. Momentum, on the other hand, is a vector quantity, and direction of motion is very important. Even though 24 ft·lb of mechanical energy has been lost in the impact, the total energy of the system is constant, and this lost mechanical energy will have been converted into heat.

The impact in this sample problem is a plastic impact or a completely inelastic impact with a coefficient of restitution equal to zero. If a small body rebounds from a very large one, such that the change of velocity of the large body is negligible, the coefficient of restitution $e = -(v_2/v_1)$. (Velocity is a vector quantity; thus v_2 will be opposite in sign to v_1.) If a ball rebounds or bounces up after being dropped onto a fixed plate, then

$$e = -\frac{v_2}{v_1} = \frac{\sqrt{2gh_2}}{\sqrt{2gh_1}} = \sqrt{\frac{h_2}{h_1}}$$

or the coefficient of restitution is equal to the square root of the height of bounce divided by the height of drop. If the impact is between two objects which do not differ greatly in mass, then

$$e = \frac{v_2 - u_2}{v_1 - u_1}$$

where

$$v_1 - u_1 = \text{relative velocity before impact}$$
$$v_2 - u_2 = \text{relative velocity after impact}$$

■ Assume that an 8-lb ball traveling with a velocity of 20 ft/s overtakes and strikes a 12-lb ball traveling in the same direction with a velocity of 8 ft/s. If the coefficient of restitution is 0.75, which of the following most nearly equals the velocity of the 12-lb ball after the impact?

(a) 8 ft/s   (b) 10 ft/s   (c) 12 ft/s   (d) 14 ft/s
(e) 16 ft/s

$$e = 0.75 = -\frac{v_2 - u_2}{20 - 8}$$

or

$$9 = u_2 - v_2$$

but also

$$m_1 v_1 + m_2 u_1 = m_1 v_2 + m_2 u_2$$

$$\frac{8}{g} \times 20 + \frac{12}{g} \times 8 = \frac{8}{g} v_2 + \frac{12}{g} u_2$$

$$32 = v_2 + 1.5u_2$$

$$9 = -v_2 + u_2$$

$$u_2 = 16.4 \text{ ft/s}$$

$$v_2 = 7.4 \text{ ft/s}$$

The correct answer is (e).

The linear momentum of any moving-mass system is the product of the mass of the whole system and the velocity of the mass center of the system, and its direction agrees with that of the velocity of the mass center.

The equations of equilibrium, $\Sigma F_x = 0$ and $\Sigma F_y = 0$, state that if the sum of the external forces acting on a system equals zero, the acceleration of the center of mass of that system equals zero. This means that if there is no external force acting on a system, the center of mass of the system remains at rest or continues its motion in a straight line at a constant speed.

Note that the forces considered are external forces; it can, therefore, be stated that the action of an internal force cannot result in a displacement of the center of mass of a system, or if no work is done upon (*not* inside) a system, the location of the center of mass of the system cannot change in a horizontal direction. As an example, a 200-lb person is standing at the rear of a 450-lb barge 18 ft in length. The front of the barge is 2 ft away from a pier, but the barge is floating freely in the water and is not tied to anything. Will the person be able to reach the pier by walking forward in the barge? Assume no frictional drag of any kind on the boat, i.e., no external force acting.

Since no external force is acting, the location of the center of mass of the system must remain constant. Then the moment of the system about point A must also remain constant (Fig. 3-3):

$$M_A = -11 \times 450 - 20 \times 200 = -8950 \text{ ft·lb}$$

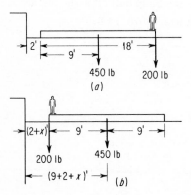

Figure 3-3

The moment of the system about point A will be the same after the person has walked the 18 ft to the front end of the barge, so the barge must move outward x ft:

$$-8950 = -450(11 + x) - 200(2 + x)$$

Thus $x = 5.54$ ft, or the barge will move 5.54 ft away from the pier. The person would then be 7.54 ft from the pier—a long reach; *or,* assuming that the individual moves and then the system (barge and person) moves, the 200-lb person moves 18 ft, so the 650-lb system must move (200/650) × 18 = 5.54 ft. Or, we could state that person impulse equals barge impulse; that is, $200(18 - x) = 450x$, giving $x = 5.54$ ft.

3-6 CONSERVATION OF ENERGY

Another important relationship, which has been mentioned before, is that of the conservation of energy. The work done on a system must equal the change in energy of the system plus the energy lost in the process (e.g., energy lost in overcoming friction).

Take the case of an object which slides down a chute onto a horizontal table, as shown in Fig. 3-4.

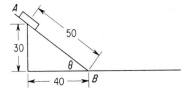

Figure 3-4

- Assume that the initial horizontal velocity is the same as that which is gained in sliding from A to B. Which of the following most nearly equals the distance S along the horizontal plane that the object will slide before coming to rest, if the coefficient of friction is 0.3?

(*a*) 25 ft   (*b*) 37 ft   (*c*) 48 ft   (*d*) 54 ft   (*e*) 60 ft

First find the answer using the velocity and acceleration relationships. The force parallel to the surface of the plane tending to cause the object to slide is $0.60W$. The resisting frictional force f is equal to the normal force $(0.80W)$ times the coefficient of friction, which gives $f = 0.24W$. The resultant force causing the object to accelerate down the plane is then $0.60W - 0.24W = 0.36W$. Since $F = ma$, the acceleration is $a = 0.36W/(W/g) = 0.36g$. The velocity at point B is then found to equal $\sqrt{2 \times 0.36g \times 50} = 6\sqrt{g}$, from the relationship $v^2 = v_0^2 + 2as$. The force acting on the object as it slides horizontally is equal to $-0.3W$, giving an acceleration of $-0.3g$. Since the final velocity will be zero,

$$S = \frac{v_0^2}{-2a} = \frac{36g}{0.6g} = 60 \text{ ft}$$

A second method of determining the desired distance is by means of the theory of the conservation of energy. In sliding down the plane the block will lose $30W$ ft·lb of potential energy. This must all be dissipated in work done in overcoming friction. The frictional work will equal $0.24W \times 50$ while sliding down the plane and $0.30W \times S$ while sliding along the horizontal. Then

$$30W = 0.24W \times 50 + 0.30W \times S$$

and $$S = \frac{30 - 12}{0.30} = 60 \text{ ft}$$

The correct answer is (*e*).

3-7 D'ALEMBERT'S PRINCIPLE

The subject of dynamics can be considerably simplified by the application of D'Alembert's principle. By this means problems in dynamics can be considered as problems in statics and can be treated by the methods of statics. D'Alembert's principle is based on the use of a fictitious "inertial force," which is equal and opposite to the force causing the acceleration.

An example is illustrated in Fig. 3-5. The block is pulled by a cord which goes over a frictionless, inertialess pulley and is then attached to a 25-lb weight, which is allowed to fall. The forces on both weights, including the inertial forces, are added, and the problem is seen to reduce to a problem in statics which can be solved by means of the relationship that the summation of all forces acting must equal zero.

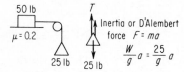

Figure 3-5

The force acting downward is the weight of the block, or 25 lb. This is balanced by two forces upward, the tension in the cord T and the D'Alembert or inertia force, which equals mass times acceleration, giving $T = 25 - (25/g)a$. The forces acting on the 50-lb block are added next, and since the tension in the cord is the same over its entire length, the force acting to cause motion of the 50-lb weight is equal to T. This is balanced by two resisting forces, the frictional force—$50 \times 0.2 = 10$ lb—and the D'Alembert force—$(50/g)a$, which gives $T = (50/g)a + 10$. Since $T = T$, $25 - (25/g)a = (50/g)a + 10$; $a = 0.2g$, and $T = 20$ lb.

3-8 ROTATIONAL MOTION

The concepts of linear velocity, displacement, and acceleration all have their exact counterparts in rotational motion. So, too, have the concepts of force and mass.

Linear concepts	Rotational concepts
Displacement s, ft	θ, rad
Velocity v, ft/s	ω, rad/s
Acceleration a, ft/s²	α, rad/s²
Force, lb	Torque, ft·lb
Mass $= W/g$, slugs	$I =$ moment of inertia, slug·ft²

The corresponding relationships are

$$s = v_0 t + \tfrac{1}{2}at^2 \qquad \theta = \omega_0 t + \tfrac{1}{2}\alpha t^2$$

$$v = v_0 + at \qquad \omega = \omega_0 + \alpha t$$

$$F = ma \qquad T = I\alpha$$

$$\mathrm{KE} = \tfrac{1}{2}mv^2 \qquad \mathrm{KE} = \tfrac{1}{2}I\omega^2$$

The relationships $T = I\alpha$ and $\mathrm{KE} = \tfrac{1}{2}I\omega^2$ are somewhat different, and the moment of inertia I is also different.

The "moment of inertia" I in rotational motion corresponds to the mass in linear motion and equals $\int r^2\, dm$.

The quantity "radius of gyration" of a system is sometimes given. This is the distance from the axis of rotation, or neutral axis, at which all the mass or area could be concentrated to give the same moment of inertia. The radius of gyration is usually indicated by k rather than r, giving $I = k^2 m$ or $I = k^2 A$.

The moment of inertia about an axis in the plane of an area is called the "plane moment of inertia"; the moment of inertia of an area about an axis perpendicular to the area is termed the "polar moment of inertia."

3-9 POLAR MOMENT OF INERTIA

The polar moment of inertia of an area with respect to any axis is equal to the sum of the moments of inertia of the area with respect to any two rectangular axes in the plane of the area which intersect on the given polar axis. Thus

$$J_z = \Sigma r^2\, \Delta A = \Sigma(x^2 + y^2)\, \Delta A$$

where J_z = polar moment of inertia about the perpendicular axis 0:

$$J_z = \int x^2\, dA + \int y^2\, dA = I_x + I_y$$

Application of this principle aids in simplifying calculation of some of the moments of inertia of the different areas. As an example, determine the moment of inertia of a circular section about the Y axis (Fig. 3-6):

$$I_y = \int_{-r}^{r} x^2\, dA = \int_{-r}^{r} x^2(2y\, dx) = \int_{-r}^{r} (r^2 - y^2)(2y\, dx)$$

which becomes involved.

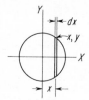

Figure 3-6

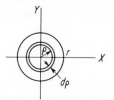

Figure 3-7

If we utilize the relationship that $J_z = I_y + I_x = 2I_y$, we can determine I_y much more easily (Fig. 3-7):

$$J_z = \int_0^r \rho^2 \, dA \qquad dA = 2\pi\rho \, d\rho$$

$$J_z = \int_0^r 2\pi\rho^3 \, d\rho = \frac{\pi r^4}{2}$$

3-10 PARALLEL-AXIS THEOREM

Calculation of the moment of inertia about an axis different from the moment of inertia about the centroid of an area can be done by means of the "parallel-axis" theorem. As an example, determine the moment of inertia of an area about an axis X-X given that the moment of inertia about an axis 0-0 through the centroid of the area equals I_0. Axis X-X is parallel with axis 0-0 (Fig. 3-8). Thus

$$I_x = \int_{-r}^r (s - r)^2 \, dA = \int_{-r}^r r^2 \, dA - \int_{-r}^r 2rs \, dA + \int_{-r}^r s^2 \, dA$$

$$= \int_{-r}^r r^2 \, dA - 2s \int_{-r}^r r \, dA + s^2 \int_{-r}^r dA = I_0 + As^2$$

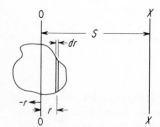

Figure 3-8

3-11 CENTRIPETAL FORCE

It can be shown that an object which is moving in a circular path has an acceleration toward the center of the circle of motion equal to the product of the square of the angular velocity times the radius.

Since we have an acceleration, if we also have a mass then we must have a force from $F = ma$, where the force equals $m \times \omega^2 r$ or $m(v^2/r)$. This force is exerted toward the center and is termed the "centripetal" force. It is balanced by an equal and opposite outward force termed a "centrifugal" force, centrifugal meaning "fleeing a center."

Two points should be noted in connection with centrifugal force. The first is that the forces and accelerations discussed are instantaneous; even if the path of the body is not a true circle, it will have an instantaneous center of curvature, and there will thus be an instantaneous centripetal acceleration (and an instantaneous centripetal force) toward this center of curvature which can be calculated from the relationships given. If the object is moving in essentially a straight line, then the distance to the center of curvature will be so large that the term v^2/r will be so small as to be negligible. The second point of importance is that if the body moving in a curved path is too large to be considered as a point, then the mass should be considered as concentrated at the center of mass and the location of the center of mass will determine the values of the radius of curvature and the tangential velocity.

▪ Which of the following most nearly equals the superelevation (slope of roadway) which would be required on a highway for 60 mi/h traffic on a 2500-ft-radius curve so that a 200-lb driver would exert no sideward force on the car seat?

(*a*) 2.5°   (*b*) 3.5°   (*c*) 4.5°   (*d*) 4.8°   (*e*) 5.5°

First, draw a figure showing the forces acting (Fig. 3-9). The centrifugal force equals $\omega^2 rm$, or $(v^2/r) \times m$, and the force of gravity acting downward equals $W = mg$:

$$\tan^{-1} \frac{(v^2/r)m}{mg} = \tan^{-1} \frac{v^2}{rg}$$

$$60 \text{ mi/h} = \frac{60 \times 5{,}280}{60 \times 60} = 88 \text{ ft/s}$$

$$\alpha = \tan^{-1} \frac{88^2}{2500\,g} = 5.5°$$

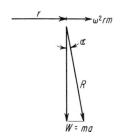

Figure 3-9

The superelevation of the roadway should be 5.5°. The correct answer is (e).

The maximum force which the driver would exert on the seat can also be calculated:

$$R = \sqrt{[(v^2/r)m]^2 + (mg)^2} = m\sqrt{9.6 + 1036.8}$$

$$= 32.35m = \frac{200}{g} 32.35 = 201 \text{ lb}$$

which is the force exerted by the driver perpendicular to the seat.

3-12 SATELLITE

▪ If the moon circles the earth once every 28 days, which of the following most nearly equals the approximate distance between the surface of the earth and the surface of the moon?

(a) 222,000 mi   (b) 234,000 mi   (c) 239,000 mi
(d) 244,000 mi   (e) 248,000 mi

The earth-moon system is in equilibrium; therefore, the forces acting on the moon are balanced. That is, the force acting to hold the moon away from the earth is exactly equal and opposite to the force attracting it toward the earth.

Simplify the problem slightly by making two assumptions: (1) the orbit of the moon about the earth is circular; (2) no influence is exerted on the moon by any members of our solar system other than the earth. These assumptions will affect the accuracy of our answer only slightly.

The two forces acting on the moon are (1) the centrifugal force outward and (2) the gravitational force toward the earth.

The centrifugal force equals $\omega^2 rm$, where individual quantities are as previously discussed. The speed of rotation is once in 28 days, so

$$\omega = \frac{2\pi}{28 \times 24 \times 3600} = 2.60 \times 10^{-6} \text{ rad/s}$$

The radius of rotation r is the distance between the center of the moon and the center of the earth. The centrifugal force is

$$F_c = 6.75 \times 10^{-12} \, rm_m$$

The gravitational attraction between two masses is equal to the gravitational constant times the product of the two masses divided by the square of the distance between the two centers of mass. This equates to

$$F_g = K \frac{m_m m_e}{r^2}$$

where

K = gravitational constant
m_m = mass of the moon
m_e = mass of the earth
r = distance between the two centers of mass

Gravitational force acting on an object at the earth's surface equals mg, where g is the acceleration of gravity. The radius of the earth is approximately 4000 mi:

$$F_g = m_0 g = m_0 32.2 = K \frac{m_e m_0}{(4000 \times 5280)^2}$$

$$K = \frac{1.44 \times 10^{16}}{m_e}$$

$$F_g = \frac{1.44 \times 10^{16}}{m_e} \frac{m_e m_m}{r^2} = \frac{1.44 \times 10^{16} m_m}{r^2}$$

The two forces F_g and F_c are equal. Equating these gives

$$\frac{1.44 \times 10^{16} m_m}{r^2} = 6.57 \times 10^{-12} r m_m$$

$$r = 1.286 \times 10^9 \text{ ft, or } 244{,}000 \text{ mi}$$

Subtract from this the radius of the earth (4000 mi) and the radius of the moon (1080 mi) to obtain 238,920 mi as the distance between the surfaces of the earth and the moon. This is very close to the reported value of 238,857 mi for the mean distance between the earth and the moon. The correct answer is (c).

■ An armature is keyed to a shaft 4 in in diameter, which rests transversely on two parallel steel rails inclined to the horizontal with a slope of 1:12. The shaft rolls, without slipping and without rolling friction, a distance of 6 ft along the rails from rest in 1 min. Which of the following most nearly equals the radius of gyration of the armature and its shaft about the longitudinal axis?

(a) 4.9 ft   (b) 4.7 ft   (c) 4.5 ft   (d) 4.3 ft   (e) 4.1 ft

Make a sketch (Fig. 3-10). The energy possessed by the armature at the end of 1 min equals $\frac{1}{2}mv^2 + \frac{1}{2}I\omega^2$:

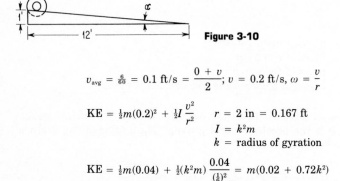

Figure 3-10

$$v_{\text{avg}} = \tfrac{6}{60} = 0.1 \text{ ft/s} = \frac{0 + v}{2}; \, v = 0.2 \text{ ft/s}, \, \omega = \frac{v}{r}$$

$$\text{KE} = \tfrac{1}{2}m(0.2)^2 + \tfrac{1}{2}I\frac{v^2}{r^2} \qquad \begin{aligned} r &= 2 \text{ in} = 0.167 \text{ ft} \\ I &= k^2 m \\ k &= \text{radius of gyration} \end{aligned}$$

$$\text{KE} = \tfrac{1}{2}m(0.04) + \tfrac{1}{2}(k^2 m)\frac{0.04}{(\frac{1}{6})^2} = m(0.02 + 0.72k^2)$$

This increase in kinetic energy must equal the decrease in potential energy, since no work is done against friction, $\alpha = \tan^{-1}(\frac{1}{12})$, giving $\sin \alpha = 0.083$.

$$\varDelta\text{PE} = W \sin \alpha \times 6 = mg \times 0.083 \times 6 = 16.05m$$
$$16.05m = m(0.02 + 0.72k^2)$$
$$k = \sqrt{\frac{16.05 - 0.02}{0.72}} = 4.72 \text{ ft}$$

The correct answer is (b).

The same result can be obtained by determining the acceleration of the armature (Fig. 3-11):

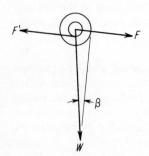

Figure 3-11

$$s = v_0t + \tfrac{1}{2}at^2$$

$$6 = 0 + \tfrac{1}{2}a(60)^2 \qquad a = \tfrac{1}{300} \text{ ft/s}^2$$

$$F = W \sin \beta = 0.083W$$

$$T_q = I\alpha = I\frac{a}{r} = F'r = Fr$$

since there is no slipping and $F' = F$. We further obtain

$$Fr = I\frac{a}{r} = k^2m\frac{a}{r}$$

$$mg \sin \beta r = k^2m\frac{a}{r}$$

$$k^2 = \frac{0.083gr^2}{a} = \frac{0.083g(\tfrac{1}{8})^2}{\tfrac{1}{300}} = 22.3$$

$$k = 4.72 \text{ ft radius of gyration}$$

A word might be in order here about the summation of the forces acting on the armature. The force F (Fig. 3-12) is just equal to the frictional

Figure 3-12

force f and is balanced by it. There is, then, no excess force to cause purely linear acceleration of the armature as a whole, and the total acceleration is angular.

If F exceeded f and the armature slipped as well as rolled, then there would be a linear acceleration in addition to that resulting from the angular acceleration.

■ A 64.4-lb cylinder of radius $R = 2$ ft is pushed by a moving bulldozer as shown in Fig. 3-13. The coefficient of friction at points A and B

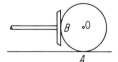

Figure 3-13

68   **Dynamics**

is 0.20. Which of the following most nearly equals the maximum acceleration of the bulldozer if the cylinder is to roll without slipping?

(*a*) 28.6 ft/s²   (*b*) 29.6 ft/s²   (*c*) 30.8 ft/s²   (*d*) 31.6 ft/s²
(*e*) 32.2 ft/s²

Refer to the sketch in Fig. 3-13. The maximum frictional force at point *A* will equal 0.20*W*. The net torque acting on the cylinder will equal $f_A R - f_B R$, so if $f_B > f_A$ (Fig. 3-14), the cylinder will not roll, and if $f_B < f_A$, the cylinder will roll. If $f_B = f_A$, then the inertial force, or D'Alembert's

f_A **Figure 3-14**

force, must equal $W = mg = ma$, so if the cylinder is to move to the right without rotating, the acceleration must be greater than *g*, and if the cylinder is to roll without sliding, the acceleration must be less than *g*. The maximum permissible acceleration of the bulldozer for the cylinder to roll without slipping would thus equal 32.2 ft/s². The correct answer is (*e*).

3-13 SPRINGS

The energy required to compress a spring a distance *dx* equals *F dx*, where the force will be equal to *kx*; *k* is the spring constant, and *x* is the distance the spring has been compressed. The total work to compress the spring a distance *S* will then equal $\int_0^S kx \, dx$, which equals $\frac{1}{2}kS^2$.

- A mass with a weight of 25 lb falls a distance of 5 ft onto a helical compression spring which has a spring constant of 20 lb/in. Which of the following most nearly equals the velocity of the mass after it has compressed the spring 8 in?

  (*a*) 15.1 ft/s   (*b*) 14.9 ft/s   (*c*) 14.7 ft/s   (*d*) 14.5 ft/s
  (*e*) 14.3 ft/s

The fall of the weight will be 5 ft plus 8 in when it has compressed the spring 8 in. At that time the spring will have absorbed energy equal to $\frac{1}{2}kS^2 = \frac{1}{2}(20 \text{ lb/in}) \times (8 \text{ in})^2 = 640 \text{ lb·in}$, or 53.3 lb·ft. The change in potential energy equals

$$Wh = 25(5 + 0.67) = \Delta\text{PE} = 142 \text{ ft·lb}$$

Part of this energy has been absorbed by the spring; the remainder, $142 - 53.3 = 88.7$ ft·lb, must remain as kinetic energy of the mass $= \frac{1}{2}mv^2$:

$$v = \sqrt{\frac{88.7 \times 2}{m}} = \sqrt{\frac{88.7 \times 2g}{25}} = 15.1 \text{ ft/s}$$

The correct answer is (a).

3-14 SIMPLE HARMONIC MOTION

Simple harmonic motion (SHM) is defined as the motion of a point in a straight line such that the acceleration of the point is proportional to the distance of the point from its equilibrium position, or $a = -kx = d^2x/dt^2$.

Simple harmonic motion can be shown to be the same motion to which a suspended mass is subjected by the force of a spring (Fig. 3-15). The

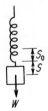

S_0
S
W **Figure 3-15**

static position of the weight is that position in which the spring is stretched the distance S_0. Then $S_0 k = W$, where k is the spring constant. If the weight is pulled down farther, a distance S, then the sum of the forces acting on the weight will equal

$$k(S_0 + S) - W = kS$$

Since the distance S is down and the force is up, $F = -kS$, giving $ma = -kS$, or $a = (-k/m)S$, where k/m is a constant, and we have the defining relationship for simple harmonic motion.

The velocity and time relationships for simple harmonic motion are usually determined by means of a "reference circle."

It can be shown that if a point moves with constant speed in a circular path, the motion of the projection of the point on a diameter of this reference circle is a simple harmonic motion (Fig. 3-16):

$$x = r \cos \theta = r \cos \omega t$$

Figure 3-16

where ω = a constant:

$$v_p = \frac{dx}{dt} = -\omega r \sin \omega t = -\omega y$$

$$a_p = \frac{dv_p}{dt} = -\omega^2 r \cos \omega t = -\omega^2 x$$

This last equation is the equation defining simple harmonic motion. The time for P to make one complete oscillation is the same as the time for the point M to make one complete revolution, or $T = 2\pi/\omega$, and the frequency equals $1/T = \omega/2\pi$.

- A truck body lowers by 5 in when a load of 25,000 lb is placed on it. Which of the following most nearly equals the frequency of vibration if the total spring-borne weight is 40,000 lb?

   (a) 0.9/s   (b) 1.0/s   (c) 1.1/s   (d) 1.2/s   (e) 1.3/s

The frequency = $1/T = \omega/2\pi$, and the acceleration of the weight $a = -\omega^2 x$, but $F = ma$, and the force acting on the mass $= -kx$, so $-kx = -\omega^2 mx$, and $\omega = \sqrt{k/m} = \sqrt{kg/W}$; then

$$f = \frac{\omega}{2\pi} = \frac{1}{2\pi} \sqrt{\frac{kg}{W}}$$

The spring constant $k = 25{,}000 \text{ lb}/5 \text{ in} = 5000 \text{ lb/in}$:

$$f = \frac{1}{2\pi} \sqrt{\frac{5000 \text{ lb/in} \times 12 \text{ in/ft} \times 32.2 \text{ ft/s}^2}{40{,}000 \text{ lb}}} = 1.105 \text{ oscillations/s}$$

The restoring force acting on the truck body equals $-kx$, where x is the distance the spring is compressed past the neutral point (Fig. 3-17). The correct answer is (c).

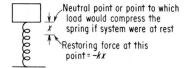

Neutral point or point to which
load would compress the
spring if system were at rest

Restoring force at this
point = $-kx$

Figure 3-17

▪ If the lever in the system shown in Fig. 3-18 is assumed to be weightless, which of the following most nearly equals the frequency with

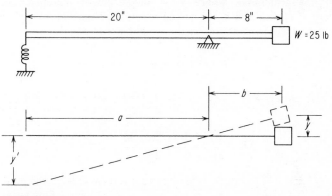

Figure 3-18

which the weight would move up and down after it has been displaced vertically? (The spring constant equals 5 lb/ft.)

(a) 0.9/s (b) 1.0/s (c) 1.1/s (d) 1.2/s (e) 1.3/s

This is an example of undamped free vibration with a single degree of freedom. For small displacements the system will vibrate in conformance with the relationship for simple harmonic motion in the same way as the suspended mass shown in Fig. 3-18. Thus the motion will satisfy the differential equation

$$\frac{d^2y}{dt^2} + \frac{k}{m}y = 0$$

The general solution of this equation gives $\omega^2 = k/m$, where ω is measured in radians per second. A circular function repeats itself in 2π rad, so one cycle of vibratory motion is completed when $\omega\tau = 2\pi$, where τ is the period of motion, or the time required for a single cycle. The frequency, in cycles per second, would equal $1/\tau$, or $f = \omega 2\pi$.

In addition, $d^2y/dt^2 = $ acceleration. From Newton's second law, $F = ma$, we obtain $a = F/m$. From this $d^2y/dt^2 = F/m$, which, in turn, gives

$$\frac{F}{m} + \frac{k}{m}y = 0 \quad \text{or} \quad F = -ky$$

where the minus sign is explained by the fact that the force is opposite in sense to the displacement of y.

The force acting on the weight (see Fig. 3-18) would equal $-(a/b)(k_{sp}y')$ (for small displacements), where $y' = $ displacement (extension or compression) of spring and $k_{sp} = $ spring constant. Then $b/y = a/y'$ so $y' = (a/b)y$, and the force acting on the weight would equal

$$F = -\left(\frac{a}{b}\right)^2 k_{sp}y$$

But force also equals $F = -ky$, so $k = (a/b)^2 k_{sp}$, $\omega^2 = k/m$, and $f = \omega/2\pi$.

Then
$$f = \frac{\sqrt{k/m}}{2\pi} = \frac{\sqrt{(a/b)^2 k_{sp}/m}}{2\pi}$$

$$\left(\frac{a}{b}\right)^2 = \left(\frac{20}{8}\right)^2 = 6.25 \qquad m = 25/32.2 = 0.776 \text{ slugs}$$

$$f = \frac{\sqrt{6.25 \times 5/0.776}}{2\pi} = 1.01 \text{ cycles per second}$$

The correct answer is (b).

Sample Problems

3-1 A chain hangs over a smooth peg, with 8 ft of its length on one side and 10 ft on the other. If the force of friction is equal to the weight of one foot of the chain, which of the following most nearly equals the time for the chain to slide off?

(a) 3.6 s   (b) 1.4 s   (c) 4.8 s   (d) 1.9 s   (e) 2.6 s

3-2 A falling weight of 1000 lb is used to drive a pile into the ground. If the weight of the pile is 800 lb and the weight is dropped 20 ft, which of the following most nearly equals the depth of penetration of the pile, assuming an average resistance to penetration of 30,000 lb? Assume the impact between weight and pile to be perfectly inelastic.

(*a*) 4.7 in   (*b*) 7.2 in   (*c*) 3.1 in   (*d*) 6.3 in   (*e*) 5.7 in

3-3 A horizontal force is applied to a block moving in a horizontal guide. When the block is at a distance x from the origin, the force is given by the equation $F = x^3 - x$. Which of the following most nearly equals the work that is done by moving the block from 1 ft to the left of the origin to 1 ft to the right of the origin?

(*a*) 6 ft·lb   (*b*) 4 ft·lb   (*c*) 2 ft·lb   (*d*) 0   (*e*) 3 ft·lb

3-4 In the system shown in Fig. 3P-4, $W = 20$ lb and is observed to vibrate 10 times in 15 s. Disregard the weight of the lever arm and find which of the following most nearly equals the stiffness coefficient of the spring S.

(*a*) 1.85 lb/ft   (*b*) 2.72 lb/ft   (*c*) 3.87 lb/ft   (*d*) 4.11 lb/ft
(*e*) 5.24 lb/ft

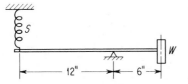

Figure 3P-4

3-5 An unbalanced force in pounds that varies with time in seconds ($F = 800 - 120t$) acts on a body with a weight of 3220 lb for 5 s. If the body has an initial velocity of 10 ft/s in the direction of the unbalanced force, which of the following most nearly equals the final velocity that the body would have?

(*a*) 15 ft/s   (*b*) 20 ft/s   (*c*) 25 ft/s   (*d*) 30 ft/s   (*e*) 35 ft/s

3-6 A wheel 12 ft in diameter weighing 64.4 lb has a radius of gyration of 9 in. If it starts from rest and rolls without slipping down a 30° plane, which of the following most nearly equals its speed when it has moved 20 ft?

(*a*) 10.2 ft/s   (*b*) 15.7 ft/s   (*c*) 20.3 ft/s   (*d*) 26.2 ft/s
(*e*) 30.2 ft/s

3-7 A rectangular door swings horizontally, is 3 ft wide, and weighs 60 lb uniformly distributed. The door is supported by frictionless hinges installed along one of its vertical sides. It is controlled by a spring which exerts on it a torque proportional to the angle through which it is turned from its closed position. If the door is opened 90° and released, the torque due to the spring is then 3 lb·ft. Which of the following most nearly equals the angular velocity at the time it has swung back and is passing through the 45° position?

(*a*) 0.8 rad/s   (*b*) 0.7 rad/s   (*c*) 0.9 rad/s   (*d*) 0.6 rad/s
(*e*) 1.0 rad/s

3-8 A body is moving in a straight line according to the law $S = \frac{1}{4}t^4 - 2t^3 + 4t^2$ where S = distance traveled in time t. Which of the following most nearly equals the time interval during which it is moving backward?
(a) 0 to 2 s   (b) 3 to 5 s   (c) 1 to 3 s   (d) 2 to 4 s   (e) 4 to 6 s

3-9 A locomotive is traveling around a curve of mean radius 1000 ft at a speed of 40 mi/h. The superelevation of the track is such that there is no side thrust on the flanges of the wheels at that speed. If the gauge of the railway track is 5 ft and the height of the center of gravity of the locomotive is 4 ft above rail level, which of the following most nearly equals the greatest speed at which the locomotive could travel around the curve without actually turning over?
(a) 110 mi/h   (b) 97 mi/h   (c) 88 mi/h   (d) 75 mi/h
(e) 68 mi/h

3-10 A belt 8 in wide and $\frac{1}{4}$ in thick drives a pulley 12 in in diameter at 1650 r/min. The maximum tension the belt may have is 200 lb/in^2. If the slack side has three-fourths the tension of the tight side, which of the following most nearly equals the horsepower that may be transmitted?
(a) 14.2 hp   (b) 16.7 hp   (c) 18.6 hp   (d) 17.2 hp   (e) 15.7 hp

3-11 A hoisting cable has a safe working strength of 40,000 lb/in^2 of cross-sectional area. Which of the following most nearly equals the greatest weight that can be raised with an acceleration of 8.05 ft/s^2 by a cable 1.5 in in diameter? Disregard friction.
(a) 57,000 lb   (b) 68,000 lb   (c) 42,000 lb   (d) 87,000 lb
(e) 71,000 lb

3-12 A kite is 120 ft high with 150 ft of cord out. If the kite moves horizontally 4 mi/h directly away from the boy flying it, which of the following most nearly equals the rate at which the cord is being paid out in feet per second?
(a) 6.8 ft/s   (b) 7.8 ft/s   (c) 3.5 ft/s   (d) 2.8 ft/s   (e) 4.7 ft/s

3-13 A bullet leaves a gun muzzle at a velocity of 2700 ft/s at an angle of 45° with the horizontal. Which of the following most nearly equals the maximum distance that the bullet will travel horizontally, measured along the same elevation as the gun muzzle? Disregard air resistance.
(a) 214,000 ft   (b) 247,000 ft   (c) 237,000 ft   (d) 227,000 ft
(e) 254,000 ft

3-14 An airplane has a true air speed of 240 mi/h and is heading true north. A 40 mi/h head wind is blowing 45° from true north from the northeast. Which of the following most nearly equals the true ground speed?
(a) 252 mi/h   (b) 206 mi/h   (c) 241 mi/h   (d) 227 mi/h
(e) 214 mi/h

3-15 The rifling in a 30-caliber rifle causes the bullet to turn 1 r per 10 in of barrel length. If the muzzle velocity of the bullet is 2900 ft/s, which of the following most nearly equals the speed at which the bullet is rotating at the instant that it leaves the muzzle?
(a) 208,000 r/min   (b) 372,000 r/min   (c) 196,000 r/min
(d) 247,000 r/min   (e) 276,000 r/min

3-16 A faster-moving body is directly approaching a slower-moving body traveling in the same direction. The speed of the faster body is 60 mi/h and the speed of the slower body is 35 mi/h. At the instant when the separation of the two bodies is 100 ft, the faster body is given a constant deceleration. Which of the following most nearly equals the acceleration so that the two bodies just touch at the instant of impact with no shock?
(a) $-g$ (b) $-0.2g$ (c) $-0.5g$ (d) $-0.4g$ (e) $-0.6g$

3-17 A parachutist is falling at a speed of 176 ft/s when the parachute opens. If the air resistance is $Wv^2/256$ lb, where W is the total weight of the person and the parachute, which of the following most nearly equals the time needed to reduce the speed of the parachutist to 20 ft/s?
(a) 1.5 s   (b) 2.3 s   (c) 0.9 s   (d) 0.5 s   (e) 1.8 s

3-18 At the beginning of the drive, a golf ball has a velocity of 170 mi/h. If the club remains in contact with the ball for $\frac{1}{25}$ s, which of the following most nearly equals the average force on the ball? The weight of the ball is 1.62 oz.
(a) 20.3 lb   (b) 32.2 lb   (c) 19.6 lb   (d) 15.4 lb   (e) 17.9 lb

3-19 An artificial satellite is circling the earth at a radius of 4500 mi. Assume that the pull of gravitation is the same at that elevation as at the earth's surface. Which of the following most nearly equals the speed required to maintain the satellite at that radius?
(a) 24 r/day   (b) 6 r/day   (c) 12 r/day   (d) 30 r/day
(e) 16 r/day

3-20 The fire door shown in Fig. 3P-20 is suspended from a horizontal track by means of wheels at points A and B. The door weighs 200 lb. The wheel at A has rusted in its bearing and slides, rather than rolls, on the track, and the coefficient of friction is one-third. The wheel at B rolls freely, and friction is negligible. Which of the following most nearly equals the force P necessary to give the door an acceleration of 8 ft/s?
(a) 107 lb   (b) 82 lb   (c) 97 lb   (d) 156 lb   (e) 74 lb

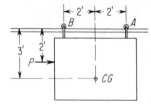

Figure 3P-20

3-21 A truck of dimensions and weights shown in Fig. 3P-21 is rolling across a 30-ft bridge. Which of the following most nearly equals the maximum moment?
(a) 134 ton·ft   (b) 256 ton·ft   (c) 106 ton·ft   (d) 168 ton·ft
(e) 187 ton·ft

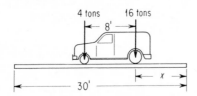

Figure 3P-21

3-22 A balloon is ascending vertically at a uniform rate for 1 min. A stone falls from it and reaches the ground in 5 s. Which of the following most nearly equals the height from which the stone fell?

(a) 371.0 ft   (b) 371.2 ft   (c) 371.4 ft   (d) 371.6 ft   (e) 371.8 ft

3-23 A body takes just twice as long to slide down a plane at 30° to the horizontal as it would if the plane were smooth. Which of the following most nearly equals the coefficient of friction?

(a) 0.433   (b) 0.500   (c) 0.667   (d) 0.245   (e) 0.317

3-24 Which of the following most nearly equals the distance from the center of a phonograph record turning at 78 r/min that a pickle can lie without being thrown off, if the coefficient of friction is 0.30?

(a) 3.0 in   (b) 2.7 in   (c) 1.9 in   (d) 1.7 in   (e) 1.4 in

3-25 Block A (Fig. 3P-25) weighs 200 lb and is placed 5 ft from the edge of the table. A string passing over a smooth pulley connects blocks A and B as shown. Block B weighs 50 lb and is 2 ft from the floor. The coefficient of sliding friction for block A is 0.15. Block B is released from rest in this position. Which of the following most nearly equals the distance from the edge of the table at which block B will come to rest?

(a) 3.2 ft   (b) falls off   (c) 2.6 ft   (d) 1.1 ft   (e) 1.9 ft

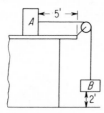

Figure 3P-25

3-26 A cast-aluminum drum, 2 ft in diameter and 1 ft in length is supported by an axial shaft on smooth bearings. A light cord wrapped around the drum is pulled with a constant force of 20 lb until 15 ft of cord is unwrapped and then released. Which of the following most nearly equals the speed of the drum when the pull on the cord has ceased?

(a) 98.6 r/min   (b) 80.5 r/min   (c) 78.0 r/min   (d) 67.2 r/min
(e) 50.7 r/min

3-27 A 50-ton car moving at a speed of 3 mi/h strikes a bumping post equipped with a 40,000 lb/in spring. Assuming that all the shock is absorbed by the spring, which of the following most nearly equals the amount that the spring will be compressed?

(*a*) 3.75 in (*b*) 5.82 in (*c*) 4.91 in (*d*) 4.25 in (*e*) 4.08 in

3-28 An automobile weighing 3700 lb approaches a curve on a concrete-paved highway. The radius of the curve is 1000 ft. If the coefficient of friction between the tires and the pavement is 0.15, which of the following most nearly equals the transverse slope of the pavement required to keep the resultant forces perpendicular to the pavement if the car is moving at a speed of 60 mi/h? Assume that the pavement is 24 ft wide and that the distance between treads is 6 ft.

(*a*) 6.2° (*b*) 8.7° (*c*) 10.2° (*d*) 12.8° (*e*) 13.5°

3-29 An elastic body weighing 100 lb is moving with a velocity of 20 ft/s and overtakes another elastic body weighing 150 lb and moving in the same direction with a velocity of 15 ft/s. Which of the following most nearly equals the loss in kinetic energy which results from the impact?

(*a*) 17.8 ft·lb (*b*) 23.4 ft·lb (*c*) 26.7 ft·lb (*d*) 29.8 ft·lb
(*e*) 19.2 ft·lb

3-30 A body weighing 200 lb starts from rest at the top of a plane which makes an angle of 60° with the horizontal. After sliding 8 ft 4 in down the plane, the body strikes a 100 lb/in spring. The coefficient of friction is 0.25. The body remains in contact with the plane throughout. Which of the following most nearly equals the compression of the spring?

(*a*) 18.76 in (*b*) 17.32 in (*c*) 19.46 in (*d*) 15.27 in
(*e*) 14.96 in

3-31 A 100-lb wheel 18 in in diameter, and rotating at 150 r/min in stationary bearings is brought to rest by pressing a brake shoe radially against its rim with a force of 20 lb. If the radius of gyration of the wheel is 7 in and the coefficient of friction between the brake shoe and the wheel rim is steady at 0.25, which of the following most nearly equals the number of revolutions the wheel will make before coming to rest?

(*a*) 10.2 (*b*) 8.7 (*c*) 6.4 (*d*) 5.5 (*e*) 4.9

3-32 A solid cylinder 4 ft in diameter and weighing 12,880 lb is rolling at a linear speed of 40 ft/s. Which of the following most nearly equals the distance that it will roll up a 10 percent grade before coming to a stop?

(*a*) 373 ft (*b*) 375 ft (*c*) 381 ft (*d*) 369 ft (*e*) 383 ft

3-33 The friction surface on a friction disk type of clutch has an outside diameter of 10 in and an inside diameter of 3 in. The coefficient of friction for the surfaces in contact may be assumed to equal 0.30. Which of the following most nearly equals the pressure which must be applied to the plate if the clutch is to transmit 100 hp at 3300 r/min?

(*a*) 22 lb/in² (*b*) 25 lb/in² (*c*) 27 lb/in² (*d*) 31 lb/in² (*e*) 29 lb/in²

3-34 For the conditions shown in Fig. 3P-34, which of the following most nearly equals the weight W in pounds if the acceleration of the 100-lb body is 10 ft/s² and the coefficient of sliding friction is 0.5? The pulleys are assumed to be frictionless.

(*a*) 417 lb   (*b*) 387 lb   (*c*) 325 lb   (*d*) 318 lb   (*e*) 307 lb

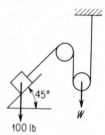

100 lb   *W*   **Figure 3P-34**

3-35 A motor car has a maximum speed of 70 mi/h on the level, and the engine is then transmitting 60 hp to the road wheels. If the tractive effort at the wheels is assumed to remain constant while the resistance to motion varies as the square of the speed, which of the following most nearly equals the slope (l in x) up which the car can maintain a speed of 50 mi/h if the weight of the car is 4000 lb?

(*a*) 0.068 ft/ft   (*b*) 0.078 ft/ft   (*c*) 0.082 ft/ft   (*d*) 0.072 ft/ft
(*e*) 0.062 ft/ft

3-36 To measure the muzzle velocity of a bullet fired from a rifle, two cardboard disks mounted on a long axle are rotated at a constant speed. The rifle is mounted parallel to the axle. Upon firing, the bullet passes through each of the disks in turn. In a test, the disks were 3.5 ft apart on the axle, the speed of rotation was 3000 r/min, and the bullet hole in the farther disk was displaced 18° with respect to the hole in the nearer disk. Which of the following most nearly equals the muzzle velocity?

(*a*) 2500 ft/s   (*b*) 2700 ft/s   (*c*) 3100 ft/s   (*d*) 3300 ft/s
(*e*) 3500 ft/s

3-37 A 3-oz bullet is fired horizontally into an 18-lb block suspended vertically so that it can swing as a pendulum. If the center of gravity of the block is caused to rise 8 in, which of the following most nearly equals the speed of the bullet?

(*a*) 635 ft/s   (*b*) 722 ft/s   (*c*) 839 ft/s   (*d*) 921 ft/s   (*e*) 587 ft/s

3-38 Referring to Fig. 3P-38, CD is turning clockwise about D at 60 r/min. Which of the following most nearly equals the tangential velocity of point B for the position shown?

(*a*) 10 ft/s   (*b*) 11 ft/s   (*c*) 12 ft/s   (*d*) 13 ft/s   (*e*) 14 ft/s

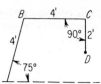

Figure 3P-38

3-39 A steel plate 30 in in diameter and 4 in thick has six 3-in-diameter holes equally spaced on a 20-in-diameter circle (Fig. 3P-39). The disk is rotating about an axis through its center at 300 r/min. Which of the following most nearly equals the kinetic energy stored in the disk?
(a) 8750 ft·lb   (b) 9470 ft·lb   (c) 7380 ft·lb   (d) 8280 ft·lb
(e) 9110 ft·lb

Figure 3P-39

3-40 A 16-lb weight is fastened to the end of a rope wound around a cylindrical drum which is free to turn in a frictionless bearing about an axis through its center (Fig. 3P-40). The drum is 1 ft in diameter and has a moment of inertia of 1.875 slug·ft². Which of the following most nearly equals the kinetic energy of the cylinder when the weight has fallen 16 ft?
(a) 180 ft·lb   (b) 210 ft·lb   (c) 240 ft·lb   (d) 275 ft·lb
(e) 292 ft·lb

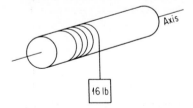

Figure 3P-40

3-41 A loaded elevator weighing 10,000 lb strikes a safety device after a 2-s free fall down the shaft. If the safety device applies a braking force of 15,000 lb, which of the following most nearly equals the total distance traveled, including the fall and deceleration to a stop?
(a) 193 ft   (b) 178 ft   (c) 207 ft   (d) 162 ft   (e) 198 ft

3-42 A drop hammer of 1 ton deadweight capacity is propelled downward by a 12-in-diameter air cylinder. At 100 lb/in^2 air pressure, which of the following most nearly equals the impact velocity if the stroke is 28 in?
(a) 18.2 ft/s   (b) 27.9 ft/s   (c) 38.2 ft/s   (d) 31.6 ft/s
(e) 29.8 ft/s

3-43 A 10-hp 1750 r/min motor is connected directly to a 30 lb·ft torque brake. The rotating system has a total inertia (WR^2) of 10 lb·ft^2. Which of the following most nearly equals the length of time that it will take for the system to stop if the brake is set at the instant the motor is shut off?
(a) 3.6 s   (b) 4.2 s   (c) 1.9 s   (d) 2.2 s   (e) 0.8 s

3-44 A train traveling at 75 mi/h receives a warning that it is approaching a "slow" signal, and the (locomotive) engineer immediately reduces speed at a uniform rate of 1.8 ft/s^2 to achieve a speed of 15 mi/h as he reaches the slow signal. The slow order covers a stretch of 5 mi of track that is being reconstructed. Upon reaching the end of the slow order, the engineer increases speed at such a uniform rate that the train is traveling 85 mi/h 90 s later. Which of the following most nearly equals the length of time that the train will have to travel at 85 mi/h in order to make up the time lost by not running continuously at 75 mi/h?
(a) 1.7 h   (b) 1.9 h   (c) 2.1 h   (d) 2.3 h   (e) 2.5 h

3-45 A 30-ton freight car with a velocity of 4 mi/h in coupling collides with a stationary 20-ton car. Assume that 25 percent of the energy is lost in impact and failure to couple. Which of the following most nearly equals the velocity of the 30-ton car after the collision?
(a) 2.1 ft/s   (b) 2.7 ft/s   (c) 1.9 ft/s   (d) 2.6 ft/s   (e) 1.6 ft/s

3-46 Which of the following most nearly equals the minimum speed that a ball starting at point P must have to loop the loop in the vertical track shown in Fig. 3P-46 without leaving the track? Assume frictionless contact.
(a) 34.3 ft/s   (b) 35.6 ft/s   (c) 36.8 ft/s   (d) 38.1 ft/s
(e) 39.1 ft/s

Figure 3P-46

3-47 Utilizing the same data as given in problem 3-46, which of the following most nearly equals the minimum speed if friction acts between the ball and the vertical circular track (i.e., if the ball rolls)?
(a) 34.3 ft/s   (b) 35.6 ft/s   (c) 36.8 ft/s   (d) 38.1 ft/s
(e) 39.1 ft/s

3-48 A mass with a weight of 25 lb falls a distance of 5 ft onto a helical compression spring which has a spring constant of 20 lb/in. Which of the following

most nearly equals the velocity of the mass after it has compressed the
spring 8 in?
(*a*) 14.7 ft/s   (*b*) 15.1 ft/s   (*c*) 16.3 ft/s   (*d*) 17.2 ft/s
(*e*) 15.6 ft/s

3-49 A circular cylinder as shown in Fig. 3P-49 rolls on a horizontal floor. At a
given instant the angular acceleration equals 8 rad/s² and the angular
velocity equals 6 rad/s. Assume that there is no slipping and that points *A*
and *C* are on a horizontal line. Which of the following most nearly equals
the acceleration of point *A* at this instant?
(*a*) 114 ft/s²   (*b*) 134 ft/s²   (*c*) 141 ft/s²   (*d*) 127 ft/s²
(*e*) 121 ft/s²

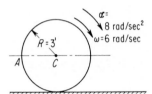

Figure 3P-49

3-50 A child weighing 100 lb is jumping up and down on a pogo stick which has
a 100 lb/in spring. The child jumps so that the lower end of the pogo stick
is 3 ft off the ground at maximum height. At the bottom of the cycle, the
child pushes down with 25 lb more than is required to stop the downward
motion in order to overcome friction and continue jumping. Which of the
following most nearly equals the number of inches that the spring is compressed at the bottom of the cycle?
(*a*) 6.0 in   (*b*) 7.5 in   (*c*) 9.8 in   (*d*) 10.2 in   (*e*) 12.1 in

3-51 A cylinder (*A*) weighing 100 lb is placed in a box (*B*) (Fig. 3P-51), the bottom
of which is inclined 30° to the horizontal. The box is accelerated to the right
at 20 ft/s². Which of the following most nearly equals the force F_1?
(*a*) 4.72 lb   (*b*) 6.48 lb   (*c*) 7.92 lb   (*d*) 5.14 lb   (*e*) 3.86 lb

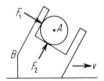

Figure 3P-51

3-52 The wheel *A* in Fig. 3P-52 is a solid cylinder weighing 1000 lb, with a
diameter of 8 ft. It is free to turn on a frictionless axle. It is desired to
arrange a brake as shown, by means of which the speed of the wheel may
be reduced from 120 r/min to zero in 10 s. Which of the following most nearly

82 Dynamics

equals the required ratio b/c of the available force, assuming that $P = 100$ lb and the coefficient of brake friction $= 0.25$?

(a) 2.125　　(b) 1.375　　(c) 2.250　　(d) 3.062　　(e) 1.875

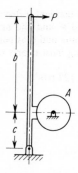

Figure 3P-52

3-53 Two weights, A and B, are suspended by a cord over a pulley as shown in Fig. 3P-53. Assume that the pulley and the cord are weightless and frictionless and that the weights are permitted to shift from the positions indicated by gravity. If A weighs 500 lb and B weighs 750 lb, which of the following most nearly equals the velocity of A when one weight hits the floor?

(a) 8.1 ft/s　　(b) 7.6 ft/s　　(c) 10.2 ft/s　　(d) 11.3 ft/s　　(e) 9.8 ft/s

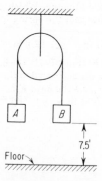

Figure 3P-53

3-54 One end of a ladder 50 ft long is leaning against a perpendicular wall standing on a horizontal plane. Suppose that the foot of the ladder is to be pulled away from the wall at the rate of 3 ft/min. Which of the following most nearly equals the rate of speed at which the top of the ladder is descending when the foot is 14 ft from the wall?

(a) 0.624 ft/min　　(b) 0.750 ft/min　　(c) 0.875 ft/min
(d) 1.025 ft/min　　(e) 0.982 ft/min

Multiple-Choice Sample Quiz

For each question select the correct answer from the five given possibilities. For practice, try to complete the 10 problems below in 12 minutes or less.

3M-1 A 16-lb weight and an 8-lb weight resting on a horizontal frictionless surface are connected by a cord, A, and are pulled along the surface with a uniform acceleration of 4 ft/s² by a second cord, B, attached to the 16-lb weight. The tension in cord A is closest to:
(a) 4 lb   (d) 1 lb
(b) 3 lb   (e) 2 lb
(c) 32 lb

3M-2 An impulse is the product of:
(a) force and displacement
(b) force and time
(c) force and velocity
(d) mass and acceleration
(e) mass and velocity

3M-3 A crane is lowering a 2-ft-diameter ball which weighs 1000 lb on the end of a weightless cable at a uniform velocity of 12 ft/s. Assuming no friction, the total tension in the cable is:
(a) 750 lb   (d) 900 lb
(b) 800 lb   (e) 1000 lb
(c) 850 lb

3M-4 The kinetic energy of a 2400-lb automobile traveling at 100 ft/s is closest to:
(a) 3.75×10^5 ft·lb
(b) 7.5×10^5 ft·lb
(c) 24×10^6 ft·lb
(d) 12×10^6 ft·lb
(e) 3.75×10^6 ft·lb

3M-5 A 10-in-diameter pulley is belt-driven with a net torque of 250 ft·lb. The ratio of tensions in the tight to slack sides of the belt is 4:1. What is the maximum tension in the belt?
(a) 250 lb   (d) 500 lb
(b) 83 lb   (e) 333 lb
(c) 800 lb

3M-6 The second derivative of distance with respect to time is:
(a) speed
(b) velocity
(c) force
(d) distance
(e) acceleration

3M-7 Centrifugal force *is not* a function of:
   (*a*) angular velocity
   (*b*) radius
   (*c*) mass
   (*d*) moment of inertia
   (*e*) any of these

3M-8 Kinetic energy has units of:
   (*a*) ft·lb   (*d*) lb·ft/s^2
   (*b*) ft·lb/s   (*e*) Btu/h
   (*c*) lb/in^2

3M-9 A newton is the force required to give:
   (*a*) 1 kg an acceleration of 1 m/s^2
   (*b*) 1 kg a velocity of 1 m/s
   (*c*) 1 kg a velocity of 1 cm/s
   (*d*) 1 g an acceleration of 1 cm/s^2
   (*e*) 1 g a velocity of 1 cm/s

3M-10 A flower pot is inadvertently dropped out of a second-story window 14 ft above the pavement. What is its velocity when it strikes the ground?
   (*a*) 30.0 ft/s   (*d*) 33.6 ft/s
   (*b*) 32.2 ft/s   (*e*) 27.8 ft/s
   (*c*) 28.1 ft/s

4

MECHANICS OF MATERIALS

The subject of mechanics of materials is an expansion of the more general subject of mechanics; it includes study of the effects of the elastic properties of structural materials. All materials deform when subjected to a stress. Fortunately, most of the important engineering materials exhibit unit deformations, or strains, which are very nearly proportional to the applied stresses within the working limits of the materials. This fact, expressed in the statement of Hooke's law that stress and strain are proportional, is the basis of the study of mechanics of materials.

4-1 EXAMINATION COVERAGE

Mechanics of materials is covered in the morning session and is also one of the required subjects in the afternoon session as a part of the general subject of engineering mechanics. In addition, it is one of the five optional subjects as mechanics of materials.

The subdivisions of the major subject which may be covered, and which should be reviewed, include the following:

Stress and strain

Tension

Shear, both transverse and parallel to the axis of a beam

Combined stresses

Beams

Columns

Composite sections
Uniaxial loading
Torsion
Bending
Combined loading
Thermal stress
Shear and moment diagrams

4-2 MATERIAL PROPERTIES

The ratio of stress (in pounds per square inch) to strain (in inches per inch), called "Young's modulus" or the "modulus of elasticity," is constant for a given material up to the proportional limit. The modulus of elasticity, the proportional limit, the yield point, etc., are all determined experimentally for a given material. Curves illustrating these properties for some of the common engineering materials are shown in Fig. 4-1.

When a metal deforms slowly under the long-time application of a stress which is well below its yield strength, the deformation is known as "creep." A typical curve showing deformation of a material vs. time under load is given in Fig. 4-2.

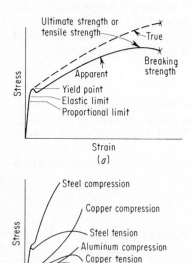

Figure 4-1 Stress-strain diagrams. (a) Stress-strain diagram for mild steel; (b) Typical stress-strain curves.

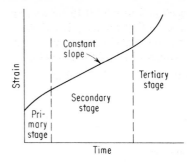

Figure 4-2 Idealized creep curve.

"Fatigue" is the term used to describe the failure or rupture of a metal part under repeated application of a load which is well below the permissible load as calculated from ordinary static stress considerations. Figure 4-3 shows some curves of stress vs. number of cycles to cause failure for a few engineering materials. For most steels the endurance limit, or fatigue stress, is that stress which will withstand 10 million cycles without breaking. For nonferrous metals it may be necessary to flex the metal through many more than 10 million cycles to determine the limiting stress at which rupture will occur, and some metals may fail to show any endurance limit at all.

Notch sensitivity, or the stress concentration factor, is also extremely important in the design of members subject to cyclic loading. Any discontinuity or change of section, e.g., a hole, a groove, a notch, or a bend, is a stress raiser.

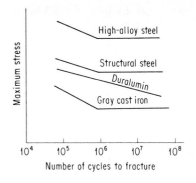

Figure 4-3 Typical S-N (stress-cycle) diagrams.

4-3 STRESS AND STRAIN

The general relationship of $E = S/\delta$ is all that is necessary to solve many of the simpler problems in mechanics of materials.

▪ A steel rod 10 ft long has a cross-sectional area of 0.40 in² and is carrying an axial load of 6000 lb. Assume that the modulus of elasticity for this steel is 30,000,000 lb/in². The total elongation produced in the rod by this load most nearly equals which of the following?

(a) 0.040 in   (b) 0.050 in   (c) 0.060 in   (d) 0.070 in
(e) 0.080 in

For a 6000-lb load the stress would be

$$\frac{6000}{0.40 \text{ in}^2} = 15{,}000 \text{ lb/in}^2$$

$$\delta = \frac{S}{E} = \frac{15{,}000}{30{,}000{,}000} = 5 \times 10^{-4} \text{ in/in}$$

The total elongation would then equal

$$(10 \times 12) \text{ in} \times (5 \times 10^{-4}) \text{ in/in} = 0.060 \text{ in}$$

The correct answer is (c).

▪ Two parallel wires 6 in apart and 2 ft long, one of copper and one of steel, are used to support a load $P = 28$ lb (see Fig. 4-4). The cross-sectional area of the copper wire is twice the area of the steel wire. The distance x from the steel wire to the point of application of the load should most nearly equal which of the following distances so that the wires will remain equal in length? (Given: $E_s = 30 \times 10^6$ lb/in² and $E_c = 10 \times 10^6$ lb/in².)

(a) 3.0 in   (b) 2.8 in   (c) 2.6 in   (d) 2.4 in   (e) 2.3 in

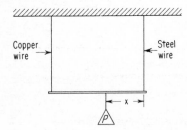

Copper wire

Steel wire

x

P

Figure 4-4

Since $\delta = S/E$ and δ_s is to equal δ_c, $S_s/E_s = S_c/E_c$, which means that the stress in the steel wire must be three times the stress in the copper wire for the lengths of the two to remain equal. The load on the steel wire will equal $S_s A_s$. Applying the ratios given, $S_s = 3S_c$ and $A_s = \frac{1}{2}A_c$, the steel wire must support $\frac{3}{2}$ as much load as the copper wire for the lengths of the two to remain the same. From $\Sigma F_y = 0$, we find that the copper wire will have to support $\frac{2}{5}$ of P and the steel wire $\frac{3}{5}$ of P. Taking moments about S gives $6 \times \frac{2}{5}P = Px$, or $x = 2.4$ in. The correct answer is (d).

- The lower chord in a panel of a bridge truss consists of three eyebars, each 8 by 1 in in cross section. The two outside bars are each 20 ft 0 in long, center to center of pinholes, and the middle bar is 20 ft $\frac{1}{8}$ in long. The total tension in the three bars is 360,000 lb. Disregard the bending of the pins, and assume that they are parallel to each other and fit loosely in the pinholes before a load is applied. If the modulus of elasticity $E = 30 \times 10^6$ lb/in^2, which of the following most nearly equals the maximum stress in the bars?

  (a) 4600 lb/in^2 (b) 30,400 lb/in^2 (c) 18,600 lb/in^2
  (d) 20,200 lb/in^2 (e) 36,800 lb/in^2

First, draw a figure (Fig. 4-5). Since the two outside bars are identical, the elongation in both will be the same. The inner bar is $\frac{1}{8}$ in longer, so the total elongation of it will be $\frac{1}{8}$ in less. The stress in the two outer bars will be $E\delta_1$, and the stress in the inner bar will be $E\delta_2$. The total load will equal the sum of the products of the stresses times the areas stressed, or

$$360,000 = (8 \times 1)E\delta_1 + (8 \times 1)E\delta_1 + (8 \times 1)E\delta_2$$

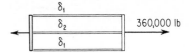

Figure 4-5

which gives $360,000/8E = 2\delta_1 + \delta_2$, where δ_1 and δ_2 are in inches per inch or feet per foot. Because of the different lengths of the bars and since the total elongation equals length times δ_1, we have

$$20 + 20\delta_1 = 20.0104 + 20.0104\delta_2 \qquad (20.0104 \text{ ft} = 20 \text{ ft } \tfrac{1}{8} \text{ in})$$

which gives $\delta_1 = \delta_2 + 0.00052$. We then have

$$\frac{360,000}{8 \times 30 \times 10^6} = 0.0015 = 2\delta_1 + \delta_2$$

and $\delta_1 = 0.000673$ ft/ft or in/in; $\delta_2 = 0.000153$ ft/ft or in/in. The stress in the two outer bars is $S = E\delta_1 = 0.000673 \times 30 \times 10^6 = 20{,}200$ lb/in², and the stress in the inner bar is $0.000153 \times 30 \times 10^6 = 4600$ lb/in². The correct answer is (d).

4-4 THERMAL STRESS

When the effects of temperature (thermal expansion) are included, the picture is changed a little. In most cases it will simplify the calculations if strain due to stress and the strain due to temperature are calculated separately and are then added algebraically to obtain the net or final result. This can best be illustrated by means of examples.

■ A 3-ft-long copper bar having a circular cross section 1 in in diameter is arranged as shown in Fig. 4-6 with a 0.001-in gap between its end and the rigid wall at room temperature. If the temperature increases 60°F, the stress in the rod will most nearly equal which of the values listed below? Assume that the coefficient of thermal expansion for copper is 9.3×10^{-6} in/(in)(°F), the modulus of elasticity is 17×10^6 lb/in², and the bar does not buckle.

  (a) 9000 lb/in² (b) 7500 lb/in² (c) 10,400 lb/in²
  (d) 11,200 lb/in² (e) 8200 lb/in²

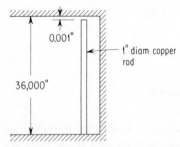

Figure 4-6

The unrestrained increase in length due to the increase in temperature would have been

$$[9.3 \times 10^{-6} \text{ in/(in)(°F)}] \times 60°F \times 36 \text{ in} = 0.0201 \text{ in}$$

The bar can expand 0.001 in and then will be restrained. This is the same (stresswise) as if it had been allowed to increase 0.0201 in and was then compressed 0.0191 in. The resulting stress would be $S = E\delta$:

$$S = 17 \times 10^6 \text{ lb/in}^2 \times 0.0191 \text{ in}/36 \text{ in} = 9020 \text{ lb/in}^2$$

The correct answer is (a).

- A steel liner is assembled in an aluminum pump housing by heating the housing and cooling the liner in dry ice. At 70°F before assembly, the liner outside diameter is 3.508 in and the housing inside diameter is 3.500 in. After assembly and inspection, several units were rejected because of poor liners. It is desired to salvage the housing by heating the complete unit to a temperature which would cause a difference (clearance) in diameter of 0.002 in between liner and housing and permit free removal of the liner. If the coefficient of thermal expansion is 0.0000065 in/(in)(°F) for steel and 0.000016 in/(in)(°F) for aluminum, which of the following most nearly equals the temperature to which the assembly would have to be heated?

  (a) 300°F (b) 320°F (c) 400°F (d) 385°F (e) 370°F

At 70°F the interference is 0.008 in. A clearance of 0.002 in is desired, so the expansion of the aluminum housing must be 0.010 in more than the expansion of the steel liner, or the unit expansion of the aluminum must be $0.010/3.5 = 0.00286$ in/in greater than that of the steel. The differential expansion is $0.000016 - 0.0000065 = 9.5 \times 10^{-6}$ in/(in)(°F). The required temperature increase would then be

$$\frac{0.00286 \text{ in/in}}{9.5 \times 10^{-6} \text{ in/(in)(°F)}} = 301°F$$

so the assembly would have to be heated to a temperature of 371°F. The correct answer is (e).

4-5 TORSION

Pure torsional loading (couple only, no bending) produces a shearing stress in a shaft with the magnitude of the stress, in any cross section, being proportional to the distance from the center of the shaft (Fig. 4-7). The polar moment of inertia of the area is ordinarily represented by the symbol J. This gives the relationship that torque $= (S_s/r)J$, and the maximum stress in the shaft due to the applied torque is $S_s = rT/J$. The

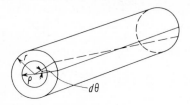

Figure 4-7

stress at any other distance ρ from the center of the shaft equals

$$\frac{\rho}{r}\frac{rT}{J} = \frac{\rho T}{J}$$

Within the proportional limit, $\delta_s = S_s/E_s$, which gives the relationship $S_s = E_s\delta_s = (E_s r\theta)/l$, since, from Fig. 4-8, $\delta_s = r\theta/l$. This also gives $T = (E_s J\theta)/l$.

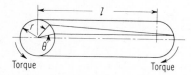

Torque   Torque   **Figure 4-8**

■ A solid round shaft $3\frac{1}{2}$ in in diameter transmits 100 hp at 200 r/min. Which of the following most nearly equals the increase in the stress in the shaft if the shaft were made hollow with a $1\frac{1}{2}$ in ID?

(a) 1050 lb/in²   (b) 970 lb/in²   (c) 840 lb/in²
(d) 520 lb/in²   (e) 130 lb/in²

The power transmitted by a shaft equals $2\pi NT/33{,}000$ (Fig. 4-9), which means $T = (33{,}000 \times 100)/(2\pi \times 200) = 2630$ ft·lb, or 31,500 in·lb. The polar moment of inertia of a circle is $(\pi r^4)/2$, so $S_s = (r \text{ in} \times T \text{ in·lb})/(J \text{ in}^4) = 2T/(\pi r^3)$:

$$S_s = \frac{2 \times 31{,}500}{\pi(1.75)^3} = 3740 \text{ lb/in}^2$$

Figure 4-9 Torque $= Tr = T$ ft·lb; constant velocity, so peripheral speed $= 2\pi N$ ft/min ($N = $ r/min); power $=$ distance $\times$ force/time $= 2\pi NT$ ft·lb/min.

For a hollow shaft of the same outer diameter but with an inner diameter of $1\frac{1}{2}$ in, we obtain

$$J = \frac{\pi(r_{\text{OD}})^4}{2} - \frac{\pi(r_{\text{ID}})^4}{2} = \frac{\pi}{2}(1.75^4 - 0.75^4) = 14.25 \text{ in}^4$$

$$S_s = \frac{1.75 \times 31{,}500}{14.25} = 3870 \text{ lb/in}^2 \qquad \text{stress in outer fiber}$$

Increase of stress $= 3870 - 3740 = 130 \text{ lb/in}^2$. The correct answer is (e).

- Referring to Fig. 4-10, which of the values listed below most nearly equals the torque M_B which should be applied to the shaft combination shown in the figure to produce the same maximum unit shear stress in both parts? Disregard any effects of stress concentration.

  (a) 5900 in·lb (b) 6200 in·lb (c) 4700 in·lb
  (d) 5200 in·lb (e) 5600 in·lb

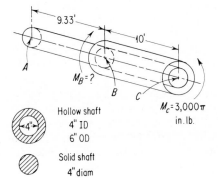

Hollow shaft
4" ID
6" OD

Solid shaft
4" diam

$M_c = 3{,}000\pi$
in. lb.

Figure 4-10

The torque $M_c = 3000\pi$ in·lb. The stress in the outer fibers of the shaft BC may be found from the equation

$$S_s = \frac{rT}{J} = \frac{3 \times 3000\pi}{(\pi/2)(3^4 - 2^4)} = 277 \text{ lb/in}^2$$

To produce the same stress in the solid portion of the shaft would require a torque of

$$T = \frac{JS_s}{r} = \frac{(\pi/2)(2^4)277}{2} = 1110\pi \text{ in·lb}$$

so torque in the amount of $3000\pi - 1110\pi$, or 1890π in·lb, must be taken off at B, giving $M_B = 1890\pi$ in·lb, which equals 5938 in·lb. The correct answer is (a).

▪ For the given applied torque, which of the following most nearly equals the angle of twist of the length BC in Fig. 4-10?

    (a) 0.0009°   (b) 0.05°   (c) 0.007°   (d) 0.08°   (e) 0.02°

The angle of twist in length BC may be determined from the relationship $T = (E_s J\theta)/l$, or

$$\theta = \frac{Tl}{JE_s} = \frac{3000\pi(10 \times 12)}{(\pi/2)(3^4 - 2^4)12 \times 10^6} \frac{\text{in·lb} \times \text{in}}{\text{in}^4 \times \text{lb/in}^2} = 0.000923 \text{ rad}$$

and 0.000923 rad $\times$ 57.3°/rad $= 0.0529°$. The correct answer is (b).

4-6 BEAMS

The general relationship for determining the stress in a beam due to bending is $S = (Mc)/I$, where $M =$ moment (in inch-pounds), $I =$ moment of inertia of the cross section, and $c =$ distance from the neutral axis to the outermost fiber. The stress at any other point than the outer fiber is proportional to its distance from the center and would be $S = (y/c)(Mc/I) = (My)/I$.

The transverse shear stress at any point is equal to the magnitude of the shearing force divided by the cross-sectional area at that point. It is seldom possible to determine by a glance just where the maximum shear and the maximum moment will occur in a loaded beam, so it is usually desirable to construct both the shear diagram and the moment diagram for the beam under consideration; then the points of maximum shear and maximum moment will be readily apparent.

4-7 SHEAR DIAGRAM

The shear at any section of a beam is the algebraic sum of all the external forces on *either* side of the section; it is considered positive if the segment of the beam on the left of the cross section tends to move up with respect to the segment on the right. The algebraic sum of all the forces on *both* sides of the section will, of course, equal zero, since the system is in static equilibrium. It is usually simplest to add all the shearing forces to the left of the section being considered; this sum is equal to the shearing force acting at that point. If this process is repeated for different points over the length of the beam, a complete shear diagram may be constructed.

4-8 MOMENT DIAGRAM

The moment diagram may be constructed in a similar manner; one must remember that the bending moment at any section of a beam is the algebraic sum of the moments of all the external forces on either side of the section. A positive moment is one which tends to cause the beam to be concave on the upper side (top fibers in compression).

Examples of shear and moment diagrams for a few of the common types of loading are shown in Fig. 4-11.

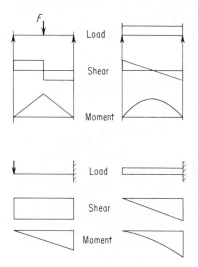

Figure 4-11 Shear and moment diagrams for some of the more common types of loading.

For combined loadings the shear and moment diagrams may be constructed by adding (algebraically) the individual shear and moment diagrams due to the individual loadings. This will give the total values for shear and moment for all parts of the beams. Each load will produce the same effect as if it had acted alone; it is unaffected by the other loads. This is illustrated in Fig. 4-12.

There are a few relationships which are of value in constructing and in checking shear and moment diagrams. One of these is that the derivative of the moment M with respect to distance is equal to the shear V, or $V = (dM)/(dx)$. This means that the slope of the moment curve at

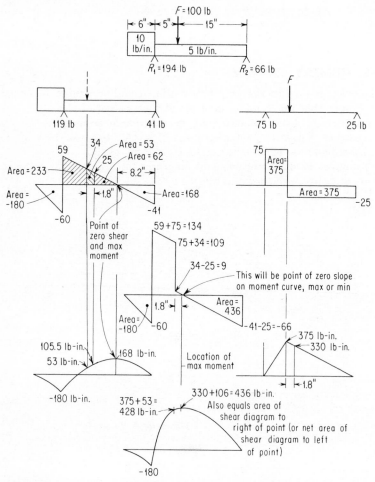

Figure 4-12

any point is equal to the magnitude of the shear at that point. The relationship may also be written

$$\int_{M_1}^{M_2} dM = \int_1^2 V\, dx \qquad \text{or} \qquad M_2 - M_1 = \int_1^2 V\, dx$$

which means that the difference between the values of the moments at points 1 and 2 is equal to the area under the shear diagram between points 1 and 2. Similarly, we have $W = (dV)/(dx)$, where W is the load at any point and $V_2 - V_1 = \int_1^2 W\, dx$. The use of these relationships is illustrated in Fig. 4-12.

- A wooden beam is made up of two timbers, one 4 by 6 in and one 4 by 8 in, as shown in Fig. 4-13. If this beam is loaded uniformly in a simple span 32 ft long, which of the following is the maximum total load that it could hold if the bending stress is limited to 1200 lb/in²?

  (*a*) 5880 lb   (*b*) 6220 lb   (*c*) 5090 lb   (*d*) 4870 lb
  (*e*) 5210 lb

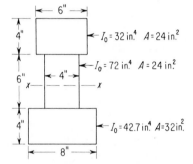

Figure 4-13

The location of the horizontal neutral axis can be obtained by taking moments about the lower edge:

$$\bar{y} = \frac{(4 \times 8) \times 2 + (6 \times 4)(4 + 3) + (4 \times 6)(4 + 6 + 2)}{(4 \times 8) + (6 \times 4) + (4 \times 6)}$$

Here, $\bar{y} = 6.5$ in from the bottom of the beam and 7.5 in from the top. Using the parallel-axis theorem and $I_{\text{rect}} = (bh^3)/12$, we obtain

$$I_{x-x} = 32 + 24 \times 5.5^2 + 72 + 24 \times 0.5^2 + 42.7 + 32 \times 4.5^2 = 1527 \text{ in}^4$$

and, further, $S = Mc/I$, $M_{\max} = (wl^2)/8$, and $c = 7.5$. Thus

$$\text{Total load} = wl = \frac{SI}{c}\frac{8}{l} = \frac{1200 \times 1527 \times 8}{7.5(32 \times 12)} = 5090 \text{ lb}$$

for a compressive stress in the topmost fibers of 1200 lb/in².

If the tensile stress (or bending stress) in the lower fibers should control (1200 lb/in² tensile stress),

$$wl = \frac{1200 \times 1527 \times 8}{6.5(32 \times 12)} = 5880 \text{ lb}$$

The correct answer is (c).

- For the beam loaded as shown in Fig. 4-14, which of the following values most nearly equals the required section modulus of the beam if the maximum stress is limited to 20,000 lb/in²?

  (a) 11.5 in³   (b) 12.3 in³   (c) 10.1 in³   (d) 14.2 in³
  (e) 13.1 in³

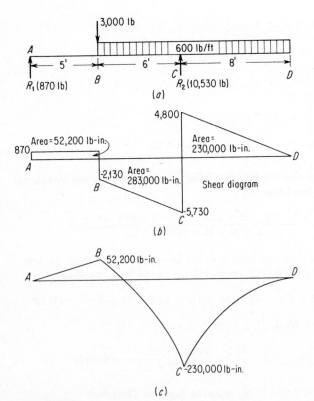

Figure 4-14

The shear and moment diagrams can be constructed as shown in Fig. 4-14. Referring to the figure, the reaction R_2 may be found by taking moments about R_1: R_2 = 10,530 lb. Taking moments about R_2 gives R_1 = 870 lb.

The shear and moment diagrams are drawn directly below the given figure. The maximum moment is $-230,000$ lb·in. Thus

$$S = \frac{Mc}{I} = \frac{M}{Z}$$

where $Z = I/c$, section modulus:

$$Z = \frac{M}{S} = \frac{230,000 \text{ lb·in}}{20,000 \text{ lb/in}^2} = 11.5 \text{ in}^3$$

The correct answer is (a).

The four important points on the beam have been labeled A, B, C, and D. Note in the shear diagram that the slope from A to B is zero since the load on the load diagram from A to B is zero. At B is a concentrated load, and the shear changes perpendicularly at that point. From B to C the load is constant and the slope angle on the shear diagram is also constant. At point C the load changes abruptly because of the reaction R_2, and the shear also changes by the same amount. From C to D the load is constant, and the slope of the shear diagram is also constant. (Remember that loads on the *left* in an *upward* direction give a positive shear and loads *down* on the *right* give a positive shear.)

Looking at the moment diagram, we see that the slope at any point equals the magnitude of the shear diagram at that point. Thus the slope is constant from A to B and makes an abrupt change at B to a steadily increasing negative slope. It makes another abrupt change at C from a large negative slope to a large positive slope which decreases to no slope (tangent to abscissa) at D.

- A cantilever beam 12 ft long is fixed at the left end and carries a uniformly distributed weight of 200 lb/lin ft and a concentrated load of 2000 lb at a point 8 ft from the right end. Which of the following most nearly equals the moment acting on the beam at a point 4 ft from the point of support?

   (a) 76,800 lb·in   (b) $-192,000$ lb·in   (c) 192,000 lb·in
   (d) $-76,800$ lb·in   (e) $-268,800$ lb·in

This problem can be worked directly from the load diagram, but for the purpose of illustration the shear and moment diagrams will be constructed and the answer drawn from the moment diagram. The shear and moment diagrams with the accompanying calculations are shown in Fig. 4-15. The required moment, as shown in the figure, is $-76,800$ lb·in. The correct answer is (d).

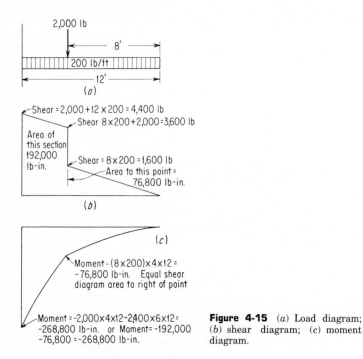

Figure 4-15 (a) Load diagram; (b) shear diagram; (c) moment diagram.

4-9   BEAM SLOPE AND DEFLECTION

The basic relationship for determining deflection or slope is $EI(d^2y/dx^2) = M$, to a close approximation. The radius of curvature $R = EI/m$.

■ An I beam 14 in deep on a 20-ft span is simply supported at the ends. It has a moment of inertia of 440 in⁴. Assume that $E = 30 \times 10^6$. Which of the following most nearly equals the load which could be suspended midway between the supports without producing a deflection of more than $\frac{1}{4}$ in? Ignore the weight of the beam.

(a) 14,200 lb (b) 15,800 lb (c) 13,600 lb (d) 12,900 lb
(e) 11,500 lb

Referring to Fig. 4-16, take the right-hand half of the beam, considering

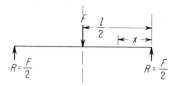

Figure 4-16   Taking right half of beam and considering center point stationary, right end free to move up.

the center point as stationary, and the right end as being free to move up. All motions are relative, so it is as valid to consider that the right end is free to move up as it is to consider the center point free to move down. From these considerations, the following relationships can be developed:

$$M_x = \frac{F}{2}x \qquad EI\frac{dy}{dx} = \int M\,dx = \int \frac{F}{2}x\,dx = \frac{Fx^2}{4} + C_1$$

$$\frac{dy}{dx} = 0 \text{ at } x = \frac{l}{2} \qquad C_1 = \frac{-Fl^2}{16}$$

$$EIy = \int \frac{F}{4}x^2\,dx - \int \frac{Fl^2}{16}\,dx = \left(\frac{Fx^3}{12}\right) - \left(\frac{Fl^2}{16}\right)x + C_2$$

$$y = 0 \text{ at } x = \frac{l}{2} \qquad C_2 = \frac{-Fl^3}{48}$$

$$y = \left(\frac{F}{EI}\right)\left(\frac{x^3}{12} - \frac{l^2x}{16} + \frac{l^3}{48}\right) \qquad y_{max} = \frac{Fl^3}{48EI} \text{ at } x = 0$$

Next, draw a figure (Fig. 4-17). Disregarding the weight of the beam, the magnitude of the desired load can be obtained as follows:

$$F = \frac{48EIy}{l^3} = \frac{48 \times 30 \times 10^6 \times 440 \times 0.25}{(20 \times 12)^3} = 11{,}460 \text{ lb}$$

The correct answer is (e).

- Which of the following most nearly equals the stress which would occur in the beam as a result of applying the calculated load at the center point of the beam (Fig. 4-17)?

  (a) 14,600 lb/in²   (b) 16,200 lb/in²   (c) 8700 lb/in²
  (d) 11,000 lb/in²   (e) 10,600 lb/in²

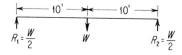

Figure 4-17

Utilizing the data from the previous example, we obtain

$$S = \frac{Mc}{I} \qquad M = \frac{11,460}{2}(10 \times 12) = 688,000 \text{ lb·in}$$

$$c = \frac{14}{2} = 7 \text{ in} \qquad S = \frac{688,000 \times 7}{440} = 10,950 \text{ lb/in}^2$$

The correct answer is (d).

- For the beam described above, which of the following most nearly equals the total load, which if uniformly distributed over the length of the beam, would produce the same deflection of $\frac{1}{4}$ in?

  (a) 18,300 lb   (b) 17,600 lb   (c) 11,500 lb   (d) 19,400 lb
  (e) 12,300 lb

For a uniformly distributed load, $y_{max} = (5wl^4)/(384EI)$, which gives

$$w = \frac{384EIy}{5l^4} = \frac{384 \times 30 \times 10^6 \times 440 \times 0.25}{5(20 \times 12)^4} = 76.4 \text{ lb/in}$$

The total uniformly distributed load would then equal

$$76.4 \text{ lb/in} \times (20 \times 12) \text{ in} = 18,300 \text{ lb}$$

The correct answer is (a).

- A simply supported beam 10 ft long carries three equal concentrated loads located 3, 5, and 7 ft from the left end. Which of the following most nearly equals the ratio of the maximum deflection to the deflection under the load located 3 ft from the left end?

  (a) 1.12   (b) 1.24   (c) 1.33   (d) 1.67   (e) 1.40

Construct a figure (Fig. 4-18). The loading is symmetrical about the center, so the maximum deflection will occur at the center. The maximum deflection will equal the sum of the deflections at the center of the beam due to each of the three individual loads. Each of these three deflections will be the same as if each load acted by itself.

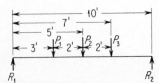

Figure 4-18

The deflection at the center due to loads P_1 and P_3 can be calculated from the equation

$$\Delta_{max} = \frac{Pa}{24EI}(3l^2 - 4a^2)$$

where $a = 36$ in and $l = 120$ in. This gives

$$\Delta_{max} = \frac{57,000P}{EI}$$

The deflection at the center of the beam due to P_2 is

$$\Delta_{max} = \frac{Pl^3}{48EI} = 36,000\frac{P}{EI}$$

The total deflection at the center of the beam would then be equal to $93,000P/EI$, which is the sum of the partial deflections due to loads P_1, P_2, and P_3.

The deflection under the load located 3 ft from the left end would also be made up of two components. The deflection due to the two equal symmetrically applied loads P_1 and P_3 would equal

$$\frac{Pa}{6EI}(3la - 4a^2) = 46,600\frac{P}{EI}$$

The deflection at the same point due to the force P_2 would equal

$$\frac{Pa}{48EI}(3l^2 - 4a^2) = 28,500\frac{P}{EI}$$

The total deflection under the left load would then equal $75,100P/EI$, and the required ratio would be 1.24. The correct answer is (b).

4-10 STATICALLY INDETERMINATE BEAMS

- The beam illustrated in Fig. 4-19 has a 10 WF 54 section. Which of the following values most nearly equals the maximum bending stress in the beam?

  (a) 17,000 lb/in^2 (b) 14,500 lb/in^2 (c) 12,300 lb/in^2
  (d) 10,400 lb/in^2 (e) 9000 lb/in^2

This is an example of a statically indeterminate beam. It is called "statically indeterminate" because the number of unknown reactions

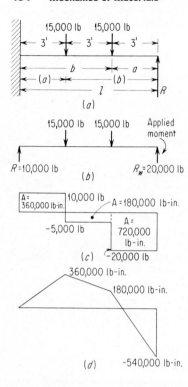

(a)

(b)

(c)

(d)

Figure 4-19 (c) Shear diagram; (d) moment diagram. The moment at the wall may also be calculated from the relationship $M = -[Fab(l + a)]/(2l^2)$, which gives for the two applied loads $-(240,000 + 300,000) = -540,000$ lb·in, which checks the values obtained by the other method.

exceeds those which may be determined by the static relationships alone. In this case there are three unknown reactions—the vertical reactions at the two ends and the couple at the left end. Because there are only vertical forces, we have only two equations based on the static relationships $\Sigma F_y = 0$ and $\Sigma M = 0$. To obtain the required answer, we resort to a handbook for a method of determining the reaction R. For a single concentrated load, beam fixed at one end, supported at the other,

$$R = \frac{Fb^2}{2l^3}(a + 2l)$$

If we apply this equation twice, once for each concentrated load, we obtain the value for R:

$$R_1 = \frac{15,000 \times (6)^2}{2 \times (9)^3}(3 + 2 \times 9) = 7780 \text{ lb}$$

$$R_2 = \frac{15,000 \times (3)^2}{2 \times 9^3}(6 + 2 \times 9) = 2220 \text{ lb}$$

Here, $R = R_1 + R_2 = 10,000$ lb, where R_1 is due to the right-hand 15,000-lb load and R_2 is due to the left of the two loads.

With this value for R we can determine the upward reaction at the wall, $R_w = 15,000 + 15,000 - 10,000 = 20,000$ lb. Having the loading of the beam, we can construct the shear and moment diagrams. These are shown in Fig. 4-19 with the loading diagram. The direction of the beam has been reversed, with the free end placed at the left to simplify the analysis.

The shear diagram may be constructed in the manner which has been discussed previously. To construct the moment diagram, start at the left end of the beam, where the moment is zero. We see from the shear diagram that the moment diagram will have a constant positive slope to the right to the point of application of the first load. Furthermore, the magnitude of the intercept of the moment curve at this point equals the algebraic sum of the areas under the shear diagram to the left of this point, which is 10,000 lb $\times$ 36 in. $= 36 \times 10^4$ lb·in. The moment curve then takes a constant negative slope to the point of application of the next load, at which point the moment equals

$$36 \times 10^4 - (5000 \times 36) = 18 \times 10^4 \text{ lb·in}$$

At this point the negative slope increases and remains constant to the end of the beam, where the moment equals

$$18 \times 10^4 - (20,000 \times 36) = -540,000 \text{ lb·in}$$

which is the maximum moment along the length of the beam. The section modulus of a 10 WF 54 beam is (from a table) 60.4 in³. Since stress $= (Mc)/I = M/Z$, where $Z = I/c$,

$$S = \frac{540,000}{60.4} = 8950 \text{ lb/in}^2$$

The correct answer is (e).

4-11 SHEAR STRESSES IN BEAMS

A loaded beam will strain in some fashion. When it does, it flexes and the longitudinal fibers are placed in tension or compression everywhere except at the neutral axis. The effect of flexure on a simple beam is shown in Fig. 4-20. If the beam were made of two parallel pieces in contact, but not fixed to one another, which were originally the same length, the ends of the top member would overlap the ends of the bottom member as shown in Fig. 4-21. The bottommost fibers of the top member

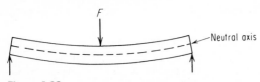

Figure 4-20

would strain in tension and elongate. The uppermost fibers of the bottom member would strain in compression and reduce in length. If the two members were joined together in such a manner that the differential strain were prevented, there would be a definite longitudinal shearing stress at the junction, the neutral axis of the beam shown in Fig. 4-20.

Figure 4-21

The longitudinal shear stress is related to the transverse (vertical in this case) shear stress by the relationship

$$S_s = \frac{VQ}{It} \quad \text{lb/in}^2 \quad \text{and} \quad \text{shear force} = \frac{VQ}{I} \quad \text{lb/in}$$

where

Q = shear flow
V = vertical shear load at the point being considered
I = transverse moment of inertia
t = length (width) of section being considered

Then

$$Q = \int y \, dA = \bar{y}A$$

where $\bar{y}$ = distance from the neutral axis of the beam to the neutral axis of area A.

Take, for example, a rectangular cross-section beam as shown in Fig. 4-22. Longitudinal shear stress will be greatest at the neutral axis.

Using the above relationship for longitudinal shear, we obtain

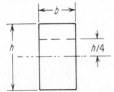

Figure 4-22

$$Q = \bar{y}A = \frac{h}{4}\frac{bh}{2} = \frac{bh^2}{8}$$

$$I = \frac{bh^3}{12} \qquad t = b$$

$$S_s = \frac{VQ}{It} = V\frac{bh^2/8}{(bh^3/12)b} = \frac{3V}{2bh}$$

The transverse shear stress equals V/bh, so the longitudinal shear stress at the neutral axis is 50 percent greater than the transverse shear stress for a rectangular beam.

For a circular beam, see Fig. 4-23.

$$Q = \bar{y}A = \frac{4r}{3\pi}\frac{\pi r^2}{2} = \frac{2r^3}{3}$$

$$I = \frac{\pi r^4}{4} \qquad t = 2r$$

$$S = \frac{VQ}{It} = V\frac{2r^3/3}{(\pi r^4/4) \times 2r} = \frac{4V}{3\pi r^2}$$

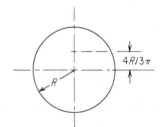

Figure 4-23

So the longitudinal shear at the neutral axis of a beam with a circular cross section is $33\frac{1}{3}$ percent greater than the transverse shear.

- Take the beam shown in Fig. 4-24. The beam is made of two pieces of a 2×6 plank (actual dimensions) fastened together with nails. If one nail is capable of withstanding a shearing force of 150 lb, what should be the spacing of the nails?

$$\bar{y} = \frac{(6 \times 2) \times 3 + (6 \times 2) \times 7}{2 \times (6 \times 2)} = 5.00 \text{ in} \qquad (\bar{y} \text{ for beam})$$

$$I = \frac{2 \times 6^3}{12} + 12 \times 2^2 + \frac{(6 \times 2^3)}{12} + 12 \times 2^2 = 136.0 \text{ in}^4$$

$$Q = \bar{y}A = 2 \times 12 = 24 \text{ in}^3$$

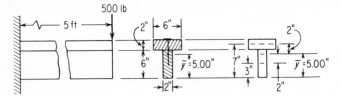

Figure 4-24

where $\bar{y}$ is the distance to neutral axis of area A. The shear force $V = $ 500 lb. Further,

$$\text{Shear force} = \frac{VQ}{I} = \frac{500 \times 24}{136} = 88.24 \text{ lb/in}$$

One nail will withstand 150 lb, so a nail would be required every $150/88.24 = 1.70$ in.

- If the crosspiece were glued in place, what would be the shear stress in the glue?

$$S_s = \frac{VQ}{It} = \frac{500 \times 24}{136 \times 2} = 44.1 \text{ lb/in}^2$$

- If the T beam were made of a 2×8 plank with pieces of 2×2 nailed to the sides, as shown in Fig. 4-25, what should be the spacing of the nails? Assume that each nail would withstand a shearing force of 150 lb.

Figure 4-25

The dimensions of the beam are the same as for the previous case, so $\bar{y}$ and I will be the same. This case is different from the previous case in that the shear force will be divided and act at two junctions instead of just one.

Calculate the shear flow acting on one sidepiece:

$$Q = \bar{y}A = 2 \times (2 \times 2) = 8 \text{ in}^3$$

$$\text{Shear force} = \frac{VQ}{I} = \frac{500 \times 8}{136} = 29.4 \text{ lb/in}$$

The required nail spacing for each of the 2- by 2-in sidepieces would equal $150/29.4 = 5.10$ in.

4-12   COMPOSITE BEAMS

The beams so far considered have all been of one homogeneous material. Many beams, however, are made of two materials, principally reinforced-concrete beams and steel-reinforced wood beams. Such beams are called "composite beams"; because of the difference in the moduli of elasticity of the two materials, they require a slightly different type of analysis than that used for homogeneous beams. One type of composite beam is shown in Fig. 4-26. It is made up of a 6- by 8-in timber 10 ft in length, with a 6-in wide strip of steel $\frac{1}{8}$ in thick attached to the bottom edge. To withstand the applied load, the beam will deflect, the upper part of the wooden portion will compress, and the lower part of the wood and the steel will stretch. The stress in any part of the beam will equal the unit deformation in that particular portion times the modulus of elasticity of the material of which that portion is made. Since the steel on the bottom of the beam deforms the same amount as the wood fibers on the bottom of the beam, the steel will be stressed in the ratio E_s/E_w times the stress in the bottommost wood fibers. Since $E_s = 30 \times 10^6$ and $E_w = 10^6$, the stress in the steel will be 30 times the stress in the wood. This means that the tensile load in the steel will be considerably greater than that in the wood, and the neutral axis will be below the geometric center of the beam.

Another way of looking at this problem is that since $F = SA$ and $S = E\delta$, the steel strip could, theoretically, be replaced by a piece of wood

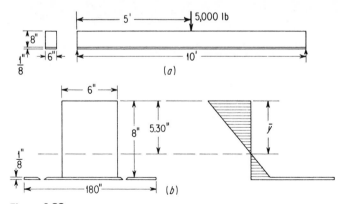

(a)

(b)

Figure 4-26

with an area equal to $E_s/E_w \times$ area of steel, or 30 times the area of the steel, but with the same thickness. The equivalent wooden beam is shown in the figure, and the internal-force diagram showing the distribution of SA over the cross section of the beam is also shown. To determine the stresses in the wood and in the steel, we need to know the distance to the neutral axis $\bar{y}$:

$$(6 \times 8) \times 4 + (\tfrac{1}{8} \times 180) \times 8\tfrac{1}{16} = (6 \times 8 + \tfrac{1}{8} \times 180)\bar{y}$$

$$\bar{y} = 5.30 \text{ in}$$

Further, $S = Mc/I$, where I is the moment of inertia of the equivalent beam about the neutral axis. Utilizing the parallel-axis theorem, we obtain

$$I_x = \frac{6 \times 8^3}{12} + 48 \times (1.30)^2 + \frac{180 \times (\tfrac{1}{8})^3}{12} + (180 \times \tfrac{1}{8})(2.70 + \tfrac{1}{16})^2$$

$$I_x = 256 + 81.8 \times 0.0293 + 171.7 = 509 \text{ in}^4$$

$$M_{max} = \frac{5000}{2} \times (5 \times 12) = 150,000 \text{ lb·in} \qquad c = 5.30$$

Maximum compressive stress in wood $S = \dfrac{150,000 \times 5.30}{509} = 1562 \text{ lb/in}^2$

The tensile stress in the bottommost fibers of the equivalent beam would be

$$\frac{150,000 \times 2.77}{509} = \frac{2.77}{5.30} \times 1562 = 816 \text{ lb/in}^2$$

and the stress in the steel would then equal $30 \times 816 = 24,500 \text{ lb/in}^2$, since $\delta_w = \delta_s$ and $E_s = 30E_w$.

The commonest type of composite beam is the reinforced-concrete beam, which can also be handled by the method described. One difference, however, is that the concrete is assumed to withstand no tensile stress and the portion of the concrete on the tension side of the neutral axis is disregarded in the stress calculations. A reinforced-concrete beam is shown in Fig. 4-27. It is required to determine the allowable concentrated load which the beam will hold if $n = 12$ and the maximum allowable stresses are 1000 lb/in^2 for the concrete and 20,000 lb/in^2 for the steel ($n = E_s/E_c$).

The equivalent beam cross section is drawn; as the concrete will take only compressive loading, the effective concrete area will extend only to the neutral axis. The location of the neutral axis is found as before:

$$10kd\frac{kd}{2} = (20 - kd) \times 24 \qquad (kd)^2 + 4.8kd - 96 = 0$$

$$kd = \frac{-4.8 + \sqrt{23 + 284}}{2} = 7.7$$

The moment of inertia is expressed as

$$I = \frac{10 \times (7.7)^3}{12} + (10 \times 7.7)\left(\frac{7.7}{2}\right)^2 + 24(20 - 7.7)^2 = 5150 \text{ in}^4$$

Note that the moment of inertia of the steel portion about its own neutral axis is disregarded since it is so small compared with the other components. Then

$$M_c = \frac{1000 \times 5150}{7.7} = 669,000 \text{ lb·in} \qquad \text{for } 1000 \text{ lb/in}^2 \text{ stress}$$

$$M_s = \frac{20,000 \times 5150}{n \times 12.3} = 698,000 \text{ lb·in} \qquad \text{for } 20,000 \text{ lb/in}^2 \text{ stress}$$

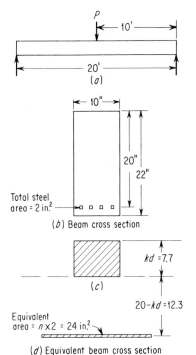

(a)

(b) Beam cross section

(c)

(d) Equivalent beam cross section   **Figure 4-27**

Compressive stress in the concrete controls. Note the use of the n in the denominator for determining the allowable moment for the steel.

$$M_{max} = \frac{P}{2}\frac{L}{2} \qquad P = \frac{4 \times 669{,}000 \text{ lb·in}}{20 \times 12 \text{ in}} = 11{,}150$$

4-13 COMPOUND STRESSES

We shall restrict ourselves here to a brief discussion of one method of determining the principal stresses and the maximum shearing stress for a condition of combined loading. The method is based on Mohr's circle.

■ Derive the equation for the maximum shearing stress in a member that is subjected to combined tension, torsion, and circumferential tension, specifically, a pipe under pressure with torsion and axial tension applied, using the diagram in Fig. 4-28 (Mohr's circle). The equation is to be derived for the maximum shearing stress from the information in the diagram in terms of $S_{s,max}$ = maximum shearing stress; S_s = stress due to torsion, S_x = stress due to pure tension, and S_y = stress due to circumferential tension. In addition, draw a stress diagram for a unit area of the external pipe surface.

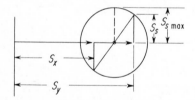

Figure 4-28

The problem as stated requires only a relatively simple application of Mohr's circle. The subject can be examined in a more general sense to obtain maximum benefit from the example in describing the application of the principles of Mohr's circle.

4-14 MOHR'S CIRCLE

First determine the stresses acting at a section on the outer surface of the pipe. Given: S_x = axial load divided by the cross-sectional area, S_y = $Pd/2t$ (assuming a thin-walled pipe), and S_s = Tc/J. Having determined these stresses, we can construct Mohr's circle for the given case. The general procedure is shown by means of Fig. 4-29. First lay out the coordinate axes. From the origin, point 0, lay off a distance $(S_x + S_y)/2$

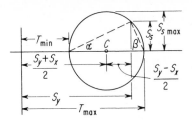

Figure 4-29

along the abscissa. This gives the location of C, the center of the circle. Note here that $S_x + S_y$ is the algebraic sum of the two perpendicular stresses; if one of these stresses had been compressive, we should have had the arithmetic difference instead of the arithmetic sum. At a distance from the origin equal to S_y draw a vertical line, and at a distance equal to S_s above the abscissa mark a point on this line. This point lies on the circumference of the circle. Now draw the circle.

The maximum shearing stress is equal to the radius of the circle. From the figure, the trigonometric relationship is

$$S_{s,\max} = \sqrt{S_s^2 + \left(\frac{S_y - S_x}{2}\right)^2}$$

The principal stresses can also be quickly obtained and are shown on

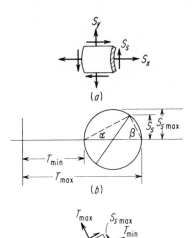

Figure 4-30 (a) Section of pipe surface showing applied stresses; (b) minimum and maximum stresses and angles α and β; (c) location of minimum and maximum stresses.

the figure as T_{min} and T_{max}. The directions of these principal stresses are also readily obtained from the figure. The lines drawn from the ends of the horizontal diameter to the construction point on the circumference give these angles, and the principal stresses act perpendicularly to these lines; i.e., T_{min} acts at an angle of β and T_{max} acts at an angle of α with the direction of S_y. The maximum shearing stress acts on a plane at an angle of 45° to these two principal planes.

The required stress diagram for a unit area of the external pipe surface shows the directions of the different stresses and would ordinarily be drawn before Mohr's circle was constructed. The diagram is shown in Fig. 4-30.

4-15 RIVETED AND WELDED JOINTS

Riveted joints can be lumped into two general classes—those in which the line of action of the applied load passes through the centroid of the rivet group and those which are subject to eccentric loading.

The simpler case, in which the line of action of the applied load passes through the centroid of the rivet group may be illustrated by referring to the rivet section illustrated in Fig. 4-31. What feature of the joint would limit the applied load, and what is the maximum safe load? The allowable load could be limited by tension in the plates, bearing stresses in the plates, or shearing stress in the bolts.

The controlling area in tension is the same for the left and right members and equals $\frac{3}{4}(5\frac{1}{2} - 2 \times \frac{3}{4}) = 3$ in². The maximum allowable load

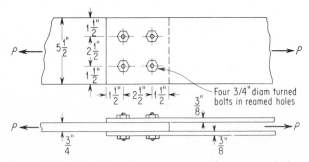

Figure 4-31 Allowable unit stresses, in pounds per square inch.
Tension, structural steel, net section 20,000
Shearing, turned bolts in reamed holes 15,000
Bearing, turned bolts in reamed holes:
    Double shear 40,000
    Single shear 32,000

from tension considerations then equals $3 \times 20,000 = 60,000$ lb. The bearing area in single shear equals $4 \times 2(\frac{3}{4} \times \frac{3}{8}) = 2\frac{1}{4}$ in². This is the same as the bearing area in double shear, so the maximum allowable load as limited by bearing equals $2\frac{1}{4} \times 32,000 = 72,000$ lb.

The shear area equals eight times the cross-sectional area of one bolt, or $8 \times \pi/4 \times (\frac{3}{4})^2 = 3.53$ in². The maximum allowable load as limited by shearing stress is then $3.53 \times 15,000 = 53,000$ lb. The allowable load is then limited by shear and equals 53,000 lb.

In this case each of the four rivets has been assumed to withstand one-quarter of the applied load; the analysis is applicable only when the line of action of the applied force passes through the centroid of the rivet group. When the load is applied eccentrically, there will be a torque loading, as well as a portion of the applied load to be withstood by each rivet, and since different rivets will ordinarily be at different distances from the centroid of the rivet group, the stresses in the rivets will vary.

- If the maximum allowable unit shearing stress in the rivets in the riveted joint shown in Fig. 4-32 is 13,000 lb/in², which of the following values most nearly equals the maximum allowable load P? Assume that bearing stress does not govern. All rivets are $\frac{3}{4}$ in in diameter.

  (a) 13,200 lb   (b) 12,800 lb   (c) 12,300 lb   (d) 11,900 lb
  (e) 11,500 lb

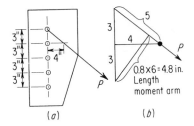

(a)   (b)   **Figure 4-32**

First, determine the centroid of the rivet group. This is the centroid of the rivet areas. In this case the rivet group is perfectly symmetrical about the center rivet, so the centroid of the group is at the center of the center rivet.

Next, calculate the length of the moment arm D; $\alpha = 37°$ and $D = 6 \cos \alpha = 6 \times 0.8 = 4.8$ in.

As the eccentric force P is applied to the plate, the plate will turn and the rivets will deform slightly, since stress is proportional to strain and there can be no stress without a corresponding strain. Assuming that all rivets fill their holes completely, we know that the deformation of rivet 1 due to the twisting of the plate will equal $\frac{6}{3}$ times the deformation of rivet 2 (Fig. 4-33). This means that the stress in rivet 1 due to the torque loading will be twice the stress in rivet 2. Because of the symmetry of the joint, the stresses in rivets 1 and 5 due to resisting the applied torque will be equal, as will the stresses in rivets 2 and 4. The stress in rivet 3 due to the torque will equal zero. Since all rivets are the same size, the resisting forces will be proportional to the strains (or the stresses). This gives us $4.8P = 6 \times 2 \times TF_1 + 3 \times 2 \times TF_2$, where TF_1 and TF_2 are the forces exerted by rivets 1 and 2 resisting the applied torque. Since the deformation of rivet 1 is just twice that of rivet 2, TF_1 is twice TF_2. This gives $4.8P = 15TF_1$, or $TF_1 = 0.32P$.

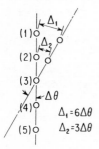

$$\Delta_1 = 6\Delta\theta$$
$$\Delta_2 = 3\Delta\theta$$

Figure 4-33

To this must be added, vectorially, $P/5$, since each rivet must withstand one-fifth of the applied force directly. The two resisting force vectors are shown in Fig. 4-34. The sum of the resisting horizontal force components is $0.32P + 0.12P = 0.44P$, and the vertical force component is $0.16P$. The total shearing force acting on rivets 1 and 5 is then $P\sqrt{0.44^2 + 0.16^2} = 0.468P$.

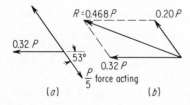

Figure 4-34

Since the total permissible shearing force in any one rivet is $13,000 \times \pi/4 \times (\frac{3}{4})^2 = 5740$ lb, the maximum allowable value of $P = 5740/0.468 = 12,300$ lb. The correct answer is (c).

The majority of the features discussed above also apply to welded joints. First, the minimum shearing stress will occur when the line of action of the force passes through the centroid of the rivet or weld area. If the load is applied eccentrically, it will provide a moment about the centroid and the centroidal point will act as a pivot point about which the riveted or welded joint will tend to rotate. Stress and strain are proportional; the greatest stress due to the applied moment will occur in the rivet or portion of weld farthest from the centroid, since the greatest strain will occur at that point.

- A 3- by 3- by $\frac{1}{2}$-in angle is to be welded to each side of a gusset plate, as shown in Fig. 4-35. The weld beads are to have the same thickness as the angle legs (see Fig. 4-35). Which of the following is the required length of weld at the corner of the angle iron to safely resist an applied tensile load of 88,000 lb? The allowable shearing stress in the weld material is 11,300 lb/in^2.

(a) 11 in   (b) 15 in   (c) 18 in   (d) 13 in   (e) 22 in

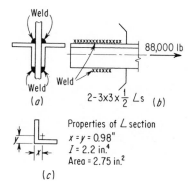

$2\text{-}3\text{x}3\text{x}\frac{1}{2} \angle \text{s}$ (b)

Properties of $\angle$ section
$x = y = 0.98"$
$I = 2.2$ in.^{4}
Area = 2.75 in.^{2}

(c)

Figure 4-35

The shear area of a welded joint is taken as the smallest section of the joint, and the fillet weld is made the same thickness as the plate being welded. This is shown in Fig. 4-36. The effective thickness of the welds would then be $\frac{1}{2} \sin 45° = 0.354$ in. The line of action of the force passes through the centroids of the angles; it must also pass through the centroid

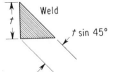

Figure 4-36

of the weld area to provide a joint free from any torque effects. The welds will then be stressed in pure shear; since the moments of the two weld areas about the centroid must be equal in order to have no moment resulting from unequal strain we have

$$0.98W_u = 2.02W_L \quad \text{or} \quad W_u = 2.06W_L$$

From stress considerations, 88,000 lb = 11,300 lb/in² × effective weld area. Then

$$\text{Effective weld area} = 7.8 \text{ in}^2 = (W_u + W_L)(0.500 \times 0.707)$$
$$W_u + W_L = 22 \text{ in} \quad W_L = 7.24 \quad W_u = 14.76 \text{ in}$$

Make the welds $7\frac{1}{2}$ and 15 in long. The correct answer is (b).

If the line of action of the force should not pass through the centroid of the weld area, the maximum stress will equal the vector sum of two component forces, as was the case for the eccentrically loaded riveted joint.

Sample Problems

4-1   See Fig. 4P-1. A cast-iron punch-press frame is reinforced with a $\frac{1}{4}$- by 6-in steel plate on the inner face, as shown. Allowable stresses, in pounds per square inch, are:

Cast-iron compression	=	15,000
Cast-iron tension	=	5,000
Steel tension	=	20,000
E_i	=	18,000,000
E_s	=	30,000,000

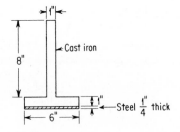

Figure 4P-1

The maximum safe bending moment for the frame most nearly equals which of the following?

(a) 313,000 in·lb   (b) 260,000 in·lb   (c) 566,000 in·lb
(d) 430,000 in·lb   (e) 515,000 in·lb

4-2 Three wires, each having a cross-sectional area of 0.20 in² and the same unstressed length of 200 in at 60°F, hang side by side in the same plane. The outer wires are copper. The middle wire, equidistant from each of the others, is steel. Given: $E_s = 30 \times 10^6$ lb/in² and $E_c = 17 \times 10^6$ lb/in². If a weight of 1000 lb is gradually picked up by the three wires, the part of the weight carried by the steel wire would most nearly equal which of the following? The coefficient of thermal expansion for steel is 6.5×10^{-6} and for copper, 9.3×10^{-6} in/(in)(°F).

(*a*) 470 lb   (*b*) 500 lb   (*c*) 560 lb   (*d*) 610 lb   (*e*) 390 lb

4-3 Referring to the data in problem 4-2, which of the following most nearly equals the temperature at which the entire load would be carried by the steel wire?

(*a*) 90°F   (*b*) 110°F   (*c*) 120°F   (*d*) 130°F   (*e*) 140°F

4-4 The maximum tensile stress at section *A*-*A* of the part shown in Fig. 4P-4 most nearly equals which of the following?

(*a*) 5000 lb/in²   (*b*) 45,000 lb/in²   (*c*) 38,000 lb/in²
(*d*) 50,000 lb/in²   (*e*) 48,000 lb/in²

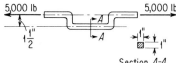

Section *A*-*A*   **Figure 4P-4**

4-5 Given a cantilever beam 8 ft long with a concentrated load of 1000 lb at the free end and a uniform load of 100 lb/ft along its length, which of the following would most nearly equal the moment at the center point of the beam?

(*a*) 57,600 lb·in   (*b*) 62,100 lb·in   (*c*) 48,700 lb·in
(*d*) 39,200 lb·in   (*e*) 35,700 lb·in

4-6 The maximum moment for a 22-ft-long beam loaded as shown in Fig. 4P-6 would most nearly equal which of the following?

(*a*) 410,000 lb·in   (*b*) 320,000 lb·in   (*c*) 502,000 lb·in
(*d*) 408,000 lb·in   (*e*) 363,000 lb·in

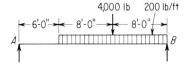

Figure 4P-6

4-7 A mine hoist weighing 5 tons empty is descending at a speed of 20 mi/h when the hoist mechanism jams and stops suddenly. If the cable is steel with a cross-sectional area of 1.00 in² and the cage is 5000 ft below the hoisting drum when the drum jams, which of the following most nearly

equals the stress produced in the cable (E for hoist cable $= 12 \times 10^6$ lb/in²; neglect the weight of the cable)?
(*a*) 47,100 lb/in² (*b*) 37,200 lb/in² (*c*) 28,900 lb/in²
(*d*) 50,600 lb/in² (*e*) 41,400 lb/in²

4-8 The distance y to the neutral axis for the reinforced-concrete beam shown in Fig. 4P-8 most nearly equals which of the following?
(*a*) 6.0 in (*b*) 6.4 in (*c*) 6.7 in (*d*) 7.1 in (*e*) 7.3 in

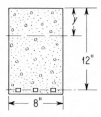

Figure 4P-8

4-9 A select oak timber finish sized to 6 × 8 in is 12 ft long and is employed as a rigidly embedded cantilever beam with a load. Given $P = 100$ lb as shown in Fig. 4P-9. If $E = 1.8 \times 10^6$ lb/in², which of the following most nearly equals the deflection at the midpoint of the beam?
(*a*) 0.22 in (*b*) 0.14 in (*c*) 0.18 in (*d*) 0.10 in (*e*) 0.07 in

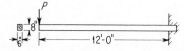

Figure 4P-9

4-10 A steel column is an H section 50 ft long and supports a concentric load of 350,000 lb. The column section has the following dimensions: $A = 49.09$ in²; $I_{1-1} = 790.2$ in⁴; $d_{1-1} = 15\frac{1}{8}$ in; $d_{2-2} = 15\frac{1}{8}$ in; $I_{2-2} = 2020.8$ in⁴. Employing the American Institute of Steel construction method, we obtain

$$\frac{P}{A} = \frac{18,000}{1 + (1/18,000)(l/r)^2}$$

Which of the following most nearly equals the allowable column load?
(*a*) 412,000 lb (*b*) 394,000 lb (*c*) 278,000 lb (*d*) 312,000 lb
(*e*) 364,000 lb

4-11 The maximum moment for the beam loaded as shown in Fig. 4P-11 most nearly equals which of the following?

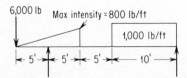

Figure 4P-11

(a) 416,000 lb·in   (b) 356,000 lb·in   (c) 380,000 lb·in
(d) 342,000 lb·in   (e) 312,000 lb·in

4-12 The load deflection curve for a spring is shown in Fig. 4P-12. When the load
reaches 150 lb the spring is fully closed and no further deflection results.
The load deflection curve can be represented by the equation

$$d = 4 - \frac{(L - 200)^2}{10,000}$$

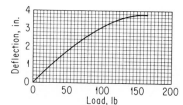

Figure 4P-12

where d is deflection in inches and L is load in pounds. Which of the following
most nearly equals the energy stored in the spring when the load equals
300 lb?

(a) 450 in·lb   (b) 325 in·lb   (c) 275 in·lb   (d) 225 in·lb
(e) 215 in·lb

4-13 A homogeneous steel beam with an allowable stress of 20,000 lb/in² is loaded
as shown in Fig. 4P-13. Which of the following most nearly equals the
required section modulus?

(a) 0.96 in³   (b) 0.84 in³   (c) 0.79 in³   (d) 0.68 in³   (e) 1.12 in³

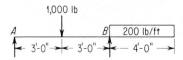

Figure 4P-13

4-14 A beam 40 ft long is supported 10 ft from the left end and 5 ft from the right
end. The beam is loaded with a concentrated load of 6000 lb at the left end
and a uniform load of 600 lb/ft from the left end to the right end. Which of
the following most nearly equals the maximum moment?

(a) 7200 ft·lb   (b) 84,000 ft·lb   (c) −75,000 ft·lb
(d) −82,000 ft·lb   (e) −90,000 ft·lb

4-15 A 6-in by 12-in by 14-ft timber is supported at the left end and also at 4 ft
from the right end. It has a load of 2000 lb at the right end, a load of 2000
lb 3 ft from the left end, and a load of 800 lb/ft between the supports. Which
of the following most nearly equals the maximum moment in the beam?

(a) −8000 lb·ft   (b) −12,200 lb·ft   (c) 10,200 lb·ft
(d) 8200 lb·ft   (e) 9400 lb·ft

4-16 A beam 40 ft long rests on simple supports at the ends and has the following loads: dead load, 300 lb/lin ft; uniform live load, 700 lb/ft; concentrated live load, 10,000 lb. Which of the following most nearly equals the maximum bending moment that can be developed at a point 16 ft from the left support?
(*a*) 312,000 lb·ft   (*b*) 288,000 lb·ft   (*c*) 322,000 lb·ft
(*d*) 317,000 lb·ft   (*e*) 306,000 lb·ft

4-17 A steel wire accurately graduated in feet was used to measure the depth of a deep well. The cross-sectional area of the wire was 0.03 in², and a 250-lb weight was attached to the wire to lower it into the well. According to the graduated scale, the measurement read 2000 ft. A steel bar 1 in² in cross-sectional area and 1 yd long weighs 10 lb. Which of the following most nearly equals the true depth of the well?
(*a*) 2000.8 ft   (*b*) 2001.2 ft   (*c*) 2000.00 ft   (*d*) 2003.1 ft
(*e*) 2002.1 ft

4-18 Which of the following most nearly equals the correct diameter of a steel shaft to transmit 40 hp at a speed of 240 r/min if the allowable working unit stress in the shaft is $S = 9000$ lb/in²?
(*a*) 1.00 in   (*b*) $1\frac{1}{4}$ in   (*c*) $1\frac{1}{2}$ in   (*d*) $1\frac{3}{8}$ in   (*e*) $1\frac{7}{8}$ in

4-19 Concrete is to be mixed with a proportion by volume of 1:2:3 (1 cement, 2 sand, 3 aggregate, and 6 gal of water per sack of cement). The cement has 51.3 percent voids and weighs 94 lb/ft³, the sand has 35.4 percent voids and weighs 111 lb/ft³, the gravel has 32.0 percent voids and weighs 108 lb/ft³. Which of the following most nearly equals the weight of sand required to make 1 yd³ of concrete?
(*a*) 800 lb   (*b*) 1100 lb   (*c*) 1420 lb   (*d*) 1300 lb   (*e*) 1510 lb

4-20 A beam 20 ft long carries a uniform load of 2000 lb/lin ft, including its own weight, on two supports 14 ft apart. The right end of the beam cantilevers 2 ft; the left end cantilevers 4 ft. Which of the following most nearly equals the maximum moment in the beam?
(*a*) −16,000 lb·ft   (*b*) 18,000 lb·ft   (*c*) 22,000 lb·ft
(*d*) 39,000 lb·ft   (*e*) 28,000 lb·ft

4-21 Which of the following most nearly equals the uniform load w (in pounds per foot) which will cause a simple beam 10 ft long to deflect 0.3 in if it is also supported at midspan by a spring whose modulus $k = 30,000$ lb/in? Assume that $I = 100$ in⁴, $E = 3 \times 10^7$ lb/in², and $d = 10$ in. At no load the spring just touches the beam (see Fig. 4P-21).
(*a*) 4210 lb/ft   (*b*) 5440 lb/ft   (*c*) 6010 lb/ft   (*d*) 4860 lb/ft
(*e*) 5840 lb/ft

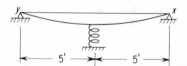

Figure 4P-21

4-22 A 2-in-diameter shaft transmits 30 hp at 200 r/min. A pulley which delivers the entire amount of power is keyed to the shaft by a $\frac{3}{8}$- by $\frac{1}{2}$- by 3-in flat key. The shaft is in pure torsion. Which of the following most nearly equals the shear stress in the key?

(a) 10,200 lb/in²   (b) 9400 lb/in²   (c) 8400 lb/in²   (d) 8700 lb/in²
(e) 7800 lb/in²

4-23 The bending moment under the 5000-lb load of the beam shown in Fig. 4P-23 most nearly equals which of the following?

(a) 11,200 lb·ft   (b) 14,600 lb·ft   (c) 9800 lb·ft   (d) 14,700 lb·ft
(e) 12,100 lb·ft

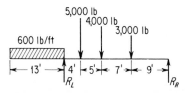

Figure 4P-23

4-24 Which of the following most nearly equals the maximum tensile unit stress for the loaded beam shown in Fig. 4P-24?

(a) 12,000 lb/in²   (b) 13,500 lb/in²   (c) 12,800 lb/in²
(d) 10,200 lb/in²   (e) 11,700 lb/in²

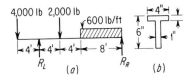

Figure 4P-24

4-25 Which of the moments listed below most nearly equals the maximum moment in the following described beam? The beam has a 24-ft span and carries a load of 24,000 lb spread uniformly over its entire length. It has an additional load of 12 tons spread uniformly over a section that starts 6 ft from the left support and extends 8 ft to the right.

(a) −124,000 lb·ft   (b) 146,000 lb·ft   (c) 188,000 lb·ft
(d) 200,000 lb·ft   (e) −162,000 lb·ft

4-26 A steel gear is to be asembled on a steel motor shaft by heating the gear and cooling the shaft with dry ice. At 70°F the shaft measures 7.000 in in diameter and the hole in the gear measures 6.996 in in diameter. Which of the following most nearly equals the temperature to which the shaft must be cooled so that the gear may be assembled on the shaft with 0.005-in clearance when the gear is warmed to 140°F? [The coefficient of linear thermal expansion for steel equals 0.0000065 in/(in)(°F).]

(a) 0°F   (b) −10°F   (c) −42°F   (d) −51°F   (e) −58°F

4-27 Which of the following most nearly equals the maximum moment for the beam shown in Fig. 4P-27?

(a) 9050 ft·lb   (b) 10,080 ft·lb   (c) 8400 ft·lb   (d) 9450 ft·lb
(e) 9850 ft·lb

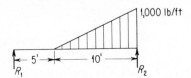

Figure 4P-27

4-28 A machine part is loaded as a cantilever beam. Two possibilities of design are being considered which would utilize different materials and different cross sections as shown in Fig. 4P-28. The length of each would be 6 in. Which of the following would most nearly equal the ratio of the deflection of the rectangular beam to the circular beam?
(a) 0.98   (b) 1.00   (c) 1.08   (d) 1.16   (e) 1.25

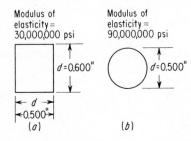

Figure 4P-28

4-29 A rigid horizontal bar 10 ft long, of uniform cross section and weighing 300 lb, is supported by two vertical wires, one of steel and one of copper. Each wire is 5 ft long and $\frac{1}{100}$ in^2 in cross section. The copper wire is attached 1 ft from one end of the bar, and the steel wire at such a distance X from the same end that both wires stretch by the same amount. Which of the following most nearly equals the distance X? (Young's modulus for steel $= 30 \times 10^6$ lb/in^2 and for copper $= 15 \times 10^6$ lb/in^2.)
(a) 6.0 ft   (b) 6.5 ft   (c) 7.0 ft   (d) 7.1 ft   (e) 6.8 ft

4-30 A block of steel would increase in volume by 330 ppm when heated from 20 to 30°C at constant presure. Which of the following most nearly equals the increase in the pressure which would have to be applied to the block to prevent it from expanding when its temperature was thus raised? (The compressibility of steel is 4.75×10^{-8} in^2/lb.)
(a) 7000 lb/in^2 (b) 8000 lb/in^2 (c) 10,000 lb/in^2
(d) 12,000 lb/in^2 (e) 11,000 lb/in^2

4-31 A weight of 3500 lb is supported by one piece of material A and two pieces of material B, as shown in Fig. 4P-31. Each of the three pieces has a cross section of 2 in^2 and a height of 6 in. The weight is symmetrically placed on the supporting blocks. Young's modulus for material A is 3×10^7 lb/in^2 and for material B is 2×10^7 lb/in^2. The stress in the piece of material A would most nearly equal which of the following?
(a) 700 lb/in^2 (b) 750 lb/in^2 (c) 800 lb/in^2 (d) 850 lb/in^2
(e) 925 lb/in^2

Nat. Assoc. of Black
Consulting Eng.
6406 Georgia Ave. N.W.
Washington, D.C. 20012

Nat. Society of Black
Eng.
344 Commerce St.
Alexandria, VA 22314
(703) 549-2207

Piedmont Business League
P.O. Box 10048
Greenville, SC 29609
(803) 242 1050

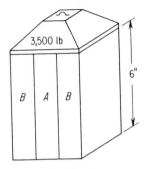

Figure 4P-31

4-32 A solid circular shaft has a uniform diameter of 2 in and is 10 ft long. At its midpoint 65 hp is delivered to the shaft by means of a belt passing over a pulley. This power is used to drive two machines, one at the left end of the shaft consuming 25 hp and one at the right end consuming the remaining 40 hp. If the shaft turns at 200 r/min, which of the following most nearly equals the relative angle of twist between the two ends of the shaft?
(a) 1.1°   (b) 2.0°   (c) 0.5°   (d) 0.7°   (e) 0.9°

4-33 An eccentrically loaded riveted joint is shown in Fig. 4P-33. The applied load is 18,000 lb acting with an eccentricity of 8 in from the geometric center of the group of six $\frac{3}{4}$-in rivets. The maximum shearing stress in the rivets most nearly equals which of the following?
(a) 13,800 lb/in^2 (b) 15,700 lb/in^2 (c) 17,900 lb/in^2
(d) 18,700 lb/in^2 (e) 19,000 lb/in^2

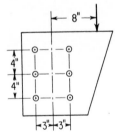

Figure 4P-33

4-34 A rectangular wooden beam is loaded as shown in Fig. 4P-34. The maximum shearing stress parallel to the gain most nearly equals which of the following?

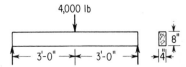

Figure 4P-34

(a) 85 lb/in^2 (b) 120 lb/in^2 (c) 94 lb/in^2 (d) 132 lb/in^2
(e) 147 lb/in^2

4-35 A 4-in-diameter shaft transmits 1000 hp at 2000 r/min. If the same power
is to be transmitted at 10,000 r/min, which of the following shaft diameters
is closest to the diameter required to maintain the same maximum shearing
stress?
(a) 3.5 in   (b) 3.1 in   (c) 2.9 in   (d) 2.75 in   (e) 2.3 in

Multiple-Choice Sample Quiz

For each question select the most nearly correct answer from the five
given possibilities. For practice, try to complete the following 10 problems
in 12 minutes or less.

4M-1 If a simple coil spring is compressed by an axial load, which of the following
is true? (The spring is not compressed sufficiently to bring the sides of the
coils into contact.)
(a) The maximum stress occurs at the center of the spring wire.
(b) The spring wire is in torsion.
(c) If the diameter of the spring is increased while the wire size and
number of turns are kept constant, the force necessary to compress
the spring a given amount will be increased.
(d) If the number of turns in the spring is increased, the spring will be
harder to compress with load.
(e) None of these statements are correct.

4M-2 If a homogeneous simple wooden beam fails under load with a crack parallel to the grain at the end supports, the failure was most likely caused
by excessive:
(a) compressive stresses
(b) bearing stresses
(c) shearing stresses
(d) tension stresses
(e) none of these

4M-3 The term $I = bd^3/12$ refers to:
(a) radius of gyration
(b) section modulus
(c) instantaneous center
(d) moment of inertia
(e) product of inertia

4M-4 In Euler's formula $P/A = \pi^2 E/(L/r)^2$, the L/r is commonly known as the:
(a) slenderness ratio
(b) proportional limit
(c) stiffness ratio
(d) rigidity factor
(e) Poisson's ratio

4M-5 In a uniformly loaded simple beam, the maximum vertical shearing force occurs:
(a) at either end support
(b) at the section of maximum moment
(c) at the center of the span
(d) on the bottom fiber of the beam
(e) on the top fiber of the beam

4M-6 If a simple beam is loaded with a uniform and a concentrated load, which of the following statements is true?
(a) The maximum bending moment is indeterminate.
(b) The area of the shear diagram to the left of a section represents the bending moment at that section to some scale.
(c) The maximum shear, as represented in the shear diagram, is located at the same section as the maximum moment.
(d) The maximum bending moment always occurs at the support which carries the most load.
(e) None of these statements are correct.

4M-7 The moment curve for a simple beam with a concentrated load at midspan takes the shape of a:
(a) triangle
(b) semicircle
(c) semiellipse
(d) parabola
(e) rectangle

4M-8 In a loaded horizontal, homogeneous, rectangular beam the horizontal shearing stress at any cross section ($\perp$) perpendicular to the neutral axis:
(a) is a constant across that section
(b) has a maximum value of $\frac{3}{2}$ the average vertical shearing stress on the cross section at the neutral axis
(c) is equal to zero
(d) can be neglected if the beam is short
(e) has a maximum value at the outer fibers of the beam

4M-9 A steel test specimen is $\frac{5}{8}$ in ϕ at root of thread. If it is to be stressed to 50,000 lb/in^2 tension, what load must be applied?
(a) 12,790 lb   (d) 25,600 lb
(b) 15,340 lb   (e) 31,250 lb
(c) 16,320 lb

4M-10 What load must be applied to a 1-in round steel bar 8 ft long ($E = 30,000,000$ lb/in^2) to stretch the bar 0.05 in?
(a) 7200 lb   (d) 12,300 lb
(b) 9850 lb   (e) 15,000 lb
(c) 8600 lb

5

FLUID MECHANICS

5-1 EXAMINATION COVERAGE

Fluid mechanics is covered in the morning session and is one of the five optional subjects in the afternoon session.

The subdivisions of the major subject which may be covered, and which should be reviewed, include the following:

Fluid properties

Fluid statics

Impulse and momentum

Dimensional analysis

Open-channel flow

Pipe flow

Flow measurement

Similitude

Compressible flow

Hydraulics

5-2 FLUID PROPERTIES

A fluid is defined as a substance which deforms continuously when subjected to a shear stress, no matter how small that stress may be. The term "fluid" includes both gases and liquids. A gas is considered to be compressible, and a liquid is considered to be incompressible. This is true only on a relative basis, since liquids are compressible, and this compressibility can be important, especially at high pressures. The compressibility of a liquid, or its modulus of elasticity in compression, is known as its "bulk modulus" K, where $K = -[dP/(dV/V)]$. For water, a rough value of the compressibility factor is 300,000 lb/in² and for petroleum oil, 200,000 lb/in². The bulk modulus of a liquid will change with both temperature and pressure.

A "perfect" gas will expand to occupy all the space in which it is contained. A liquid, under static conditions, will flow under the influence of gravity until its surface is smooth and horizontal. A real gas is also acted on by gravity.

A fluid cannot resist a continuous change of shape under the influence of a shear force. The resistance to the rate of deformation is known as "viscosity." The dynamic, or absolute, is usually indicated by the symbol μ (mu) and is defined physically by Newton's law of viscosity—$\tau = \mu (dv/dy)$, where the units are pounds per square foot, equal to μ times feet per second per foot, giving μ the units of pound-seconds per square foot or dyne-seconds per square centimeter (newton-seconds per square meter). Viscosity is also given as kinematic viscosity ν (nu), where $\nu = \mu/\rho$ and has the units of square feet per second or square centimeters per second.

5-3 HYDROSTATICS

One definition of the liquid state is that "a liquid in equilibrium offers no resistance to change of shape." From this it follows that the resultant of all the forces acting on a particle of liquid at its surface must be perpendicular to the tangent to the surface at that point. An example is shown in the determination of the shape of the surface of the fluid in an upright cylindrical container rotating about its central axis (Fig. 5-1).

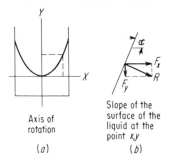

Axis of rotation

(a)

Slope of the surface of the liquid at the point x,y

(b)   **Figure 5-1**

First sketch a section through the center of the cylinder. To determine the required curve of the surface, select any particle of fluid on this surface and add the forces acting on this particular particle. The term F_x equals the force on the particle due to the rotation about the axis and equals the centripetal force, or

$$F_x = \omega^2 rm \qquad F_y = \text{gravitational force} = mg$$

The slope of the surface of the liquid at the point (x, y) is perpendicular to the slope of the resultant R of the forces acting on the particle of liquid; $\tan \alpha = F_x/F_y$, but also $\tan \alpha = dy/dx$ at this particular point, so

$$\frac{dy}{dx} = \frac{\omega^2 rm}{mg}$$

where ω and g are constants, $r = x$, and $dy = (\omega^2/g)\, x\, dx$, or the equation for the curve of the surface is $y = (\omega^2/2g)x^2$ plus a constant of integration. If the axes are selected as shown in the figure, $y = 0$ for $x = 0$ and the constant of integration is zero, the final equation becomes $y = (\omega^2/2g)x^2$, which will be recognized as the equation of a parabola. The curve of the surface of the liquid is then a paraboloid of revolution.

▪ A container partially filled with water and having a total weight of 10 lb is sliding down a 30° inclined plane. The coefficient of sliding friction $\mu = 0.30$. The angle of the surface of the water, measured from the horizontal would equal which of the following?

    (a) 30.0°   (b) 18.2°   (c) 13.3°   (d) 15.2°   (e) 6.4°

Draw a figure showing the different forces as vectors (Fig. 5-2). Then calculate the acceleration acting on the container and the liquid:

$$f = N\mu = W \cos 30° \times 0.30 = 0.260W = 2.60 \text{ lb}$$
$$P = W \sin 30° = 5.00 \text{ lb}$$

The net force acting on the container to cause acceleration is 2.40 lb:

$$a = \frac{F}{m} = \frac{2.40}{10/g} = 0.240g = 7.73 \text{ ft/s}^2$$

Next, take a particle of liquid at the surface and sum the forces acting on it—these will be the force of gravity acting down and an inertial force of ma acting at an angle of 30° as shown in the figure. The angle of the surface of the water in the container would then equal $\tan^{-1}(6.69/28.3) = 13.30°$. The correct answer is (c).

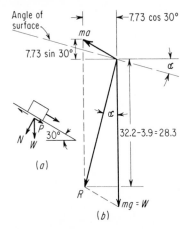

Figure 5-2

5-4 FLUID PRESSURE

The pressure at any point in a liquid is equal to the pressure on the surface of the liquid plus the weight of the column of liquid of unit cross-sectional area extending vertically from the point to the surface of the liquid. The pressure is the same at all points at the same level within a static, continuous, single-liquid system, and the pressure at a particular point is the same regardless of the direction in which the pressure is measured.

■ Carbon tetrachloride flows in pipe A in Fig. 5-3 and water flows in pipe B. The two pipes are connected by a differential gauge which contains oil (0.80 specific gravity) and pipe A is 10 ft above pipe B.

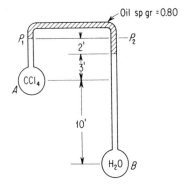

Figure 5-3

If the specific gravity of the carbon tetrachloride is 1.5 and the other dimensions are as shown, the difference in pressure (in feet of water) between pipes A and B most nearly equals which of the following?

(a) 10.0 ft   (b) 6.9 ft   (c) 8.7 ft   (d) 7.1 ft   (e) 8.1 ft

The notations P_1 and P_2 are added to the figure, and since they are at the same elevation in the same fluid system, $P_1 = P_2$:

$$P_A = P_1 + 5 \text{ ft of CCl}_4 \qquad P_B = P_2 + 2 \text{ ft of oil} + 13 \text{ ft of H}_2\text{O}$$

$$P_1 = P_A - 5 \text{ ft of CCl}_4 \qquad P_2 = P_B - 2 \text{ ft of oil} - 13 \text{ ft of H}_2\text{O}$$

$$P_1 = P_A - 7.5 \text{ ft of H}_2\text{O} \qquad P_2 = P_B - 1.60 \text{ ft of H}_2\text{O} - 13 \text{ ft of H}_2\text{O}$$

$$P_A - 7.5 \text{ ft of H}_2\text{O} = P_B - 14.60 \text{ ft of H}_2\text{O}$$

$$P_B = P_A + 7.1 \text{ ft of H}_2\text{O}$$

The correct answer is (d).

Utilizing the same reasoning, we can calculate the total force on any submerged area. The total force will equal the summation of the pres-

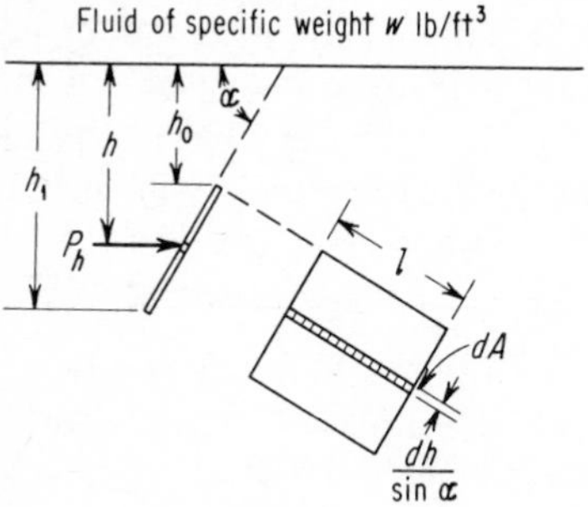

Figure 5-4

sures at depths h, $h + \Delta h$, $h + 2\Delta h$, ... times the areas at depths h, $h + \Delta h$, $h + 2\Delta h$, ... (see Fig. 5-4). The force on $dA = P_h dA = (P_1 + wh) \, dA$. The force acting on the surface is

$$F = \int_{h_0}^{h_1} (P_1 + wh) \frac{l \, dh}{\sin \alpha}$$

For the rectangular surface shown,

$$F = \frac{l}{\sin \alpha} \left[P_1(h_1 - h_0) + \frac{w}{2} (h_1^2 - h_0^2) \right]$$

Ordinarily, P_1 is only atmospheric pressure, which will act on both sides of a gate or dam subjected to hydraulic pressure, so it can be omitted. Thus

$$F = \frac{lw}{2 \sin \alpha} (h_1{}^2 - h_0{}^2) = wA \frac{h_1 + h_0}{2}$$

since $A = l(h_1 - h_0)/\sin \alpha$, $w(h_1 + h_0)/2$ is the average pressure acting on the area.

Utilizing the above, we can also derive a more general relationship for the hydrostatic force acting on any submerged surface. First, put down the equation for the force acting on a surface but leave out the effect of atmospheric pressure. This leaves

$$F = \int_{h_0}^{h_1} wh \frac{l \, dh}{\sin \alpha} = \int_{h_0}^{h_1} w \frac{h \, dA}{\sin \alpha}$$

This gives $F = w\bar{y}A/\sin \alpha$, where $\bar{y}$ is the vertical distance from the surface of the liquid to the centroid of the submerged area.

The hydrostatic force can be resisted by a single applied force acting against the surface on the side opposite the liquid. This single force must be located at a point about which there will be no moment due to the hydrostatic force; this point is called the "center of pressure."

Locating the center of pressure is similar to locating the center of gravity of a surface. The center of pressure might be termed the "center of force," since it is that point on the submerged surface about which there is no resultant moment due to the pressure forces acting on the surface. The term m is the distance from the surface of the liquid to the center of pressure:

$$m = \frac{\Sigma r \, \Delta F}{\Sigma \, \Delta F} = \frac{\displaystyle\int_{h_0}^{h_1} h \times hw \, dA}{\displaystyle\int_{h_0}^{h_1} hw \, dA} = \frac{\displaystyle\int_{h_0}^{h_1} h^2 \, dA}{\displaystyle\int_{h_0}^{h_1} h \, dA}$$

where

$$\int_{h_0}^{h_1} h^2 \, dA = \text{moment of inertia of the area about axis 0-0}$$

$$\int_{h_0}^{h_1} h \, dA = \bar{y} \int dA$$

Since $\bar{y} = (\int h \, dA)/\int dA$, the following general relationship results:

$$m = \frac{I}{A\bar{y}}$$

where
  m = distance from point 0 to the center of pressure
  I = moment of inertia of the area about a horizontal axis through 0
  A = area
  $\bar{y}$ = distance from 0 to the center of gravity (or centroid) of the area

- A vertical 4- by 4-ft gate is submerged with the level top 2 ft below the surface of the water. The gate is exposed to hydrostatic pressure on one side. The magnitude of the hydrostatic force acting on the gate most nearly equals which of the following?

  (a) 1570 lb   (b) 4130 lb   (c) 3290 lb   (d) 2170 lb
  (e) 3990 lb

Draw a figure and label it (Fig. 5-5). The pressure diagram will be a trapezoid, as shown. The total force acting on the gate because of the pressure of the water will then equal $[(2w + 6w)/2] \times 4$ lb/ft of width $\times$ 4 ft wide; therefore,

$$F = 4w \times 4 \times 4 = 3990 \text{ lb} \ (w \text{ for water} = 62.4 \text{ lb/ft}^3)$$

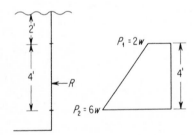

Figure 5-5

The correct answer is (e).

- For the same gate, the center of pressure will be what distance below the surface of the water?

  (a) 4.00 ft   (b) 4.33 ft   (c) 4.67 ft   (d) 5.02 ft
  (e) 4.12 ft

Note that the atmospheric pressure may be disregarded since it acts equally on both sides of the gate. The location of the center of pressure will be m ft below the surface, where $m = I/A\bar{y}$. The moment of inertia of a rectangle about a transverse axis through its center is $bh^3/12$, which equals $(4 \times 4^3)/12 = 21.3$ ft^4. The moment of inertia about the surface, 4 ft from the axis of the square, would be

$$21.3 + AS^2 = 21.3 + 16 \times 4^2 = 277.3 \text{ ft}^4$$
$$A\bar{y} = 16(2 + 2) = 64$$
$$m = 277.3/64 = 4.33 \text{ ft}$$

The correct answer is (b).

■ Flashboards placed on the crest of an ogee spillway are designed to fail when the head H is 12 ft. The boards are 4 ft wide and are supported by iron pins mounted in sockets along the crest at 10-ft intervals. Consider only hydrostatic pressure. The resisting force which must be exerted by each pin most nearly equals which of the following (see Fig. 5-6)?

(a) 13,100 lb (b) 11,900 lb (c) 14,200 lb (d) 15,400 lb
(e) 13,800 lb

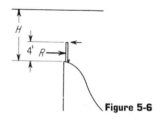

Figure 5-6

The total force acting on each foot of width will equal

$$\frac{8 + 12}{2} \times w4 \times 1 = 2500 \text{ lb/ft}$$

The location of the center of pressure $m = I/A\bar{y}$, where I is the moment of inertia of the pressurized area about an axis at the surface of the water and $\bar{y}$ is the distance below the surface to the center of gravity of the pressurized area:

$$I/\text{ft of width} = \frac{1 \times 4^3}{12} + (1 \times 4)10^2 = 405.3 \text{ ft}^4/\text{ft}$$
$$A\bar{y}/\text{ft of width} = (1 \times 4)(2 + 8) = 40$$
$$m = 405.3/40 = 10.1 \text{ ft below the surface of the water}$$

Each pin must withstand the force acting on half of the 10-ft width of flashboard on each side of it or the hydrostatic force acting on 10 ft of flashboard. Taking moments about an axis at the top of the flashboard, we find that the pins must withstand a force of

$$\frac{(2500 \text{ lb/ft} \times 10 \text{ ft})(10.1 - 8)}{4} = 13,100 \text{ lb/pin}$$

The remainder of the force of 25,000 lb/(10-ft width), or 11,900 lb/10 ft of width, must be held at the top of the flashboard. The correct answer is (a).

5-5 ARCHIMEDES'S PRINCIPLES OF BUOYANCY

The principles of buoyancy and flotation may be expressed as follows:

1. A body immersed in a fluid is buoyed up by a force equal to the weight of fluid displaced by the body.
2. A floating body displaces its own weight of the fluid in which it is floating.
3. The buoyant force acts upward through the center of gravity of the displaced volume.

   ▪ A rectangular caisson is being constructed for use in building a bridge pier. The caisson is 50 by 20 by 30 ft high (inside dimensions). The walls are 1 ft thick, and the weight is 18,000 lb. There is no bottom.

This example can be used to form the nucleus of three problems.

   ▪ Which of the following is the distance that the bottom edge would sink below the surface of the water when the caisson was launched?
   (a) caisson would sink   (b) 8.0 ft   (c) 6.0 ft   (d) 4.0 ft
   (e) 2.0 ft

The volume of the walls per foot of depth would equal $(52 \times 22) - (50 \times 20) = 144$ ft³. The volume of water displaced for each foot that the open-ended caisson sank below the surface of the water would thus equal 144 ft³, and the buoyancy per foot of caisson below the surface would equal $144 \times 62.4 = 8986$ lb/ft. The caisson would then sink $18,000/8986 = 2.00$ ft. The correct answer is (e).

- The additional weight which would be required to sink the caisson to the bottom of a 25-ft-deep lake would most nearly equal which of the following?

   (a) 207,000 lb   (b) 314,000 lb   (c) 178,000 lb
   (d) 410,000 lb   (e) 287,000 lb

The caisson would float with 2 ft submerged. To sink it another 23 ft would require displacing $23 \times 144 = 3312$ ft^3 more of water. This would require overcoming $3312 \times 62.4 = 206,670$ lb of buoyancy. The correct answer is (a).

- The amount of weight which would have to be applied to the caisson to hold it on the bottom once it had been seated and the water pumped out from the inside would most nearly equal which of the following?

   (a) 207,000 lb   (b) nothing   (c) 74,800 lb   (d) 94,300 lb
   (e) 22,500 lb

Once the caisson was seated on the bottom and the water was pumped out, there would be no buoyant force acting upward *if* (and a very big if) the bottom of the caisson were sealed tightly so that no water could seep under it. If there were seepage, then the pressure acting upward on the bottoms of the walls would range from $25 \times 62.4 = 1560$ lb/ft^2 at the outside, where the pressure would equal the static pressure at a depth of 25 ft, to zero pressure at the inside, where only atmospheric pressure would act. The average pressure acting upward on the bottoms of the walls would then equal the average pressure between 1560 and 0, or 780 lb/ft^2. The upward force acting on the caisson would thus equal 144 ft$^2 \times 780$ lb/ft$^2 = 112,300$ lb less the 18,000 lb weight of the caisson, or 94,300 lb.

This is conservative if the space inside is kept pumped out and free of internal water. If the leakage were not removed and the water level rose inside, a greater weight would be required to keep the caisson in place. The correct answer is (d).

The same principles of buoyancy are used to measure the specific gravities of liquids.

- A hydrometer bulb and stem displace 0.5 in^3 when immersed in fresh water. The stem has an area of 0.05 in^2. The distance on the stem between the graduation for 1.00 and 1.10 specific gravity would most nearly equal which of the following?

   (a) 1.10 in   (b) 0.90 in   (c) 0.85 in   (d) 0.62 in
   (e) 0.76 in

The hydrometer bulb and stem displace 0.5 in³ of water, so the bulb and stem weigh $(0.5/1728) \times 62.4$ lb. They will displace this same weight of liquid when floating in the 1.1-specific gravity liquid, so the volume of liquid displaced will equal

$$\frac{0.5}{1728} \times 62.4 \times \frac{1728}{1.1 \times 62.4} = 0.455 \text{ in}^3$$

The difference in the volume of fluid displaced (or the volume of bulb and stem immersed) will be $0.500 - 0.455 = 0.045$ in³. Since the stem has a cross-sectional area of 0.05 in², the stem will stick out of the liquid $0.045 \text{ in}^3/0.050 \text{ in}^2 = 0.90$ in higher in the 1.10-specific-gravity fluid than it did out of the 1.00-specific-gravity fluid, and the distance on the stem between the graduations for 1.00 and 1.10 specific gravity would then equal 0.90 in. The correct answer is (*b*).

- A cubical block of wood 10 cm on a side having a density of 0.5 g/cm³ floats in a container of water. Oil of density 0.8 g/cm³ is poured on the water until the top of the oil layer is 4 cm below the top of the block. The depth of the oil layer would most nearly equal which of the following?

  (*a*) 1 cm   (*b*) 2 cm   (*c*) 3 cm   (*d*) 4 cm   (*e*) 5 cm

First draw a sketch (Fig. 5-7). The mass of the block is $10 \times 10 \times 10 \times 0.5 = 500$ g. The total volume of fluid displaced equals $(10 - 4) \times 100 = 600$ cm³. The density of water is 1 g/cm³, and the weight of the block must equal the weight of the fluid displaced, so

$$100y \times 0.8 \text{ g/cm}^3 + 100(6 - y) \times 1 \text{ g/cm}^3 = 500 \text{ g}$$

$$y = 5 \text{ cm}$$

So the oil layer would be 5 cm deep. The correct answer is (*e*).

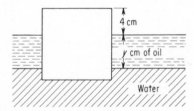

4 cm

y cm of oil

Water

Figure 5-7

- If a steel vessel 6 ft in diameter and 20 ft long is pressure-tested at 10,000 lb/in^2, which of the following most nearly equals the energy contained in the water? Ignore the energy stored due to the expansion of the vessel.

  (a) 14,500 ft·lb   (b) 20,785,000 ft·lb   (c) 13,570,000 ft·lb
  (d) 486,000 ft·lb   (e) 142,000 ft·lb

The volume of the water in the vessel equals 565.5 ft^3. The bulk modulus of water is 300,000 lb/in^2, so the compression of the water would equal $(10,000/300,000) \times 565.5 = 18.85$ ft^3, or 18.85 ft^3 additional water would have to be pumped into the vessel to increase the pressure to 10,000 lb/in^2, ignoring any expansion of the vessel. Work $= \int P\ dV$ $K = dP/(dV/V)$, or $dV = V\ dP/K$, so $\int P\ dV = 1/K \int VP\ dP = (V/K)$ $(P^2/2)$ between the limits of zero and 10,000 lb/in^2; $V = 977,163$ in^3, and the energy contained in the water alone would equal 162,860,540 in·lb, or 13,571,000 ft·lb. The correct answer is (c).

5-6 HYDRODYNAMICS

Fluids in motion are governed in large part by the same principles that were discussed in Chap. 3, "Dynamics." To analyze the characteristics of the flow of liquids, we must also, however, include the effects of viscosity; and analysis of the flow of gases must in addition include a consideration of the thermodynamic effects.

5-7 BERNOULLI'S EQUATION

Perhaps the most important fundamental relationship dealing with hydrodynamics is Bernoulli's equation. This equation is based on the concept of the conservation of energy; it states in algebraic form that the total energy at one point of a closed system is equal to the total energy at any other point in the system plus the energy lost between those two points. Bernoulli's equation may be written as

$$\frac{P_1}{w} + \frac{v_1^2}{2g} + Z_1 = \frac{P_2}{w} + \frac{v_2^2}{2g} + Z_2 + h_{L,1-2}$$

where
$\qquad P_1$ and P_2 = respective static pressures at points 1 and 2, lb/ft^2
$\qquad\qquad w$ = specific weight of the fluid flowing, lb/ft^3

v_1 and v_2 = respective velocities, ft/s

g = acceleration of gravity = 32.2 ft/s²

Z_1 and Z_2 = elevations of points 1 and 2 above some selected datum plane

$h_{L,1-2}$ = head loss measured in feet of fluid flowing between points 1 and 2

■ From the information given in the sketch (Fig. 5-8) calculate the diameter at section B in inches. The diameter at section A is 4.00 in. Kerosene is flowing at the rate of 2.5 ft³/s (specific gravity of kerosene = 0.80). Ignoring friction, the diameter at section B would most nearly equal which of the following?

(a) 3.20 in   (b) 2.75 in   (c) 1.92 in   (d) 3.10 in
(e) 2.95 in

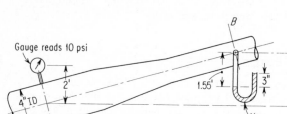

Gauge reads 10 psi

Figure 5-8

Write out Bernoulli's equation:

$$\frac{P_A}{w} + \frac{v_A{}^2}{2g} + Z_1 = \frac{P_B}{w} + \frac{v_B{}^2}{2g} + Z_2$$

Let the elevation at point A be the zero reference plane. This gives

$$Z_1 = 0 \quad \text{and} \quad Z_2 = 10 \text{ ft}$$

$$\frac{P_A}{w} = \frac{(10 \times 144) \text{ lb/ft}^2}{62.4 \times 0.80} + 2 = 28.9 + 2 = 30.9 \text{ ft}$$

$$\frac{P_B}{w} + 1.55 \text{ ft} = 3 \text{ in Hg} = \frac{3 \times 13.6}{12} \times \frac{1}{0.80} = 4.25 \text{ ft of kerosene}$$

$$\frac{P_B}{w} = 4.25 - 1.55 = 2.70 \text{ ft}$$

$$v_A = \frac{Q}{A} = \frac{2.5 \text{ ft}^3/\text{s}}{(\frac{4}{12})^2 \times \pi/4 \text{ ft}^2} = 28.7 \text{ ft/s}$$

$$\frac{v_B^2}{2g} = \frac{P_A}{w} - \frac{P_B}{w} + \frac{v_A^2}{2g} - Z_2 = 30.0 - 2.70 + \frac{28.7^2}{64.4} - 10$$

$$\frac{v_B^2}{2g} = 30.95 \qquad v_B = \sqrt{1995} = 44.7 \text{ ft/s}$$

$$A_B = \frac{Q}{v_B} = \frac{2.5}{44.7} = 0.0560 \text{ ft}^2 \qquad \text{or} \qquad 8.06 \text{ in}^2$$

$$D_B = \sqrt{8.06/0.785} = 3.20 \text{ in}$$

The correct answer is (a).

5-8 FLOW MEASUREMENT

Many flow-measuring devices are based on application of Bernoulli's principle. The ones we are most interested in are the orifice, the venturi tube, the nozzle, and the pitot tube. The orifice, the nozzle, and the venturi tube are quite similar in that all are reductions in flow area; the correlation between pressure and velocity upstream of these devices and at the sections of reduced area gives the velocity of flow through the throat and thus provides a measure of the flow rate.

None of these three devices conforms exactly to Bernoulli's equation because of fluid friction and the contraction of the jet of liquid as it flows through the section of reduced area. This leads to a correction factor called the coefficient of discharge C, which, in the case of an orifice, is the product of a velocity coefficient C_v and a coefficient of contraction C_c. For a sharp-edged orifice, $C_c = 0.62$, $C_v = 0.98$, and $C = 0.61$. This means that for a sharp-edged orifice placed in a pipe the flow through the orifice would equal $CA\sqrt{2gh}$, where h would equal $\Delta P/w$.

- If a 1-in-diameter orifice in a 4-in line has a head of 200 ft across it and the coefficient of discharge equals 0.98, the rate of flow most nearly equals which of the following?

  (a) 361 gal/min (b) 272 gal/min (c) 219 gal/min
  (d) 205 gal/min (e) 198 gal/min

$$Q = A \times 0.98 \sqrt{2g \times 200} = (\tfrac{1}{12})^2 \times \frac{\pi}{4} \times 0.98 \sqrt{64.4 \times 200}$$

$$Q = 0.607 \text{ ft}^3/\text{s} \qquad \text{or} \qquad 0.607 \times 60 \times 7.48 = 272 \text{ gal/min}$$

The correct answer is (b).

In the case of an orifice in a pipe, the fluid approaching the orifice will have a velocity which must be considered in the calculation of the orifice coefficient. If we apply Bernoulli's equation to a point upstream and at the throat of the orifice, we obtain

$$\frac{v_o^{\,2}}{2g} + \frac{P_o}{w} = \frac{v_p^{\,2}}{2g} + \frac{P_p}{w}$$

where the subscript o refers to the orifice and the subscript p refers to the pipe, since the same quantity of liquid flows through both pipe and orifice. Then

$$v_o^{\,2} = \left[1 - \left(\frac{A_o}{A_p}\right)^{\!2} \right] 2gh$$

To this must be added the effect of the previously discussed orifice coefficients, giving

$$v_o = c\sqrt{2gh}\;\frac{1}{\sqrt{1 - (A_o/A_p)^2}}$$

where $1/\sqrt{1 - (A_o/A_p)^2}$ is termed the "velocity-of-approach factor." If the diameter of the orifice is less than one-third the diameter of the pipe, the error introduced by ignoring the velocity-of-approach factor is less than 1 percent. As a result, it is usually ignored.

One other point that should be mentioned here is that much of the differential head across an orifice is recovered downstream.

Another past examination problem illustrating Bernoulli's principle was as follows.

- A 3- by 1-in venturi meter carries oil with a specific gravity of 0.86. A differential gauge connected between the pipe and the throat contains water with oil above the water in the manometer tubes. If the gauge shows a deflection of 57.6 in and the meter coefficient is 0.975, the rate of discharge will most nearly equal which of the following?

  (a) 24.2 gal/min   (b) 16.2 gal/min   (c) 18.7 gal/min
  (d) 17.1 gal/min   (e) 14.7 gal/min

First, sketch the figure. Assume that the pipe and the venturi meter are horizontal and that both are at the same elevation (Fig. 5-9). The difference in pressure between throat and body can be determined from the manometer reading:

$$P_B + 57.6 \text{ in of oil} = P_T + 57.6 \text{ in of water}$$

$$P_B - P_T = 57.6 - 57.6 \times 0.86 = 8.06 \text{ in of water}$$

$$\frac{P_B - P_T}{w} = \frac{8.06}{12} \times \frac{1}{0.86} = 0.782 \text{ ft of oil}$$

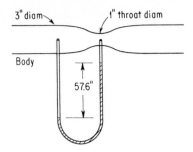

Figure 5-9

Bernoulli's equation gives

$$\frac{P_B}{w} + \frac{v_B^2}{2g} + Z_B = \frac{P_T}{w} + \frac{v_T^2}{2g} + Z_T$$

$$v_B = \frac{A_T}{A_B} v_T = \frac{v_T}{9}$$

$$\frac{P_B - P_T}{w} = \frac{v_T^2 - v_B^2}{2g} = \frac{80v_T^2}{162g}$$

$$v_T = \sqrt{\frac{0.782 \times 162g}{80}} \qquad \text{if} \quad C = 1.00$$

Actually

$$v_T = 0.975 \sqrt{\frac{0.782 \times 162g}{80}} = 6.97 \text{ ft/s}$$

$$Q = (\tfrac{1}{12})^2 \times \frac{\pi}{4} \times 6.97 = 0.0380 \text{ ft}^3/\text{s} \qquad \text{or} \qquad 17.1 \text{ gal/min}$$

The correct answer is (d).

A pitot tube is used to measure the velocity of a fluid by converting the energy due to the velocity (velocity head) to potential energy or pressure head. Referring to Bernoulli's equation, we get the relationship that $v = \sqrt{2gh}$, and since there is no flow, no flow coefficient is needed as in the previous example.

- A pitot tube having a coefficient of unity is inserted into the exact center of a long smooth pipe with an inner diameter of 1.00 in. Crude oil with a specific gravity of 0.90 and $\mu = 35 \times 10^{-5}$ lb·s/ft² is flowing through the pipe. Which of the following values most nearly equals the average velocity of the oil for the two different cases shown in Fig. 5-10?

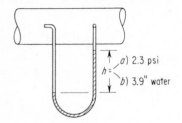

a) 2.3 psi

$h =$

b) 3.9" water

Figure 5-10

CASE 1: (a) 19.5 ft/s (b) 17.2 ft/s (c) 18.1 ft/s
(d) 16.0 ft/s (e) 15.8 ft/s
CASE 2: (a) 2.4 ft/s (b) 4.8 ft/s (c) 5.1 ft/s (d) 3.7 ft/s
(e) 3.2 ft/s

Bernoulli's equation gives us the velocity at the center of the tube. The velocity of the fluid in contact with the pitot tube is all converted into pressure head:

$$\frac{P_1}{w} + \frac{v_1^2}{2g} + Z_1 = \frac{P_2}{w} + \frac{v_2^2}{2g} + Z_2$$

where $Z_1 = Z_2$ and $v_2 = 0$. Then

$$\frac{v_1^2}{2g} = \frac{P_2 - P_1}{w}$$

For case 1:

$$\frac{P_2 - P_1}{w} = \frac{(2.3 \times 144) \text{ lb/ft}^2}{(62.4 \times 0.9) \text{ lb/ft}^3} = 5.9 \text{ ft of oil}$$

For case 2:

$$\frac{P_2 - P_1}{w} = \frac{3.9}{0.90} \times \frac{1}{12} = 0.361 \text{ ft of oil}$$

The velocity at the center of the tube would then be

$$v_a = \sqrt{5.9 \times 2g} = 19.5 \text{ ft/s}$$
$$v_b = \sqrt{0.361 \times 2g} = 4.82 \text{ ft/s}$$

5-9 AVERAGE VELOCITY IN A PIPE

The average velocity can be determined from the velocity at the center of the tube, which is the maximum velocity. In the case of viscous flow,

$v_{avg} = v_{max}/2$, and the average velocity in the case of turbulent flow is approximately equal to $0.82v_{max}$.

Turbulent flow occurs for Reynolds numbers (see below) greater than 4000; flow is ordinarily viscous for Reynolds numbers less than 2000:

$$\text{Re} = \frac{\rho D v}{\mu} = \frac{62.4 \times 0.90}{32.2} \frac{\text{lb·s}^2}{\text{ft}^4} \times \tfrac{1}{12} \text{ ft} \times \frac{1}{35 \times 10^{-5}} \frac{\text{ft}^2}{\text{lb·s}} \times v \text{ ft/s}$$

$$= 415v$$

For case 1: Re = 8100—turbulent flow

For case 2: Re = 2000—viscous flow

Average velocities will then be:

Case 1: $v_{avg} = 0.82 \times 19.5 = 16$ ft/s, Re = 6640, answer (d)

Case 2: $v_{avg} = 0.50 \times 4.82 = 2.41$ ft/s, Re = 1000, answer (a).

5-10 REYNOLDS NUMBER

The Reynolds number, abbreviated Re, is a dimensionless number equal to

$$\frac{\rho D v}{\mu}$$

where

ρ = density of the fluid flowing, slugs/ft^3

$\rho = w/g$

v = average fluid velocity in pipe, ft/s

μ = absolute viscosity in ll s/ft^2, or μ = centipoises (cP)/47,800

D = four times the hydraulic radius (D = diameter of a circular pipe)

The Reynolds number also equals Dv/v, where v is the kinematic viscosity measured in square feet per second (v = centistokes divided by 93,000).

The hydraulic radius is equal to the cross-sectional flow area divided by the wetted perimeter. For a circular pipe,

$$R_H = \frac{(\pi/4)D^2}{\pi D} = \frac{D}{4}$$

For a rectangular duct of width w and height h,

$$R_H = \frac{wh}{2w + 2h} = \frac{wh}{2(w + h)}$$

For a square of side length s, $R_H = s/4$. In calculating the hydraulic radius of an open trough or flume, it is important to remember the phrase "wetted perimeter." The hydraulic radius R_H for the trough shown in Fig. 5-11 of width w with water flowing at a depth h would equal $(wh)/(w + 2h)$, since the top is open and the "wetted perimeter" includes only the two sides and the bottom.

Figure 5-11

5-11 FRICTIONAL HEAD LOSS

The head loss due to frictional flow of fluid flowing full in a pipe or duct can be determined from the relationship (Darcy's equation)

$$h_L = f\frac{L}{D}\frac{v^2}{2g}$$

where

h_L = head loss, ft of fluid flowing
L = length of pipe or duct through which the fluid flows, ft
v = fluid velocity, ft/s
g = acceleration of gravity
D = four times the hydraulic radius (or the diameter of a circular pipe), ft
f = friction factor, which is dimensionless and depends on the value of the Reynolds number and the pipe roughness

For laminar flow, $f = 64/\text{Re}$ and f is independent of any ordinary internal pipe roughness. In the turbulent range, the value of f is somewhat dependent on the roughness of the pipe and is usually taken from a curve of f vs. Re, such as is shown in Fig. 5-12.

- What size of pipe will be required to transport 2 ft³/s of oil with a drop in pressure (frictional head loss) of 1.0 lb/in² per 1000 ft of pipe? The specific gravity of the oil is 0.85, and the absolute viscosity is 12 cP at a temperature of 70°F. Assume steel pipe.

  (*a*) 10 in   (*b*) 11 in   (*c*) 12 in   (*d*) 13 in   (*e*) 14 in

First

$$h_L = f\frac{Lv^2}{2gD}$$

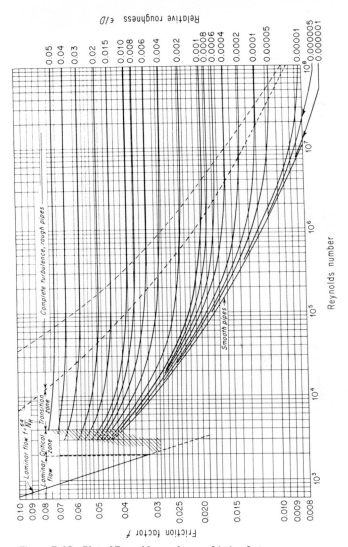

Figure 5-12 Plot of Reynolds number vs. friction factor.

Figure 5-12 (Continued)
Table of Values of Absolute Roughness ε for New Pipes

	Feet
Drawn tubing, brass, lead, glass, centrifugally spun concrete, bituminous lining, transite	0.000005
Commercial steel or wrought iron	0.00015
Welded-steel pipe	0.00015
Asphalt-dipped cast iron	0.0004
Galvanized iron	0.0005
Cast iron, average	0.000085
Wood stave	0.0006–0.003
Concrete	0.001–0.01
Riveted steel	0.003–0.03

ϵ/D for use with curves of Fig. 5-12 found by dividing the value given above for ϵ by diameter of pipe in feet.

where h_L is measured in feet of fluid flowing. Then

$$h_L = \frac{1 \times 144 \text{ lb/ft}^2}{62.4 \times 0.85 \text{ lb/ft}^3} = 2.72 \text{ ft/1000 ft}$$

$$\frac{2gh_L}{L} = f\frac{v^2}{D} = \frac{2g \times 2.72}{1000} = 0.175 = f\frac{v^2}{D}$$

Then

$$\mu = \frac{0.12 \text{ poise}}{478} = 2.50 \times 10^{-4} \text{ lb·s/ft}^2$$

$$\text{Re} = \frac{\rho Dv}{\mu} = \frac{62.4 \times 0.85}{32.2} \times \frac{1}{2.50 \times 10^{-4}} \times Dv = 6580Dv$$

$$2 \text{ ft}^3/\text{s} = Q = vA = v\frac{\pi}{4}D^2$$

giving $vD^2 = 2.55$. These three relationships reduce to

$$\text{Re} = \frac{17,500}{D} \qquad \text{and} \qquad f = 0.0269D^5$$

From Fig. 5-13, using the curve for steel pipe, $\epsilon/D = 0.00015$. If D is 1 ft, Re $= 17,500$ and $f = 0.028$, so D should be slightly larger to satisfy the requirements exactly. The theoretical inner diameter would be approximately $12\frac{1}{8}$ in, but standard pipe is actually a little less than 12 in ID (schedule 40 pipe, 11.938 in ID for 12-in nominal pipe size), and this

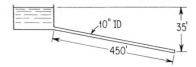

Figure 5-13

ordinarily has a $\pm\frac{1}{32}$-in tolerance. In addition, the calculated pressure drop is not absolutely accurate, and the friction factor f may vary as much as ±10 percent for commercial pipe. Considering these various factors, standard 12-in pipe would be the best selection. The 12-in-ID pipe would give

$$v = \frac{2}{0.785} = 2.55 \text{ ft/s}$$

$$\text{Re} = \frac{62.4 \times 0.85}{32.2} \frac{1 \times 2.55}{2.50 \times 10^{-4}} = 16{,}800$$

giving $f = 0.028$ and

$$h_L = 0.028 \frac{1000 \times (2.55)^2}{2g \times 1} = 2.82 \text{ ft/1000 ft}$$

$$2.82 \text{ ft} \times (62.4 \times 0.85) \text{ lb/ft}^3 \times \frac{1 \text{ ft}^2}{144 \text{ in}^2} = 1.04 \text{ lb/in}^2/1000 \text{ ft}$$

estimated pressure drop. The correct answer is (c).

- What quantity of water, in cubic feet per second, will flow in a 450-ft 10-in ID pipe if the static head is 35 ft? The friction factor may be considered essentially constant, having a value of 0.025. Disregard entrance and exit losses (Fig. 5-13). Which of the following values most nearly equals the rate of flow which would occur?

  (a) 7.2 ft³/s (b) 6.8 ft³/s (c) 6.6 ft³/s (d) 6.4 ft³/s
  (e) 6.2 ft³/s

Apply Bernoulli's equation, taking as the two points the top of the reservoir and the discharge end of the pipe:

$$\frac{P_1}{w} + \frac{v_1^2}{2g} + Z_1 = \frac{P_2}{w} + \frac{v_2^2}{2g} + Z_2 + h_{L,1-2}$$

where P_1 and P_2 are equal to atmospheric pressure and v_1 is negligible, since the area of the reservoir is many times that of the pipe cross section, and $Z_1 - Z_2 = 35.0$ ft, the difference in elevation between the two points.

Thus

$$35 = \frac{v_2^2}{2g} + h_L = 0.0155v_2^2 + h_L$$

$$h_L = f\frac{Lv^2}{2gD} = 0.025\,\frac{450v^2}{2g\frac{10}{12}} = 0.210v^2$$

$$35 = 0.0155v_2^2 + 0.210v_2^2 \qquad v = \sqrt{\frac{35}{0.226}} = 12.45 \text{ ft/s}$$

$$Q = vA = 12.45 \times (\tfrac{10}{12})^2 \times \frac{\pi}{4} = 6.80 \text{ ft}^3/\text{s}$$

The correct answer is (b).

5-12 OPEN-CHANNEL FLOW

If the flow is in a pipe which is not running full or if it is in an open channel, the flow rate is usually calculated by means of the Manning formula:

$$v = 1.486\,\frac{R^{2/3}S^{1/2}}{n}$$

where
- R = hydraulic radius
- S = slope of the channel, ft/ft
- n = Manning factor, which depends on the material of which the channel is made and the roughness of the inside of the channel (see Table 5.1)

■ A sewer made of concrete pipe (Manning coefficient 0.013) 4 ft in diameter is to carry 32.3 ft³/s when running half full. Which of the following values is closest to the minimum slope, measured in feet per 1000 ft, which would be required?

(a) 4.0   (b) 6.2   (c) 3.1   (d) 2.0   (e) 1.9

Sketch the pipe (Fig. 5-14). Using Manning's formula for flow in open

Figure 5-14

TABLE 5.1   Values of Roughness Coefficient, n (Manning Factor)

Surface	Average value of n
Wood, smooth, planed surface	0.011
Smooth metal	0.011
Finished concrete	0.012
Wood, unplaned	0.013
Good ashlar masonry or brick work	0.013
Unfinished concrete	0.014
Cast iron	0.015
Vitrified clay	0.015
Brick	0.016
Rubble masonry	0.017
Riveted steel	0.018
Firm gravel	0.020
Corrugated metal	0.022
Earth canals and rivers in good condition	0.025
Gravel	0.029
Earth canals with stones or weeds	0.035
Rough natural streams	0.050
Very weedy natural streams	0.100

channels, given above, we obtain

$$R = \frac{\text{cross-sectional area}}{\text{wetted perimeter}} = \frac{\frac{1}{4}(\pi/4)D^2}{\frac{1}{2}\pi D} = \frac{D}{4} = 1 \text{ ft}$$

$$\text{Required } v = \frac{Q}{A} = \frac{32.2}{\frac{1}{4}(\pi/4)D^2} = \frac{32.2}{6.28} = 5.13 \text{ ft/s}$$

$$5.13 = 1.486 \frac{1^{2/3}S^{1/2}}{0.013} \quad S = \left(\frac{5.13 \times 0.013}{1.486}\right)^2 = 0.00202$$

or the slope must equal 2.02 ft/1000 ft of line. The correct answer is (d).

5-13 CENTRIFUGAL PUMPS

In a centrifugal pump the fluid leaves the periphery of the impeller at a higher velocity than that at which it enters; this velocity is partly converted into pressure by the pump casing before the fluid leaves the pump through the nozzle. Thus a centrifugal pump is a device which supplies kinetic energy to a fluid (velocity head) and then converts it into potential energy (pressure head).

The pump affinity laws are expressed by the following formulas:

$$\frac{Q_1}{Q_2} = \frac{n_1}{n_2} \qquad \frac{H_1}{H_2} = \frac{n_1^2}{n_2^2} \qquad \frac{HP_1}{HP_2} = \frac{n_1^3}{n_2^3}$$

A pump must "suck" fluid into its intake, so the absolute pressure at the intake must be great enough to prevent cavitation (or "pulling apart") of the fluid. This pressure requirement is called the "net positive suction head" (NPSH). It is defined as the pressure of the fluid in feet at the inlet to the pump referred to the center line of the pump, minus the vapor pressure of the fluid at the fluid temperature in feet, plus the velocity head of the fluid at this point.

■ Several water piping systems are under consideration for use with a pump which has the operating characteristics shown in Fig. 5-15. Curves showing the operating characteristics of the piping systems under consideration are given in the lower figure. It is desired to use the piping system which will permit the pump to operate at its highest efficiency (1 gal of water weighs 8.33 lb).

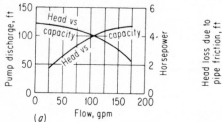 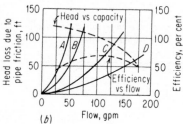

Figure 5-15 (a) Curves represent actual pump test information; (b) curves show head loss for piping systems A, B, C, and D.

Four different questions will be asked about the piping systems.

■ Which of the following systems will permit the most efficient pump operation?

(a) system A   (b) system B   (c) system C   (d) system D
(e) none of these

Plot the curve of pump efficiency vs. flow:

$$\text{Efficiency} = \frac{\text{hp out of pump}}{\text{hp input to pump}}$$

The horsepower out of the pump equals the water horsepower (Whp):

$$\text{Whp} = \frac{(\text{gal/min} \times 8.33) \text{ lb/min} \times (\text{head}) \text{ ft}}{33,000 \text{ ft·lb/min}} = 2.525 \times 10^{-4} \text{ gal/min} \times (\text{head})$$

$$\text{Efficiency} = \frac{2.525 \times 10^{-4} (\text{gal/min}) \times (\text{head})}{\text{hp input}}$$

Taking different points from the pump data curve gives the following:

Gallons per minute	Feet of head	Water horsepower	Horsepower input	Efficiency, %
25	119	0.75	1.8	42
50	115	1.45	2.5	58
75	110	2.08	3.2	65
100	102	2.58	3.8	68
125	93	2.94	4.3	69
150	78	2.95	4.5	66
175	55	2.43	4.6	53

The resulting curve of efficiency vs. flow can be plotted on the curve showing the head losses for piping systems A, B, C, and D. The efficiency curve and the head vs. capacity curve are shown dotted in the figure. Piping system C will permit the pump to operate more efficiently than any of the other systems. The correct answer is (c).

- Which of the following is closest to the quantity of flow which would be pumped through the piping system selected?

  (a) 108 gal/min   (b) 118 gal/min   (c) 130 gal/min
  (d) 137 gal/min   (e) 142 gal/min

From the curve and the chart of pump data (by interpolation), the most efficient operating point will be at 130 gal/min. The correct answer is (c).

- The efficiency of the pump at the selected operating condition would most nearly equal which of the following?

  (a) 57%   (b) 72%   (c) 61%   (d) 68%   (e) 79%

From the plotted curves and the tabulated data, it can be seen that the efficiency of the pump at the operating condition would equal 68 percent. The correct answer is (d).

- Which of the following motor sizes would you recommend so that the pump might operate over its full capacity range without over-loading of the motor?

  (a) 5 hp   (b) 6 hp   (c) 4 hp   (d) $4\frac{1}{2}$ hp   (e) 7 hp

The power requirement would equal 4.34 hp at the recommended operating condition. However, if someone opened a valve, the flow would increase and the power requirement would rise to almost 5 hp. It would be best to specify a 5-hp pump for the system. The correct answer is (a).

- A pump for the water system shown in Fig. 5-16 is to be selected from the characteristic curves shown. The pump is to service both tanks, pumping into one tank at a time. All piping is 6-in diameter, new steel (standard strength, schedule 40) with $f = 0.017$. The pump desired is to deliver more than 500 gal/min to each tank.

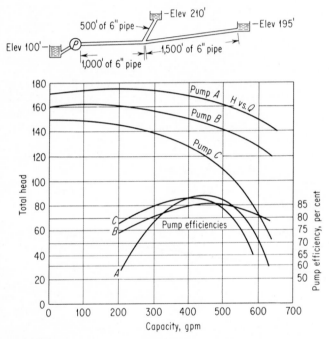

Figure 5-16

This problem can be used to illustrate two aspects of pump analysis, thus providing the nucleus for two problems.

- For the given conditions, which of the three pumps would provide the most efficient operation?

    (a) pump A (b) pump B (c) pump C

Utilizing the information given in Fig. 5-16 and the graph of the pump characteristics, and designating the points as R, F, and S for reservoir, first tank, and second tank, respectively, we can write Bernoulli's equation for each case:

$$\frac{P_R}{w} + \frac{v_R^2}{2g} + Z_R + h_P = \frac{P_F}{w} + \frac{v_F^2}{2g} + Z_F + h_{L,R-F}$$

where h_P = head supplied by pump; then

$$h_P = \frac{v_F^2}{2g} + 110 + h_{L,R-F}$$

where v_F = velocity through pipe.

The inner diameter of 6-in schedule 40 pipe is 6.065 in.

$$500 \text{ gal/min} = \frac{500}{60 \times 7.48} = 1.115 \text{ ft}^3/\text{s} \qquad A = 0.200 \text{ ft}^2$$

$$v_F = \frac{Q}{A} = \frac{1.115}{0.200} = 5.58 \text{ ft/s}$$

$$h_{L,R-F} = f\frac{L}{D}\frac{v_F^2}{2g} = 0.017 \times \frac{1500}{0.505} \times \frac{v_f^2}{2g} = 0.784 v_F^2$$

$$h_p = \frac{v_F^2}{2g} + 110 + 0.784 v_F^2 = 134.9 \text{ ft} \qquad \text{for tank 1}$$

For the second tank,

$$h_p = \frac{v_s^2}{2g} + 95 + h_{L,R-S}$$

$$h_{L,R-S} = 0.017 \times \frac{2500}{0.505} \times \frac{v^2}{2g} = 1.31 v_s^2$$

$$h_p = \frac{v_s^2}{2g} + 95 + 1.31 v_s^2 = 0.483 + 95 + 40.8 = 136.3 \text{ ft}$$

The pump would then have to supply at least 136.3 ft of head at a flow of 500 gal/min. This rules pump C out because the characteristic curve for this pump shows that it produces only about 110 ft of head at a flow rate of 500 gal/min. Either pump A or pump B could handle the operation. Pump A is more efficient at a flow rate of 500 gal/min, but to operate at this flow rate it would have to pump against a pressure of slightly over 160 ft, as shown on its characteristic curve.

The pump characteristic curve is a curve of the actual operational characteristics of a pump. Such a curve is constructed from data obtained during a test of the pump. It shows the output characteristics of a pump of a particular size and design. Note that the discharge pressure is measured in feet of fluid flowing rather than pressure. Pump A, for example, would pump against a discharge pressure of 170 ft at a flow rate of 400 gal/min. If water were being pumped, the 170 ft of head would correspond to a gauge pressure of $170 \times 0.433 = 73.6$ lb/in^2 ($0.433 = 62.4/144$ and has the units of pounds per square inch per foot of water). If oil with a specific gravity of 0.84 were being pumped, rather than the water, then the gauge would register a pressure of only $73.6 \times 0.84 = 61.8$ lb/in^2, at the flow rate of 400 gal/min.

If pump A were used in this system, the quantity of liquid handled by the pump would have to increase until the pressure required to force the larger flow through the system matched the new flow rate on the characteristic curve for pump A.

The system pressure drop to the second tank equals

$$h_p = \frac{v_s^2}{2g} + 95 + 1.31v^2 = 95 + 1.32v^2$$

The velocity will be directly proportional to the flow rate in gallons per minute, so the head loss for a 600 gal/min flow rate would equal

$$95 + (600/500)^2 \times 41.3 = 154.4 \text{ ft}$$

This is almost on the curve for pump A. At this point, 600 gal/min flow rate, the efficiency of pump A drops to 72 percent. This is the point at which pump A would operate if it were used with the given system.

Pump B would also pump more than 500 gal/min since the discharge pressure of this pump at the 500 gal/min rate is 140 ft and the system backpressure is only 136.4 ft. It would pump only a little more, however, until the flow rate and system backpressure matched a point on the curve for pump B. The operating point for the second tank would shift to approximately 515 gal/min and 139 ft of head. At this point, the efficiency would equal 84 percent. This is an appreciably better efficiency than could be obtained with pump A *in this particular system*. The correct answer is (*b*).

Two points of importance regarding centrifugal pump operation might be emphasized. The first is that the minimum power consumption by the pump occurs at pump shutoff. Power output of the pump at shutoff equals zero, since at this point the value of the integral of $P \, dV$ equals zero since dV is zero. As a word of caution, the opposite is true in the case of a positive displacement pump or a turbine vane, or regenerative-type, pump. The second point of importance is that a centrifugal pump will *always* operate at some point on its characteristic curve. It cannot operate anywhere else, and if the downstream condition of the pumping system is changed, the pump output will shift until it matches a point on the characteristic curve.

■ Under the operating conditions determined in the preceding problem, the maximum necessary horsepower output of the motor for satisfactory operation of the pumping system would most nearly equal which of the following?

(a) 22 hp   (b) 28 hp   (c) 18 hp   (d) 25 hp   (e) 31 hp

The pump would pump approximately 525 gal/min at 137 ft of head into the first tank, again with an efficiency of 84 percent.

The maximum horsepower out of the motor would be

$$\frac{525 \text{ gal/min} \times 8.34 \text{ lb/gal} \times 137 \text{ ft}}{33,000 \times 0.84 \text{ efficiency}} = 21.7 \text{ hp}$$

The correct answer is (a).

The graph of pump characteristics gives "total head" as the item listed on the ordinate. Total head would equal pressure head plus velocity head. The velocity head is the term $v_s^2/2g$. This has been included in the calculations, so the analysis is complete. However, we might check the magnitude of this factor for the 525 gal/min flow. For this rate of flow, $v^2/2g = 0.531$ ft, so the velocity head is small enough to have safely been ignored. This is usually the case in pump calculations, but the velocity head should always be checked before it is disregarded.

5-14   FLUID MOMENTUM

Impulse equals the change in momentum, or $F \, \Delta t = \Delta(Mv)$, which becomes $F \, \Delta t = M \, \Delta v$ for a mass which remains constant. This can also be recognized as an expression of Newton's second law, $F = ma$, or $F = m(dv/dt) = (m/dt) \, dv$.

■ A horizontal bend in a pipeline reduces the pipe from a 30-in diameter to an 18-in diameter while bending through an angle of 135° from its original direction. The flow rate is 10,000 gal/min, and the direction of flow is from the 30- to the 18-in pipe. If the pressure at the entrance is 60 lb/in² gauge, which of the following horizontal resultant forces would be required to hold the bend in place? Consider the elbow to be adequately supported in the vertical direction.

(a) 695 lb   (b) 10,600 lb   (c) 27,800 lb   (d) 42,700 lb
(e) 54,800 lb

Draw the figure (Fig. 5-17), and calculate the velocities and pressures in and out of the bend:

$$10,000 \text{ gal/min} \times \frac{1}{60 \times 7.48} = 22.3 \text{ ft}^3/\text{s}$$

$$\text{Area}_A = 4.91 \text{ ft}^2$$

$$\text{Area}_B = 1.77 \text{ ft}^2$$

$$v_A = \frac{22.3}{4.91} = 4.54 \text{ ft/s}$$

$$v_B = \frac{22.3}{1.77} = 12.6 \text{ ft/s}$$

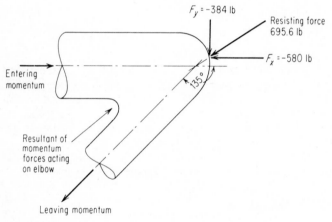

Figure 5-17

Determine the pressure at section B by means of Bernoulli's equation:

$$\frac{P_A}{w} + \frac{v_A{}^2}{2g} = \frac{P_B}{w} + \frac{v_B{}^2}{2g} \qquad P_A - P_B = \frac{w}{2g}(v_B{}^2 - v_A{}^2)$$

$$P_A - P_B = \frac{62.4}{64.4}(159 - 20.6) = 134 \text{ lb/ft}^2, \text{ or } 0.931 \text{ lb/in}^2$$

The pressure at section B is $P_B = 59 \text{ lb/in}^2$.

First determine the momentum forces acting on the elbow. Assuming water flow, 10,000 gal/min corresponds to 1388 lb/s weight flow or 43.1

slugs/s mass flow $= m/t$. The momentum force acting on the elbow would equal the mass flow rate times the change in velocity:

$$F_M = \frac{m}{t} \Delta v$$

or, in the X direction

$$F_{Mx} = \frac{m}{t} (v_{1x} - v_{2x})$$

in the Y direction

$$F_{My} = \frac{m}{t} (v_{1y} - v_{2y})$$

For this case,

$$v_{1x} = 4.54 \text{ ft/s}$$
$$v_{2x} = -12.6 \times \cos 45° = -8.91 \text{ ft/s}$$
$$v_{1y} = 0$$
$$v_{2y} = -12.6 \times \sin 45° = -8.91 \text{ ft/s}$$
$$F_{Mx} = 43.1 (4.54 + 8.91) = 580 \text{ lb}$$
$$F_{My} = 43.1 (0 + 8.91) = 384 \text{ lb}$$

Thus the fluid exerts a force on the elbow to the right equal to 580 lb and a force upward of 384 lb. To withstand these forces, the elbow must exert equal and opposite forces on the fluid. The resultant force which must be exerted on the elbow to balance the fluid momentum forces would then equal 580 lb to the left and 384 lb downward. The resulting force which must be applied externally to the elbow to withstand the fluid momentum forces would equal $\sqrt{580^2 + 384^2} = 695.6$ lb. It would act to the left at an angle equal to $\tan^{-1} (384/580) = 33.5°$ down from the horizontal (see Fig. 5-17).

Alternatively, it might be assumed that all the momentum force carried by the fluid entering the region is dissipated and that all the momentum carried by the fluid leaving the region is created. Then the net momentum force acting on the elbow in this example would equal the momentum force entering minus the momentum force leaving the elbow, bearing in mind that momentum is a vector quantity.

For the example considered:

$$\text{Entering } F_M = 43.1 \text{ slugs/s} \times 4.54 \text{ ft/s} = 196 \text{ lb}$$
$$\text{Leaving } F_M = 43.1 \text{ slugs/s} \times 12.6 \text{ ft/s} = 543 \text{ lb}$$
$$\text{Entering } F_{Mx} = 196 \text{ lb}$$
$$\text{Entering } F_{My} = 0$$
$$\text{Leaving } F_{Mx} = -543 \cos 45° = -384 \text{ lb}$$
$$\text{Leaving } F_{My} = -543 \sin 45° = -384 \text{ lb}$$

The sums of the momentum forces acting on the elbow are

$$\text{Total } F_{Mx} = 196 - (-384) = 580 \text{ lb}$$
$$\text{Total } F_{My} = 0 - (-384) = 384 \text{ lb}$$

The resultant force acting on the elbow equals 696 lb as before.

To determine the total force acting on the elbow, it is necessary to add the pressure forces to the momentum forces:

$$\text{Pressure force at } A = 30^2 \frac{\pi}{4} \times 60 = 42,400 \text{ lb} = F_A$$

$$\text{Pressure force at } B = 18^2 \frac{\pi}{4} \times 59 = 15,000 \text{ lb}$$

The force at B can be resolved into forces in the x and y directions:

$$F_{B-x} = 15,000 \times 0.707 = 10,600 \text{ lb}$$
$$F_{B-y} = 15,000 \times 0.707 = 10,600 \text{ lb}$$

The total forces add to (see Fig. 5-18)

$$F_x = F_A + F_{B-x} + \Delta M_{Fx} = 53,581 \text{ lb}$$
$$F_y = F_{B-y} + \Delta MF_y = 10,985 \text{ lb}$$
$$F_R = \sqrt{F_x^2 + F_y^2} = 54,800 \text{ lb}$$

The correct answer is (*e*).

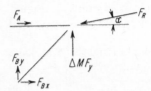

Figure 5-18

$$\alpha = \tan^{-1} \frac{F_y}{F_x} = \tan^{-1} 0.205 = 11.6°$$

5-15 JET PROPULSION

The principle of momentum force, $F = (m/dt)\,\Delta v$, has been utilized in rocketry and jet propulsion. It has also been utilized in helping to control unruly crowds.

- If a fire hose with a 1-in-diameter nozzle is supplied with water at a pressure of 100 lb/in², which of the following most nearly equals the force which would be exerted by the jet if it impinged on a flat (or slightly rounded) surface which was perpendicular to the axis of the jet?

  (a) 120 lb   (b) 151 lb   (c) 175 lb   (d) 114 lb   (e) 137 lb

The velocity of the jet can be determined by calculating the change from potential energy (100 lb/in²) to kinetic energy (velocity head). Assume that the nozzle is 98 percent efficient. Then $v = 0.98 \sqrt{2gh} = 7.86\sqrt{h}$. The head of water equivalent to 100 lb/in² equals

$$\frac{(100 \times 144)\ \text{lb/ft}^2}{62.4\ \text{lb/ft}^3} = 231\ \text{ft}$$

and the jet velocity would equal 119.5 ft/s. The flow would equal

$$119.5\ \text{ft/s} \times \frac{0.785}{144}\ \text{ft}^2 = 0.651\ \text{ft}^3/\text{s}$$

Mass flow would equal

$$0.651 \times \frac{62.4}{32.2} = 1.26\ \text{slugs/s}$$

The velocity of the jet in the axial direction would change from 119.5 ft/s to zero after the jet had impinged on the flat surface, so $\Delta v = 119.5$ ft/s. Thus

$$\text{Jet force} = 1.26 \times 119.5 = 151\ \text{lb}$$

The reaction force acting on the nozzle would equal essentially the same. That is why it takes two (or more) firefighters to hold the nozzle on a firehose. The correct answer is (b).

5-16 COMPRESSIBLE FLOW

There have been only a few references to compressible fluid flow in past examinations, and most of these could be treated satisfactorily by the application of the principles discussed in the sections on incompressible fluid flow. A general relationship to keep in mind is that the error introduced by considering (and treating) a compressible fluid as an incompressible fluid is approximately equal to one-fourth of the square of the Mach number of the fluid flowing for velocities appreciably less than the sonic velocity. In other words,

$$\text{Error} = \text{(approximately)} \frac{1}{4} \left(\frac{v}{c} \right)^2$$

where c is the velocity of sound at the existing conditions.

▪ An air-conditioning duct is designed to carry 2300 ft³ of air per minute (ft³/min) with a pressure loss of 0.4 in of water using standard air weighing 0.075 lb/ft³. During construction it was found that several right-angle turns had to be added to avoid other piping. Upon completion the duct was tested and found to have a pressure loss of 0.6 in of water at 2300 ft³/min, using air having a "density" of 0.074 lb/ft³. Which of the following most nearly equals the required increase in air horsepower?

(*a*) 0.07 hp   (*b*) 0.10 hp   (*c*) 0.04 hp   (*d*) 0.15 hp
(*e*) 0.05 hp

The pressure loss in the system will equal the friction loss

$$h_L = f \frac{Lv^2}{2gD}$$

and the losses in the elbows. The losses in the elbows can also be measured in "velocity heads." A velocity head equals $v^2/2g$. Then

$$\text{hp}_{\text{design}} = \frac{2300 \text{ ft}^3/\text{min} \times 0.4/12 \times 0.433 \times 144 \text{ lb/ft}^2}{33,000} = 0.145 \text{ hp}$$

$$\text{hp}_{\text{measured}} = \frac{2300 \text{ ft}^3/\text{min} \times 0.6/12 \times 0.433 \times 144 \text{ lb/ft}^2}{33,000} = 0.217 \text{ hp}$$

$$0.217 - 0.145 = 0.072 \text{ increase in hp}$$

The correct answer is (*a*).

Sample Problems

5-1 The force on the end of a triangular upright trough having an apex angle of 90° most nearly equals which of the following when the trough is filled with water to a height of 3 ft?

(*a*) 584 lb   (*b*) 562 lb   (*c*) 559 lb   (*d*) 548 lb   (*e*) 576 lb

5-2 From the information shown on the schematic diagram in Fig. 5P-2, the velocity of the fluid at section *B* most nearly equals which of the following?

(*a*) 52 ft/s   (*b*) 61 ft/s   (*c*) 41 ft/s   (*d*) 38 ft/s   (*e*) 48 ft/s

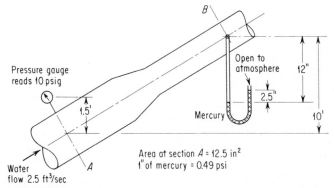

Figure 5P-2

5-3 Water flows over the triangular weir shown in Fig. 5P-3 with a head of 1 ft. The coefficient of discharge *C* is 0.60, and the temperature of the water is 140°F. Which of the following most nearly equals the flow over the weir in gallons per hour?

(*a*) 32,000   (*b*) 28,000   (*c*) 36,000   (*d*) 38,000   (*e*) 40,000

Figure 5P-3

5-4 A piece of glass weighs 125 g in air and 92 g in gasoline. The specific gravity of the gasoline most nearly equals which of the following?

(*a*) 0.54   (*b*) 0.58   (*c*) 0.66   (*d*) 0.69   (*e*) 0.72

5-5 A 21- by 12-in timber 12 ft long floats level in freshwater with 4.3 in above the water surface. One end is just touching a rock which prevents that end from sinking any deeper. Which of the following most nearly equals the

distance a 150-lb person can walk out from the supported end before the free end submerges?
(a) 7.2 ft   (b) 7.4 ft   (c) 6.8 ft   (d) 5.9 ft   (e) 5.7 ft

5-6 An 8-in ID cast-iron pipeline, 3500 ft long, conveys water from a pump to a reservoir whose water surface is 450 ft above the pump, which is pumping at the rate of 3 ft^3/s. Disregard velocity head and minor losses. Which of the following most nearly equals the gauge pressure at the discharge end of the pump?
(a) 246 lb/in^2 (b) 258 lb/in^2 (c) 235 lb/in^2 (d) 250 lb/in^2
(e) 240 lb/in^2

5-7 Water is flowing horizontally from a reservoir through a circular orifice under a head of 50 ft. The coefficient of discharge is 0.7, and the diameter of the orifice is 4 in. Which of the following values most nearly equals the rate of flow out of the orifice?
(a) 3.9 ft^3/s   (b) 2.9 ft^3/s   (c) 3.1 ft^3/s   (d) 4.2 ft^3/s
(e) 3.5 ft^3/s

5-8 A swimming pool is being cleaned with a brush fixture on the end of a 50-ft length of 2-in-diameter hose. Water is drawn through the hose by a pump which is 12 ft above the bottom of the pool. If the water in the pool is 8 ft deep, which of the following most nearly equals the suction pressure required to draw 60 gal/min through the hose? (Assume a loss of one-half of a velocity head at the brush. The kinematic viscosity of the water equals 1.2×10^{-5} ft^2/s.)
(a) -2.8 lb/in^2 (b) -3.1 lb/in^2 (c) -3.4 lb/in^2 (d) -3.7 lb/in^2
(e) -4.1 lb/in^2

5-9 Water flows from a supply tank through 80 ft of 6-in ID welded steel pipe to a hydraulic mining nozzle 2 in in diameter. The coefficient of discharge of the nozzle may be assumed to equal 0.90. The water may be assumed to have a temperature of 75°F and a kinematic viscosity of 0.00001 ft^2/s. If the rate of water flow is to be 1.00 ft^3/s, which of the following most nearly equals the distance which the nozzle must lie below the elevation of the surface of the water in the tank?
(a) 40 ft   (b) 37 ft   (c) 32 ft   (d) 45 ft   (e) 38 ft

5-10 Water falling from a height of 100 ft at the rate of 2000 ft^3/min drives a water turbine connected to an electric generator at 120 r/min. If the total resisting torque due to friction is 400 lb at 1-ft radius and the water leaves the turbine blades with a velocity of 12 ft/s, which of the following most nearly equals the horsepower developed by the turbine?
(a) 350 hp   (b) 360 hp   (c) 365 hp   (d) 372 hp   (e) 355 hp

5-11 A venturi meter with a 2-in-diameter throat is placed in a 4-in-diameter pipeline which carries fuel oil. If the differential pressure between the up-stream section and the throat equals 3.5 lb/in^2, which of the following most nearly equals the rate of oil flow? The discharge coefficient of the venturi meter is 0.97, and the specific weight of the oil is 58.7 lb/ft^3.
(a) 240 gal/min   (b) 230 gal/min   (c) 220 gal/min
(d) 235 gal/min   (e) 225 gal/min

5-12 A rectangular sluice gate, 4 ft wide and 6 ft deep, hangs in a vertical plane. It is hinged along the top (4-ft) edge. If the surface of the water is 21 ft above the top of the gate, which of the following forces applied horizontally at the bottom of the gate most nearly equals the force necessary to open it?
(a) 20,200 lb   (b) 22,800 lb   (c) 19,400 lb   (d) 18,700 lb
(e) 17,200 lb

5-13 A pump on the diagram shown in Fig. 5P-13 pumps water to an elevation of X feet above the pump. When the pump is operating, the pressure gauge A registers 39.2 lb/in². When the pump is shut off, the pressure gauge A reads 10 lb/in² because of the water in the line. A characteristic curve for the pump furnished by the manufacturer is shown. Which of the following values most nearly equals the efficiency of the pump if the input power is 1.25 hp when the discharge pressure equals 39.2 lb/in²?
(a) 71%   (b) 69%   (c) 75%   (d) 67%   (e) 73%

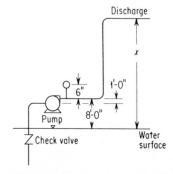 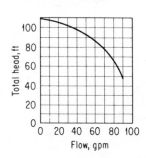

Figure 5P-13

5-14 A gravity dam composed of concrete is 45 ft high. It is 3 ft wide on top, and the base is 30 ft wide. The water side of the dam is vertical, and the other side slopes uniformly from top to bottom. The water behind the dam is 42 ft deep, and there is no pressure under the dam. The pressure at the heel of the dam most nearly equals which of the values listed below?
(a) 3010 lb/ft²   (b) 2460 lb/ft²   (c) 2020 lb/ft²   (d) 2140 lb/ft²
(e) 2290 lb/ft²

5-15 A smooth nozzle on a fire hose is 1⅛ in in diameter and discharges 250 gal/min. The force with which the stream strikes a flat surface normal to the stream placed only a short distance away most nearly equals which of the values listed below?
(a) 92 lb   (b) 87 lb   (c) 84 lb   (d) 90 lb   (e) 81 lb

5-16 A pipe 12 in in diameter carries water by gravity from a reservoir. At a point 500 ft from the reservoir and 28 ft below its surface, a pressure gauge reads 8.0 lb/in²; at a point 8500 ft from the reservoir and 285 ft below its

surface, a pressure gauge reads 61.5 lb/in^2. Which of the values listed below most nearly equals the rate of discharge from the reservoir?
(a) 3010 gal/min   (b) 2940 gal/min   (c) 3260 gal/min
(d) 3140 gal/min   (e) 3070 gal/min

5-17   The instrument readings for a pump test are shown on the schematic diagram in Fig. 5P-17. Which of the values listed below most nearly equals the efficiency of the pump under the given conditions? Assume that the pipes leading to the gauges are filled with water and that the difference between the inlet and outlet velocity heads is negligible.
(a) 67%   (b) 72%   (c) 59%   (d) 52%   (e) 61%

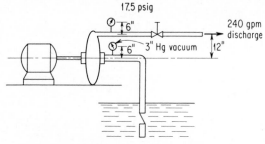

Figure 5P-17   3460 r/min at 7.89 lb·ft torque; 220 V, 21.2 A at 88 percent power factor single phase; weight of water = 8.33 lb/gal; weight of mercury = 0.49 lb/in^3.

5-18   A horizontal 36-in pipe curving through an angle of 60° contains water with an average velocity of 7.1 ft/s and a pressure of 50 lb/in^2. Which of the values listed below most nearly equals the force exerted on the bend by this discharge? (See Fig. 5P-18.)
(a) 48,900 lb   (b) 51,600 lb   (c) 50,200 lb   (d) 52,100 lb
(e) 50,900 lb

Figure 5P-18

5-19   The hinged top of a 4-ft square gate is submerged 10 ft below the surface of the water in a tank (Fig. 5P-19). Which of the values listed below most nearly equals the distance x to the center of pressure on the gate?
(a) 2.1 ft   (b) 2.4 ft   (c) 2.6 ft   (d) 2.5 ft   (e) 2.3 ft

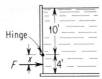

Figure 5P-19

5-20 A circular concrete culvert 8 ft in diameter and having a slope of 1 ft in 1000 ft is carrying water 6 ft in depth. The discharge, disregarding loss of head at the entrance, most nearly equals which of the values listed below? Assume a Manning coefficient of 0.013.
(a) 257 ft³/s   (b) 263 ft³/s   (c) 265 ft³/s   (d) 249 ft³/s
(e) 260 ft³/s

5-21 A jet of water is entirely reversed in direction as it hits a vane. If the vane is fixed, the velocity of the jet is 100 ft/s, the weight flowing equals 200 lb/s, and friction is such that the water leaves the vane at 0.90 times the velocity with which it strikes it, which of the values listed below most nearly equals the force exerted on the vane by the jet?
(a) 1171 lb   (b) 1179 lb   (c) 1197 lb   (d) 1181 lb   (e) 1192 lb

5-22 Which of the values listed below most nearly equals the rate of flow of water through the trough shown in Fig. 5P-22? The slope of the trough is 1.0 ft/100 ft, and the coefficient of roughness is 0.013.
(a) 591 ft³/s   (b) 597 ft³/s   (c) 594 ft³/s   (d) 601 ft³/s
(e) 587 ft³/s

Figure 5P-22

5-23 A penstock 10 ft in diameter carrying 250 ft³/s of water bends down through a 45° angle. Which of the values below most nearly equals the force F_x acting at the bend? (See Fig. 5P-23.)
(a) 1543 lb   (b) 1115 lb   (c) 2014 lb   (d) 1087 lb   (e) 1091 lb

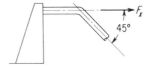

Figure 5P-23

5-24 A hydrometer weighs 50 g. The area of cross section of the stem is 1 cm². The distance on the stem between the 0.80 gradation and the 0.90 gradation most nearly equals which of the values listed below?
(a) 5.8 cm   (b) 6.2 cm   (c) 6.9 cm   (d) 7.1 cm   (e) 7.3 cm

5-25 In the venturi meter shown in Fig. 5P-25, the liquid being measured is water and the coefficient of the meter is 0.98. The rate of flow most nearly equals which of the values listed below?
(a) 1.40 ft³/s   (b) 1.37 ft³/s   (c) 1.35 ft³/s   (d) 1.42 ft³/s
(e) 1.29 ft³/s

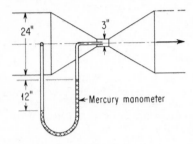

Figure 5P-25

5-26 From the data in Fig. 5P-26, the head loss from A to B due to the valves and fittings, measured in feet of water, most nearly equals which of the values listed below?
(a) 14.6 ft   (b) 13.0 ft   (c) 14.8 ft   (d) 15.1 ft   (e) 13.2 ft

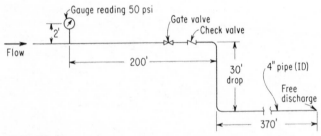

Figure 5P-26

5-27 A 30-in-diameter access hole in the side of a penstock is closed with a flat plate. The static head at the center of the access hole is 25 ft (see Fig. 5P-27). Which of the following is closest to the distance of the center of pressure below the horizontal centerline of the access hole?

Figure 5P-27

(a) 0.19 in   (b) 0.31 in   (c) 0.15 in   (d) 0.21 in   (e) 0.23 in

5-28 A standard 10-in galvanized iron pipe 15 ft long, weighing 32 lb/ft, is closed at one end by a pipe cap with a weight attached to the cap so that the pipe will float upright in a vertical orientation. The cap and weight total 25 lb and displace 0.5 ft³ of water. Which of the following is closest to the distance that the top end of the pipe will stick up above the surface of seawater of 1.025 specific gravity?
(a) 2.48 ft   (b) 3.51 ft   (c) 2.69 ft   (d) 2.98 ft   (e) 3.26 ft

5-29 A tank in the form of a frustrum of a cone, with bases horizontal and axis vertical, is 10 ft in height. The top diameter is 8 ft and the bottom diameter is 3 ft. Discharge is through a 3-in standard square orifice for which the discharge coefficient is 0.61. Which of the following is closest to the time that would be required to empty the tank?
(a) 7.21 min   (b) 6.72 min   (c) 6.49 min   (d) 6.68 min
(e) 7.04 min

5-30 A 10-in ID oil pipeline is 100 mi long and is made of welded steel. The discharge end is 500 ft above the intake. The velocity of flow is 4.0 ft/s. The total pressure in pounds per square inch gauge at the entrance to the line when the whole length is full of gasoline at 60°F and 0.68 specific gravity most nearly equals which of the following?
(a) 844   (b) 858   (c) 869   (d) 804   (e) 828

5-31 A flume of unplaned wood with a cross section as shown in Fig. 5P-31 is to carry 200 ft³/s of water. Which of the following is closest to the required slope in ft/1000 ft?
(a) 214   (b) 227   (c) 206   (d) 257   (e) 282

Figure 5P-31

5-32 A centrifugal motor-driven water pump is connected to a pipeline as shown in Fig. 5P-32. If the input to the pump motor is 1105 W, which of the following is closest to the efficiency of the motor-pump combination?

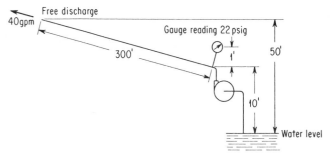

Figure 5P-32   Pipe ID = 2.067 in; pipe flow area = 3.36 in²; 7.48 gal/ft³.

(*a*) 38   (*b*) 51   (*c*) 47   (*d*) 42   (*e*) 40

5-33   A 1-in-diameter cable is used to anchor a buoy near the entrance to a harbor where the ocean current reaches a velocity of 4 mi/h. A model is made of the buoy and cable to one-tenth scale for tests in the laboratory. Using fresh water in the test, results show that for a velocity of 5 ft/s, the drag on 1 ft of model cable is 2 lb. The specific weight of seawater is 63.8 lb/ft³. If the relation between the model and the prototype can be expressed as $F/(\rho A v^2)$ = constant, where F = drag force, ρ = density of fluid, A = area (projected area perpendicular to flow), v = velocity of fluid relative to model or prototype, which of the following values is closest to the drag of prototype cable as installed in the harbor entrance?
(*a*) 31.6 lb/ft   (*b*) 34.2 lb/ft   (*c*) 28.2 lb/ft   (*d*) 26.1 lb/ft
(*e*) 29.8 lb/ft

5-34   A liquid is forced through a pipe 1 m long which has an internal diameter of 0.6 cm by a pressure difference of 1.2×10^5 dyn/cm². If 570 cm³ of liquid were discharged in 5 min, which of the following values would be closest to the viscosity of the liquid?
(*a*) 225 cP   (*b*) 187 cP   (*c*) 294 cP   (*d*) 310 cP   (*e*) 201 cP

5-35   Turpentine is stored in a tank on the second floor of a plant that manufactures paints. A pipeline is to be installed to a mixer on the first floor feeding the turpentine by gravity through an equivalent length of 55 ft of straight steel pipe. The bottom of the tank is 15 ft above the point of discharge to the mixer. Which of the following standard sizes of pipe is the minimum size which will ensure a flow rate of 10 gal/min to the mixer? Minimum temperature may be taken as 60°F, at which temperature the viscosity of turpentine is 1.6 cP and the specific gravity is 1.025.
(*a*) ⅜-in   (*b*) ½-in   (*c*) ¾-in   (*d*) 1-in   (*e*) 1¼-in

5-36   A cubical block of wood 10 cm on a side with density of 0.5 g/cm³ floats in a container of water (see Fig. 5P-36). Oil of density 0.8 g/cm³ is poured on the water until the top of the oil layer is 4 cm below the top of the block. Which of the following values is closest to the depth of the oil layer?
(*a*) 3.5 cm   (*b*) 4.2 cm   (*c*) 5.0 cm   (*d*) 5.4 cm   (*e*) 5.7 cm

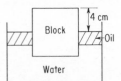

Figure 5P-36

5-37   A tank is a paraboloid of revolution with every horizontal section a circle whose radius equals the square root of the height above the bottom. There is a 1-in-diameter sharp-edged orifice in the bottom of the tank. The depth of water in the tank at the start of time is 10 ft. Which of the following values is closest to the time that will be required for the tank to empty?

(*a*) 42 min   (*b*) 51 min   (*c*) 38 min   (*d*) 35 min   (*e*) 54 min

5-38 A 12-in ID cast-iron pipe, 4000 ft in length, conveys water from a pump to a reservoir whose surface is 250 ft above the pump, which is pumping at the rate of 3 ft^3/s. If the efficiency of the pump is 89%, which of the following is the closest to the required horsepower input to the pump?
(*a*) 110 hp   (*b*) 98 hp   (*c*) 87 hp   (*d*) 103 hp   (*e*) 115 hp

5-39 A pitot tube having a coefficient of unity is inserted in the exact center of a long smooth tube of 1-in ID in which crude oil (specific gravity = 0.9 and $\mu = 35 \times 10^{-5}$ lb·s/ft^2) is flowing. If the velocity pressure measured by the pitot tube is 3.9 in of water, which of the following values is closest to the average velocity in the tube?
(*a*) 2.4 ft/s   (*b*) 3.8 ft/s   (*c*) 3.6 ft/s   (*d*) 2.9 ft/s   (*e*) 2.7 ft/s

5-40 A pitot tube having a coefficient of unity is inserted in the exact center of a long smooth tube of 1-in ID in which crude oil (specific gravity = 0.9 and $\mu = 35 \times 10^{-5}$ lb·s/ft^2) is flowing. If the velocity pressure measured by the pitot tube is 2.3 lb/in^2, which of the following values is closet to the average velocity in the tube?
(*a*) 16.2 ft/s   (*b*) 14.8 ft/s   (*c*) 15.1 ft/s   (*d*) 15.4 ft/s
(*e*) 15.8 ft/s

5-41 Water flows through 3000 lin ft of 36-in ID pipe which branches into 2000 lin ft of 18-in-diameter pipe and 2400 lin ft of 24-in-diameter pipe. These rejoin, and the water continues through 1500 ft of 30-in-diameter pipe. All pipes are horizontal, and the friction factors are 0.016 for the 36-in pipe, 0.017 for the 24-in and 30-in pipes, and 0.019 for the 18-in pipe. If the flow through the system is 60 ft^3/s, which of the following values most nearly equals the pressure drop between the beginning and the end of the system? Disregard minor losses.
(*a*) 38.8 lb/in^2 (*b*) 40.2 lb/in^2 (*c*) 41.6 lb/in^2 (*d*) 46.1 lb/in^2
(*e*) 47.2 lb/in^2

5-42 A piece of lead (specific gravity 11.3) is attached to 40 cm^3 of cork (specific gravity 0.25). When fully submerged, the combination will just float. Which of the following values most nearly equals the mass of the lead?
(*a*) 40.1 g   (*b*) 30.6 g   (*c*) 38.4 g   (*d*) 34.2 g   (*e*) 32.9 g

5-43 A weighted timber 1 ft square and 6 ft long floats upright in fresh water with 2 ft of its length exposed. Which of the following values most nearly equals the length which will project above the surface if the water is covered with a layer of oil 1 ft thick? The specific gravity of the oil is 0.8.
(*a*) 1.2 ft   (*b*) 1.4 ft   (*c*) 1.6 ft   (*d*) 1.8 ft   (*e*) 2.0 ft

5-44 The inlet and outlet of a venturi meter are each 4 in in diameter, and the throat is 3 in in diameter. The inlet velocity is 10 ft/s. The inlet has a static head of 10 ft of water. If there is no loss due to friction between the inlet and the throat of the meter, which of the following values most nearly equals the head in the throat, measured in feet of water?
(*a*) 6.4 ft   (*b*) 6.6 ft   (*c*) 6.8 ft   (*d*) 7.0 ft   (*e*) 6.2 ft

5-45 A conduit having a cross section of an equilateral triangle of sides *b* has

water flowing through it at a depth of $b/2$. Which of the following values most nearly equals the "hydraulic radius"? (See Fig. 5P-45.)
(*a*) 0.12*b*   (*b*) 0.13*b*   (*c*) 0.19*b*   (*d*) 0.17*b*   (*e*) 0.15*b*

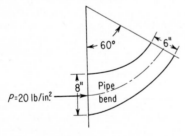

Figure 5P-45

5-46 A mass of copper, suspected of being hollow, weighs 523 g in air and 447.5 g in water. If the specific gravity of copper is 8.92, which of the following values most nearly equals the volume of the cavity?
(*a*) 14.2 cm³   (*b*) 14.7 cm³   (*c*) 15.1 cm³   (*d*) 15.8 cm³
(*e*) 16.9 cm³

5-47 A 60° pipe bend reduces from a diameter of 8 in to a diameter of 6 in (Fig. 5P-47). The static pressure at the 8-in diameter is 20 lb/in². If the flow through the elbow is 3.0 ft³/s and the fluid is water, which of the following values most nearly equals the total force acting on the bend?
(*a*) 910 lb   (*b*) 920 lb   (*c*) 930 lb   (*d*) 940 lb   (*e*) 950 lb

← 60°   ← 6"

8" Pipe
$P = 20$ lb/in² bend

Figure 5P-47

5-48 A gravity dam has a trapezoidal section 18 ft high, 2 ft wide on top, and 8 ft wide at the base. The water side is vertical. The depth of the water is 15 ft. Assume that the concrete weighs 141 lb/ft³ and that the foundation is sealed so that no water seeps under the dam. Which of the following values most nearly equals the vertical component of stress on the foundation at the upstream edge of the base?
(*a*) −277 lb/ft²   (*b*) +1863 lb/ft²   (*c*) −480 lb/ft²
(*d*) +422 lb/ft²   (*e*) +597 lb/ft²

5-49 For the condition shown in Fig. 5P-49, which of the following values most nearly equals the force F required to open the gate? The width of the gate is 4 ft, and the height is 6 ft.
(*a*) 3210 lb   (*b*) 3460 lb   (*c*) 3560 lb   (*d*) 3640 lb   (*e*) 3720 lb

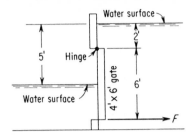

Figure 5P-49

5-50 For the arrangement shown in Fig. 5P-50, which of the following values most nearly equals the pressure difference between A and B? (Specific gravity of mercury is 13.6 and of turpentine is 0.8.)
(a) 16.2 lb/in²   (b) 17.9 lb/in²   (c) 18.7 lb/in²   (d) 19.2 lb/in²
(e) 20.1 lb/in²

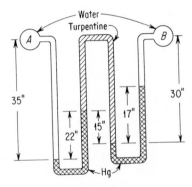

Figure 5P-50

5-51 A 36-in water main, carrying 60 ft³/s, branches at point X into two pipes, one 24 in in diameter and 1500 ft long, and the other 12 in in diameter and 3000 ft long. They join at point Y and continue as a 36-in pipe. If the friction factor f is equal to 0.022 in all pipes, which of the following values most nearly equals the rate of flow in the 12-in branch pipe?
(a) 6.10 ft³/s   (b) 7.34 ft³/s   (c) 6.65 ft³/s   (d) 7.20 ft³/s
(e) 6.42 ft³/s

5-52 The water system shown in Fig. 5P-52 is made up of cast-iron pipe. If the flow rate through the system is 2.0 ft³/s, which of the following values most nearly equals the pressure drop between A and B?
(a) 25.1 lb/in²   (b) 14.7 lb/in²   (c) 18.9 lb/in²   (d) 12.1 lb/in²
(e) 10.4 lb/in²

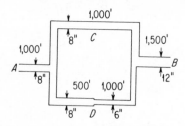

Figure 5P-52

Multiple-Choice Sample Quiz

For each question select the most nearly correct answer from the five given possibilities. For practice, try to complete the following 10 problems in 15 minutes or less.

5M-1 If absolute viscosity has the units of pounds-seconds per square foot, then kinematic viscosity has units of:
(a) centipoise
(b) British thermal units per second
(c) square feet per second-pound
(d) cubic feet per second
(e) square feet per second

5M-2 The energy of a fluid flowing at any section in a pipeline is a function of:
(a) pressure only
(b) pressure, height above a chosen datum, velocity of flow, density of fluid, viscosity of fluid, and temperature of fluid
(c) height above a chosen datum, density, internal energy, pressure, and velocity of flow
(d) velocity of flow only
(e) none of these

5M-3 If a body weighs 100 lb in air and 25 lb in fresh water, its volume in cubic feet is:
(a) 0.75   (d) 1.3
(b) 1.1   (e) 1.5
(c) 1.2

5M-4 Friction head in a pipe carrying water varies:
(a) inversely with gravity squared
(b) directly with diameter
(c) inversely with diameter
(d) directly with velocity
(e) inversely with the coefficient of friction f

5M-5 A fluid flows at a constant velocity in a pipe. The fluid completely fills the pipe, and the Reynolds number is such that the flow is just subcritical and laminar. If all other parameters remain unchanged and the viscosity of the fluid is decreased a significant amount, one would generally expect the flow to:

(a) not change   (d) increase

(b) become turbulent   (e) temporarily increase

(c) become more laminar

5M-6 A barge loaded with rocks floats in a canal lock with both the upstream and downstream gates closed. If the rocks are dumped into the canal lock water, with both the gates still in the closed position, the water level in the lock will theoretically:

(a) rise

(b) rise and then return to original level

(c) fall

(d) fall and then return to original level

(e) remain the same

5M-7 The locus of the elevations to which water will rise in a piezometer tube is termed:

(a) the phreatic line

(b) the energy gradient

(c) the hydraulic gradient

(d) the friction head

(e) critical depth

5M-8 If a liquid has laminar flow through a pipe, which of the following may be true?

(a) The Reynolds number is 35,000 or higher.

(b) A great deal of energy is lost due to turbulence.

(c) The roughness of the pipe will have no effect on the friction factor.

(d) The velocity at the center of the pipe will be the same as at the pipe wall.

(e) None of these.

5M-9 Cavitation is the result of:

(a) static pressure in a fluid becoming less than fluid vapor pressure

(b) rivets under impact load

(c) exposure of concrete to saltwater

(d) heat treatment of a low-carbon steel

(e) improper welding technique

5M-10 A cylinder of cork is floating upright in a container partially filled with water. A vacuum is applied to the container such that the air within the vessel is partially removed. The cork will:

(a) rise somewhat in the water

(b) sink somewhat in the water

(c) remain stationary

(d) turn over on its side

(e) sink to the bottom of the container

6

CHEMISTRY

A general knowledge of chemistry is presupposed; discussion here is limited to those portions of the subject which have been emphasized in past examinations and to a few of the more general fundamental relationships applicable to many engineering problems. The reader who has retained little (or less) of an introductory college chemistry course, or who had no such course, will find it of value to read a beginning text.

Chemistry is not a major subject in the fundamentals examination and is covered only in the morning session. However, a number of the principles which are ordinarily treated in chemistry also apply to the subject of thermodynamics, which is a major examination subject. Some of the principles of chemistry can also overlap into the areas of materials science and nucleonics. As a result, chemistry is well worth reviewing.

6-1 EXAMINATION COVERAGE

The subdivisions of the general subject of chemistry which may be covered in an examination include the following:

Chemical reactions

Balancing equations

Law of definite proportions

Equivalent weight

Gram mole

Avogadro's law

Freezing-point depression

Valence

Chemical formulas

Nomenclature of compounds

Reaction rate

Heat of reaction

Solubility product

Ionization

Electrolysis

Gas-phase equilibrium

Isotopes

Solutions

Acids

Metals

Bases

Hydrocarbons

Periodic table

Atomic structure

Molecular structure

Crystal structure

Bonding

Radioactivity

Gas laws

6-2 LAW OF LE CHATELIER

The law of Le Chatelier states, "When a system in equilibrium is subjected to a change (e.g., a change of temperature, concentration, or pressure), there is a shift in the point of equilibrium which tends to restore the original condition, or to relieve the strain." As an example, the combination of carbon and oxygen is exothermic and represented by the reaction $C + O_2 \rightarrow CO_2 + 97,000$ cal. If, after the reaction has reached a point of equilibrium, the system is heated, the point of equilibrium will shift to the left, using up heat and tending to lower the temperature of the system. The result will be a dissociation of some of the CO_2 which was previously formed and an increase in the concentrations of C and O_2.

An increase in the pressure acting on a system may also result in a shift in the point of equilibrium. For the reaction $N_2 + 3H_2 \rightarrow 2NH_3$, we have one volume of N_2 and three volumes of H_2 combining to form two volumes of NH_3, so an increase in pressure would cause the point of equilibrium to shift to the right. Such a shift would reduce the total equivalent volume of gas, as four volumes on the left combine to form two volumes on the right.

The shift in the point of equilibrium due to a change in concentration may be illustrated by means of the equilibrium constant, which is a constant for a given reaction at a given temperature and pressure. For a general chemical reaction of $aA + bB \rightarrow cC + dD$, where we have a molecules of A reacting with b molecules of B, etc., the equilibrium constant $K = [(C)^c(D)^d]/[(A)^a(B)^b]$. For the previously used example, then, $K = (NH_3)^2/[(N_2)(H_2)^3]$, where $(NH_3)^2$ represents the concentration of NH_3 in the final mixture, in moles per liter, squared. Since K is a constant for a given temperature, we see that an increase in the concentration of either the N_2 or the H_2 must be followed by an increase in the concentration of the NH_3.

On the other hand, a chemical reaction will go to completion when one of the products is (1) insoluble, (2) volatile, or (3) only slightly ionized.

These conditions can be illustrated by the following chemical equations:

$$NaCl + AgNO_3 \rightarrow NaNO_3 + AgCl \downarrow$$
$$NH_4Cl + NaOH + heat \rightarrow NaCl + H_2O + NH_3 \uparrow$$
$$HCl + NaOH \rightarrow NaCl + H_2O$$

In the last equation the water resulting is only slightly ionized so the reaction goes to the end.

6-3 DALTON'S LAW OF PARTIAL PRESSURES

Dalton's law of partial pressures states, "Each component of a mixture of gases exerts its pressure exactly as it would if the others were not present, and the sum of all their pressures makes up the total of the pressure exerted by the mixture of gases." The water-vapor pressure in gases collected over water is an example of a partial pressure. If oxygen is collected over water at a temperature of 64°F at a pressure of 780 mmHg, the actual volume of oxygen at this temperature and pressure will be only 764.6/780 times the apparent volume, since the vapor pressure of water at 64°F is 15.4 mmHg. This means that 15.4 mm of the measured pressure is due to water vapor, and the pressure of the oxygen is 780 − 15.4, or 764.6 mm. Then $15.4/780 \times$ total volume is water vapor, and $764.6/780 \times$ total volume is oxygen.

6-4 AVOGADRO'S LAW

Avogadro's hypothesis may be stated as: "Equal volumes of gases when subjected to the same conditions of temperature and pressure contain an equal number of molecules." This means that 1 liter of gas at 60°F and 14.7 lb/in² abs will contain the same number of molecules regardless of the type of gas occupying that space—whether the gas is chlorine, hydrogen, or a mixture of gases, such as air. The number of molecules in 1 g·mol of gas, which occupies 22.4 liters at 0°C and 1 atm of pressure, is 6.024×10^{23} and is known as "Avogadro's number."

- Which of the following most nearly equals the number of molecules of nitrogen, N_2, that would be contained in 10 liters of nitrogen saturated with water vapor at 60°C and under a total absolute pressure of 2 atm? (Absolute zero may be taken as $-273°C$, and the vapor pressure of water is 3.1 lb/in² at 60°C.)

  (a) 6.0×10^{23} (b) 4.0×10^{23} (c) 3.9×10^{24}
  (d) 3.7×10^{23} (e) 5.8×10^{14}

6-5 VAPOR PRESSURE

This introduces the concept of vapor pressure, which is the pressure (absolute) exerted by a vapor when it is in equilibrium with its liquid. The vapor pressure of a given liquid is dependent only on the temperature; it is not affected by the total pressure acting on the system nor by the amount of space above the surface of the liquid. The boiling temperature of a liquid is the temperature at which the vapor pressure is equal to the local atmospheric pressure.

The vapor pressure of the water at the given temperature is 3.1 lb/in² abs and the total pressure is $2 \times 14.7 = 29.4$ lb/in² abs. From Dalton's law of partial pressures, the pressure due to the nitrogen equals $29.4 - 3.1 = 26.3$ lb/in² abs. Application of the universal gas law gives the equivalent volume of nitrogen at 0°C and 14.7 lb/in² abs:

$$\frac{P_1 V_1}{T_1} = R = \frac{P_2 V_2}{T_2} \qquad V_2 = V_1 \frac{P_1}{P_2} \frac{T_2}{T_1}$$

$$V_2 = 10 \times \frac{26.3 \text{ lb/in}^2 \text{ abs}}{14.7 \text{ lb/in}^2 \text{ abs}} \times \frac{273 \text{ K}}{(273 + 60) \text{ K}} = 14.71 \text{ liters}$$

of pure nitrogen at 0°C and 1 atm. The total number of molecules of nitrogen contained in the original 10 liters would then equal

$$\frac{14.7}{22.4} \times 6.024 \times 10^{23} = 3.95 \times 10^{23} \text{ molecules}$$

The correct answer is (c).

▪ A tank having a capacity of 800 gal contains 500 gal of benzene at 35°F. Under these conditions the tank is sealed at atmospheric pressure. During storage the temperature of the tank rises to 100°F. Which of the following most nearly equals the gauge pressure which would exist in the tank? Assume that the air inside the tank is saturated with benzene vapor at the time that the tank was sealed and that the expansion of the tank due to the rise in temperature and pressure is negligible. The coefficient of cubical expansion for liquid benzene = 0.00077 ft³/(ft³)(°F) (7.48 gal/ft³; 760 mmHg = 1 atm).

Temperature, °F	Vapor pressure, mmHg
−9.5	10
71.2	100
109.4	200
141.9	400
163.7	600
176.2	760

(a) 4.7 lb/in² gauge   (b) 5.2 lb/in² gauge   (c) 5.6 lb/in² gauge   (d) 6.2 lb/in² gauge   (e) 5.4 lb/in² gauge

Initially the pressure in the tank is equal to atmospheric pressure at 35°F. This is the total pressure in the tank at the time the tank is closed; it is made up of the pressure of the air in the tank and the pressure of the benzene vapor. To determine the amount of the pressure due to the benzene vapor, plot a graph of vapor pressure vs. temperature from the data given. From the graph (Fig. 6-1) the vapor pressure of the benzene can be seen to equal 48 mmHg at 35°F. The pressure due to the air at the time the tank is closed is then 760 − 48 = 712 mmHg.

Three things affect the pressure: (1) the pressure of the air will increase because of the increase in the temperature, (2) the pressure of the air will increase because of the reduction in volume resulting from the expansion of the benzene, and (3) the vapor pressure will increase as shown in Fig. 6-1. Dalton's law states that the change in the pressure of the air will be the same as if the air alone were occupying the space:

Volume of air at start = $800 - 500 = 300$ gal

Final volume of air = $800 - 500(1 + 65 \times 0.00077) = 275$ gal

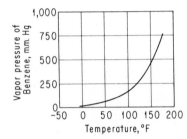

Temperature, °F   **Figure 6-1**

where 65 = change in temperature in Fahrenheit and 0.00077 = thermal coefficient of cubical expansion of the benzene. Then

$$P_2 = P_1 \frac{T_2}{T_1} \frac{V_1}{V_2}$$

The final air pressure would equal

$$P_2 = 712 \times \frac{560}{495} \times \frac{300}{275} = 880 \text{ mmHg}$$

The total pressure would then equal $880 + 170 = 1050$ mmHg, with the vapor pressure of the benzene added, and the gauge pressure would equal $1050 - 760 = 290$ mmHg, or

$$\frac{290}{760} \times 14.7 = 5.61 \text{ lb/in}^2 \text{ gauge}$$

The correct answer is (c).

6-6 SOLUTIONS

The concentration of a chemical solution may be given in a number of ways. The common methods are as follows:

$$\text{Weight percent} = \frac{\text{lb solute}}{\text{lb solvent}} \quad \text{or} \quad \frac{\text{g solute}}{\text{g solvent}}$$

(Note that this is not pounds of solute per pound of *solution*.)

$$\text{Volume percent} = \frac{\text{ft}^3 \text{ solute}}{\text{ft}^3 \text{ solvent}} \quad \text{or} \quad \frac{\text{cm}^3 \text{ solute}}{\text{cm}^3 \text{ solvent}}$$

This is commonly used, for example, to express the amount of a gas dissolved in water or other liquid. Note the following definitions of concentration and molar solution:

Concentration: may be either weight percent, mass percent (grams per gram), or volume percent.

Molar solution: the concentration of a solution may be given as the molarity (M), where the molarity of a solute equals the number of moles of solute per liter of *solution*. A 2 M (molar) solution of sodium chloride (common salt) could be made by adding water to 2 mol of NaCl [$2 \times (22.991 + 35.457) = 116.896$ g] until the volume of the solution reached 1 liter. Note that a 2 M solution would not be obtained by adding 116.896 g of NaCl to 1 liter of water, since the final volume of the solution would then be greater than 1 liter.

From the above, it can be seen that

$$\text{Moles of solute} = \text{molarity} \times \text{liters of solution}$$

- The number of grams of potassium hydroxide (KOH) contained in 750 mL of a 0.400 M solution would most nearly equal which of the following?

    (*a*) 17 g   (*b*) 13 g   (*c*) 15 g   (*d*) 11 g   (*e*) 10 g

To answer the question:

$$\text{Moles of solute} = (0.400 \ M) \times 0.750 \ \text{l} = 0.300 \ \text{mol}$$

$$0.300 \ \text{mol KOH} = 0.300(39.100 + 16.00 + 1.008) = 16.832 \ \text{g KOH}$$

The correct answer is (*a*).

Note the following definitions of molal and normal solutions:

Molal solution: Molality (m) of a solution equals the number of moles of solute dissolved in one kilogram of solvent:

$$\text{Molality} = \frac{\text{mol solute}}{\text{kg solvent}}$$

Citing the previous example, if 2 mol of NaCl (116.896 g) were dissolved in 2 kg of water, the result would be a 2 m (molal) solution. Capital M indicates molarity; lowercase m indicates molality.

Normal solution: Normality (N) of a solution is the number of gram-equivalent weights of solute contained in one liter of solution:

$$\text{Normality} = \frac{\text{gram-equivalent weights}}{\text{liters of solution}}$$

A gram-equivalent weight is that amount of an element that will combine with 8.000 g of oxygen. Thus a gram-equivalent weight would equal the atomic weight of an element divided by the ionic charge number, or the molecular weight of a compound divided by the ionic charge number. For H_2SO_4, for example, 1 g·mol would equal

$$2 \times 1.008 + 32.066 + 4 \times 16.000 = 98.082 \text{ g}$$

But since the SO_4 ion has an ionic charge of -2, the gram-equivalent weight of H_2SO_4 would equal

$$\frac{98.082}{2} = 49.041 \text{ g}$$

Similarly, K_3PO_4 has a molecular weight of 212.275, but since each molecule splits up into three K^+ ions and one PO_4^{3-} ion, the gram-equivalent weight equals

$$\frac{212.275}{3} = 70.758 \text{ g}$$

Normality is similar to molarity, but the units used to express the quantity of solute are different.

6-7   CHEMICAL REACTIONS

A chemical equation is said to be balanced when the number of each of the different kinds of atoms on one side of the equation is exactly equal to the number of the same kinds of atoms on the other side of the equation.

The weights of the different substances required for or produced by a chemical reaction will be in the same proportion as the molecular weights of the substances involved, as indicated by the balanced equation of the reaction. To illustrate this point, consider the following problem.

- Hydrogen sulfide gas, H_2S, is used to precipitate lead sulfide, PbS, in a solution of hydrochloric acid:

$$\_\_H_2S + \_\_PbCl_2 \rightarrow \_\_PbS + \_\_HCl$$

Formula weights: H = 1.008, Pb = 207.2, S = 32.06, Cl = 35.46.

a. Balance the chemical equation and state how many moles of H_2S are required to produce 1 lb of PbS.

b. How many pounds of H_2S are required to produce 1 lb of PbS?

c. How many cubic feet of H_2S gas at 70°F and atmospheric pressure are required to produce 1 lb of PbS?

The gas constant for H_2S is 45.3 ft·lb/(lb)(°R).

$$H_2S + PbCl_2 \rightarrow PbS + 2HCl$$

which shows that 1 mol of H_2S combines with 1 mol of $PbCl_2$ to give 1 mol of PbS and 2 mol of HCl.

The molecular weights can be determined from the formula weights given: $H_2S = 34$; $PbCl_2 = 278$; PbS = 239; HCl = 36.5.

The weight of H_2S to produce 1 lb of PbS can be determined by proportion: $x/34 = \frac{1}{239}$ $x = 0.142$ lb of H_2S.

The cubic feet of H_2S required can easily be determined by means of the universal gas law, $PV = WRT$, which gives

$$V = \frac{45.3 \times (460 + 70)}{14.7 \times 144} \text{ lb} \times 0.142 = 1.61 \text{ ft}^3$$

Another method of calculating the volume of gas involves the fact that 1 lb·mol of gas occupies 359 ft³ at standard conditions. Since 1 lb·mol of H_2S weighs 34.06 lb, this would give a volume of $0.142/34.06 \times 359 = 1.50$ std ft³:

$$1.50 \times \frac{460 + 70}{460 + 32} = 1.615 \text{ ft}^3$$

▪ A piece of plumber's solder weighing 3.0 g was dissolved in dilute nitric acid and then treated with dilute H_2SO_4. This precipitated the lead as $PbSO_4$, which, after washing and drying, weighed 2.93 g. The solution was then neutralized to precipitate stannic acid, which was decomposed by heating, yielding 1.27 g of SnO_2. Which of the following most nearly equals the percentage of tin that was in the solder? (Atomic weights: Pb = 207.2, Sn = 118.7, O = 16.00, S = 32.07, H = 1.00.)

(a) 66   (b) 50   (c) 47   (d) 41   (e) 33

The formula weight of $PbSO_4$ is 303.27 (207.2 + 32.07 + 64.0), which means that $(207.2/303.27) \times 2.93 = 2.00$, or that there is a total of 2 g of lead in the 2.93 g of $PbSO_4$.

The formula weight of SnO_2 is 150.7, which means that the SnO_2 contains $(118.7/150.7) \times 1.27 = 1.00$ g of tin.

The original 3 g of solder contained 2.00 g of lead and 1.00 g of tin, or 66.7 percent lead and 33.3 percent tin. The correct answer is (e).

It is sometimes necessary to convert from volumes of gas to weights.

- An Orsat analysis of flue gases yields the following volumetric analysis: CO_2 = 12.5 percent, CO = 0.5 percent, O_2 = 6.4 percent, and N_2 = 80.6 percent. (The water vapor would have condensed and thus would not be a constituent of the flue gas.) Which of the following most nearly equals the specific weight of the flue gas in pounds per mole? (The atomic weights are: C = 12, O = 16, and N = 14.)

  (a) 44   (b) 32   (c) 34   (d) 30   (e) 28

The molecular weight of CO_2 is 44, so $R = 1544/44 = 35.09$; thus

$$V = \frac{RT}{P} = \frac{35.09 \times (460 + 32)}{14.7 \times 144} = 8.16 \text{ ft}^3/\text{lb}$$

at standard conditions, or $w = 0.1227 \text{ lb/ft}^3$.

For each cubic foot of flue gas there will be 0.125 ft^3 of CO_2, which will weigh $0.125 \text{ ft}^3 \times 0.1227 \text{ lb/ft}^3 = 0.01535 \text{ lb}$. Similarly, at standard conditions there will be

$$0.005 \text{ ft}^3 \times 0.0781 \text{ lb/ft}^3 = 0.00039 \text{ lb CO/ft}^3 \text{ flue gas}$$

$$0.064 \text{ ft}^3 \times 0.0893 \text{ lb/ft}^3 = 0.00572 \text{ lb O}_2/\text{ft}^3 \text{ flue gas}$$

$$0.806 \text{ ft}^3 \times 0.0781 \text{ lb/ft}^3 = 0.0630 \text{ lb N}_2/\text{ft}^3 \text{ flue gas}$$

The weight of 1 ft^3 of flue gas at 1 atm of pressure and 32°F is then 0.08446 lb, and the weight percentages of the different constituent gases are CO_2 = 18.1 percent; CO = 0.5 percent; O_2 = 6.8 percent; N_2 = 74.6 percent.

The weight percentage may be calculated more easily by determining the weight of each gas in a mole volume of 359 ft^3. Since the volume percent would also be the mole percent, we would have, for 1 mol of flue gas:

Gas	Volume, %	Molecular weight	Flue gas, lb/mol	Weight, %
CO_2	12.5	44	5.50	18.2
CO	0.5	28	0.14	0.44
O_2	6.4	32	2.05	6.76
N_2	80.6	28	22.57	74.6
			30.26	

The correct answer is (d).

6-8 COMBUSTION

Combustion need not necessarily be an oxidation; many substances burn with intense heat without combining with oxygen (e.g., the reaction in an atomic hydrogen welding torch which produces a very high temperature, $H + H \rightarrow H_2 +$ heat). Oxidation is the commonest type of burning, however, and is the one most frequently encountered.

The principles of combustion analysis can be illustrated with the aid of a multifaceted problem.

- One molecular weight of methyl alcohol, CH_4O, is burned in 10 percent excess air. Write the combustion equation and determine the following:
  - *a.* Pounds of air required
  - *b.* The respective partial pressure of each of the products for a total pressure of 15 lb/in^2 abs, assuming that the water vapor has not condensed
  - *c.* Volume occupied by the products at 240°F and 15 lb/in^2 abs
  - *d.* Volume of air at 15 lb/in^2 abs and 60°F needed to burn 1 ton of methyl alcohol per hour under the above conditions

First, write the equation:

$$\_CH_4O + \_O_2 \rightarrow \_CO_2 + \_H_2O$$

Since we are burning with 10 percent excess air, we assume that the end products will be carbon dioxide and water vapor. Only the oxygen of the air will enter into the reaction; the nitrogen will not. Balancing the equation gives

$$2CH_4O + 3O_2 \rightarrow 2CO_2 + 4H_2O$$

One pound mole of CH_4O weighs 32 lb and 32 lb of methyl alcohol combines with 48 lb of oxygen for complete combustion. Ten percent excess air gives a total of 52.8 lb of oxygen. Since air consists of 23.2 percent oxygen by weight, a total of

$$\frac{52.8}{0.232} = 228 \text{ lb}$$

of air would be required.

The products (of the combustion) would include CO_2, H_2O, the excess O_2, and the N_2, which took no part in the process. The complete equation, assuming the air to consist of 23.2 percent O_2 and 76.8 percent N_2 by weight (21 percent O_2 and 79 percent N_2 by volume), would be

$$CH_4O + O_2 + N_2 \rightarrow CO_2 + H_2O + N_2 + O_2$$

We are disregarding the approximately 1 percent of CO_2 and inert gases present in the air, but this will not give any appreciable error.

The weights of the different components would be

$$CH_4O + O_2 \quad\quad + N_2 \quad\quad \rightarrow CO_2 \;+ H_2O \;+ N_2 \quad\quad + O_2$$
$$32 \text{ lb} + 52.8 \text{ lb} + 175.2 \text{ lb} \rightarrow 44 \text{ lb} + 36 \text{ lb} + 175.2 \text{ lb} + 4.8 \text{ lb}$$

The partial pressures will be in the same ratios as the volumes. From $PV = RT$, the specific volume of the CO_2 is

$$V = \frac{RT}{P} = \frac{1544}{44} \times (460 + 240) \times \frac{1}{15 \times 144} = 11.37 \text{ ft}^3/\text{lb} \quad\quad \text{for } CO_2$$

For H_2O, $V = 27.8$ ft^3/lb; for N_2, $V = 17.85$ ft^3/lb; for O_2, $V = 15.6$ ft^3/lb.

The volume occupied at 240°F and 15 lb/in^2 abs would then equal

$$44 \times 11.37 + 36 \times 27.8 + 175.2 \times 17.85 + 4.8 \times 15.60 = 4715 \text{ ft}^3$$

The answer to part b would then be, for CO_2,

$$P = \frac{500}{4715} \times 15 = 1.59 \text{ lb/in}^2 \text{ abs}$$

for H_2O, $P = 3.18$ lb/in^2 abs; for N_2, $P = 10$ lb/in^2 abs; and for O_2, $P = 0.24$ lb/in^2 abs.

Since it requires 228 lb of air to burn 32 lb of methyl alcohol, it would require $\frac{2000}{32} \times 228 = 14{,}250$ lb of air to burn 1 ton. At 15 lb/in^2 abs and 60°F, the specific volume of air would be

$$V = \frac{53.3 \times 520}{15 \times 144} = 12.83 \text{ ft}^3/\text{lb}$$

The required volume of air would equal $14{,}250 \times 12.83 = 183{,}000$ ft^3/h.

We could also have calculated these answers by using mole volumes:

Gas	Weight, lb	Molecular weight	Moles	Volume, %	Partial pressure, lb/in^2 abs
CO_2	44	44	1	10.63	1.59
H_2O	36	18	2	21.26	3.19
N_2	175.2	28	6.26	66.52	9.98
O_2	4.8	32	0.15	1.59	0.24
			9.41	100.00	15.00

$$9.41 \times 359 \times \frac{14.7}{15} \times \frac{460 + 240}{460 + 32} = 4710 \text{ ft}^3$$

We know that 1 mol of CH_4O requires 1.5 mol of O_2 for complete combustion. Air is 21 percent O_2 by volume, so this would mean $1.5 \times 0.79/0.21 = 5.64$ mol of N_2. Adding 10 percent gives 1.65 mol of O_2 plus 6.21 mol of N_2, or a total of 7.86 mol of air, which would occupy

$$7.86 \times 359 \times \frac{14.7}{15} \times \frac{520}{492} = 2920 \text{ ft}^3 \text{ air/mol } CH_4O$$

Finally, 1 ton of CH_4O contains $\frac{2000}{32} = 62.5$ mol of CH_4O:

$$62.5 \times 2920 = 182{,}500 \text{ ft}^3/\text{h}$$

6-9 WATER SOFTENING

The degree of hardness is usually expressed in terms of the equivalent amount of $CaCO_3$ (calcium carbonate) in parts per million (ppm) by weight.

Hardness is caused by the presence in the water of bivalent ions of calcium, magnesium, or iron. These ions form insoluble precipitates when the water is boiled or when soap is added. Softening of the hard water may be accomplished by chemical precipitation of the undesirable ions or by replacing them with sodium ions which form soluble compounds.

■ A water-treating plant processes 2000 gal of raw water per day. Analysis shows that the raw water contains 60 ppm (parts per million by weight) of calcium bicarbonate, $Ca(HCO_3)_2$. In passing through the zeolite softening process, the calcium ions are exchanged for the sodium ions, so that the softened water contains sodium bicarbonate, $NaHCO_3$, instead of $Ca(HCO_3)_2$. Which of the following most nearly equals the parts per million of $NaHCO_3$ contained in the softened water?

(a) 60   (b) 62   (c) 64   (d) 66   (e) 58

Substitute Z for the zeolite radical and write the balanced chemical equation. The formula for sodium zeolite is Na_2Z. Formula weights are: $Na = 23.0$, $H = 1.0$, $C = 12.0$, $Ca = 40.1$, and $O = 16.0$.

The equation of the process is $Na_2Z + Ca(HCO_3)_2 \rightarrow CaZ + 2NaHCO_3$. The formula weight of $Ca(HCO_3)_2$ is 162.1, and that of $NaHCO_3$ is 84, so 162.1 lb of calcium bicarbonate will form 168 lb of sodium bicarbonate. Then, 60 ppm of calcium bicarbonate would give $60 \times (168/162.1) = 62.2$ ppm of sodium bicarbonate. The correct answer is (b).

6-10 ELECTROCHEMISTRY

There is a direct relationship between a chemical balance and an electrical balance. Chemical ions are created by the addition, or subtraction, of electrons from atoms or molecules in exact numbers. A hydrogen ion, H^+, is created by the removal of one electron from a hydrogen atom. A sulfate ion, SO_4^{2-}, is created by the addition of two electrons to an SO_4 molecule.

One gram-mole of a substance contains one Avogadro number of molecules. One gram-atom of an element contains one Avogadro number of atoms. The Avogadro number is 6.024×10^{23}.

A faraday equals 96,500 coulombs (C) and is the amount of electricity that will deposit one gram-equivalent weight of a substance at an electrode. One gram-equivalent weight of an oxidizing agent is that number of grams of the substance that picks up one Avogadro number of electrons. One gram-equivalent weight of a reducing agent is that amount of the substance, in grams, that gives up one Avogadro number of electrons. Thus, one gram-equivalent of any oxidizing agent will react exactly with one gram-equivalent of any reducing agent.

It would, then, require one Avogadro number of electrons (1 faraday), or 96,500 C, to deposit one gram-equivalent weight of a positively ionized substance at the cathode of an electrolytic cell, or one gram-equivalent weight of a negatively ionized substance at the anode of an electrolytic cell.

One coulomb equals one ampere-second.

- Which of the following most nearly equals the number of ampere-seconds (coulombs) which would be required to produce 100 std ft^3 of chlorine from an NaCl solution?

  (a) 6.024×10^{23} (b) 96,500   (c) 26.44×10^{10}
  (d) 13.08×10^8 (e) 24.32×10^6

Chlorine has an atomic weight of 35.457 and a molecular weight of 70.914. The weight equivalent of 100 std ft^3 would equal

$$W = \frac{PV}{RT} = \frac{(14.7 \times 144) \times 100}{(1544/70.914)(460 + 32)} = 19.75 \text{ lb}$$

where 19.75 lb corresponds to $19.75 \times 454 = 8966$ g, which equals $8966/70.914 = 126$ g·mol or 252 gram-equivalent weights of chlorine, since a chlorine atom has a valence of -1 and there are two atoms per molecule.

The release of this amount of chlorine gas, Cl_2, at the anode of the electrolytic cell would require 252 faradays, or 252 × 96,500 = 24.32 × 10⁶ C or, 24.32 × 10⁶ A·s. This would require a current of 200 A for 33.8 h. The correct answer is (e).

The chlorine would be produced from a solution of sodium chloride. The NaCl would ionize into Na⁺ and Cl⁻. The anode reaction would be

$$2Cl^- \rightarrow Cl_2 \uparrow + 2e^-$$

The sodium ions would be simultaneously attracted to the cathode. But there would be H⁺ ions in the solution as well. And less electric energy is required to deposit hydrogen than is required to deposit sodium. So hydrogen would be given off at the cathode and the sodium ions would remain in solution.

The reaction at the cathode would be

$$2HOH + 2e^- \rightarrow H_2 \uparrow + 2OH^-$$

A total of 252 gram-equivalent weights would be released, or 252 × 1.008 = 254 g of hydrogen gas.

Sample Problems

6-1 A 5.82-g silver coin is dissolved in nitric acid. When sodium chloride is added to the solution, all the silver is precipitated as AgCl. The AgCl precipitate weighs 7.20 g. The percentage of silver in the coin most nearly equals which of the following? (Atomic weights: H = 1.008, N = 14.008, O = 16.000, Ag = 107.88, and Cl = 35.457.)
(a) 78%   (b) 93%   (c) 82%   (d) 88%   (e) 85%

6-2 In the electric-furnace method for producing phosphorous, the raw materials are phosphate rock, silica, and coke. The products are calcium silicate, phosphorous, and carbon monoxide. The chemical equation is of the following form:

$$?Ca_3(PO_4)_2 + ?SiO_2 + ?C = ?CaSiO_3 + 2P + ?CO$$

Which of the following most nearly equals the number of tons of $Ca_3(PO_4)_2$ required to produce a ton of phosphorous? The atomic weight of P = 30.98; the formula weight of $Ca_3(PO_4)_2$ = 310.2.
(a) 3.5   (b) 4.0   (c) 5.0   (d) 5.5   (e) 6.0

6-3 Which of the following most nearly equals the number of pounds of $PbSO_4$ which would be formed when a lead storage battery supplied 200 A·h of its capacity? (Atomic weights: Pb = 207.2, S = 32, and O = 16; 1 faraday of charge = 96,494 C.)
(a) 2.48 lb   (b) 3.65 lb   (c) 2.74 lb   (d) 3.02 lb   (e) 2.25 lb

6-4 Caustic soda (NaOH) is an important commercial chemical. It is often prepared by the reaction of soda ash, Na_2CO_3, with slaked lime, $Ca(OH)_2$. Which of the following most nearly equals the weight of lime, CaO, which would be required to treat 11.023 lb of soda ash with slaked lime? Assume complete reaction. Use approximate atomic weights: Ca = 40.1, C = 12, Na = 23, O = 16, and H = 1.

(a) 7.2 lb   (b) 6.8 lb   (c) 6.4 lb   (d) 5.8 lb   (e) 5.6 lb

6-5 In a certain chemical process two liquids enter a mixing chamber (Fig. 6P-5) and are thoroughly mixed before being discharged at 80°F and 50 gal/min. Liquid 1 enters at 140°F and has a specific heat of 10 Btu/(gal)(°F). Liquid 2 enters at 65°F and has a specific heat of 8.33 Btu/(gal)(°F). Assume that there is no chemical reaction between liquids 1 and 2 and that there is no heat lost or gained to the system. Which of the following would most nearly equal the specific heat of the mixed liquid?

(a) 9.2 Btu(gal)(°F)   (b) 8.6 Btu(gal)(°F)   (c) 8.4 Btu(gal)(°F)
(d) 8.2 Btu(gal)(°F)   (e) 7.9 Btu(gal)(°F)

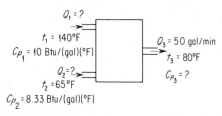

Figure 6P-5

6-6 A producer gas has the following percentage composition by volume at standard conditions of temperature and pressure: CO_2 = 5.8, CO = 19.8, H_2 = 15.1, CH_4 = 1.3, O_2 = 1.3, and N_2 = 56.7. Which of the following most nearly equals the volume of dry air which would be required at the same conditions of temperature and pressure to produce complete combustion of 100 ft³ of this gas?

(a) 62 ft³   (b) 66 ft³   (c) 74 ft³   (d) 81 ft³   (e) 89 ft³

6-7 Sulfuric acid, H_2SO_4, of 93 percent by weight purity is obtained from zinc sulfide, ZnS, ore of 50 percent by weight purity. Which of the following most nearly equals the amount of air at standard conditions of 1 atm and 32°F which will be required per ton of ore?

$$2ZnS + 3O_2 = 2ZnO + 2SO_2$$

$$2SO_2 + O_2 = 2SO_3$$

$$SO_3 + H_2O = H_2SO_4$$

(a) 22,000 ft³   (b) 28,000 ft³   (c) 36,000 ft³   (d) 38,000 ft³
(e) 42,000 ft³

6-8 Which of the following most nearly equals the weight of calcium carbonate ($CaCO_3$) which will react with excess hydrochloric acid (HCl) to produce 10 liters of carbon dioxide, CO_2, at 25°C and 770 mmHg? (Atomic weights: Ca = 40, C = 12, O = 16, and Cl = 35.)

(a) 41 g   (b) 39 g   (c) 37 g   (d) 35 g   (e) 32 g

6-9 Iron pyrite (FeS_2) is roasted in the presence of air so that sulfur dioxide (SO_2) is driven off in the form of a gas. The SO_2 is collected and combined with water to form sulfuric acid, H_2SO_4. Which of the following most nearly equals the amount of 60 percent sulfuric acid (60 percent H_2SO_4 and 40 percent H_2O) which can be produced per ton of iron pyrite, assuming no losses in the process?

(a) 3400 lb   (b) 3700 lb   (c) 4600 lb   (d) 5500 lb   (e) 6200 lb

6-10 Pure silver (Ag) is combined with nitric acid, HNO_3, to yield silver nitrate, $AgNO_3$; nitric oxide, NO; and water, H_2O. Which of the following most nearly equals the amount of silver that would be required to liberate 100 ft^3 of nitric oxide at 60°F and 1 atm? The gas constant for nitric oxide is 49.5 ft·lb/(lb)(°R). (Formula weights: Ag = 107.9, N = 14.0, H = 1.01, and O = 16.0.)

(a) 72 lb   (b) 75 lb   (c) 79 lb   (d) 83 lb   (e) 89 lb

6-11 Soda ash, Na_2CO_3, is produced commercially by the Solvay process, which may be represented by the following chemical equations:

$$CO_2 + NH_3 + NaCl + H_2O \rightarrow NaHCO_3 + NH_4Cl$$

$$2NaHCO_3 + heat \rightarrow Na_2CO_3 + H_2O + CO_2$$

Which of the following most nearly equals the amount of salt, NaCl, required to produce 2.2 kg of soda ash? Use the following atomic weights: Na = 23, C = 12, O = 16, and Cl = 35.5.

(a) 1900 g   (b) 2000 g   (c) 2100 g   (d) 2200 g   (e) 2400 g

6-12 Limestone containing 60 percent by weight of $CaCO_3$ and 40 percent by weight of $MgCO_3$, when heated in a kiln, undergoes the following reactions:

$$CaCO_3 \rightarrow CaO + CO_2$$

$$MgCO_3 \rightarrow MgO + CO_2$$

If 1 lb·mol of gas occupies 380 ft^3 at the conditions of production, which of the following most nearly equals the volume of CO_2 which would be produced from 1 ton of limestone? The atomic weights are: Ca = 40.1, Mg = 24.3, C = 12.0, and O = 16.0.

(a) 8200 ft^3 (b) 8400 ft^3 (c) 8800 ft^3 (d) 9200 ft^3 (e) 9500 ft^3

6-13 If a current of 1 A passes through a solution of NaCl for 1 h, which of the following most nearly equals the amount of material which would form at the cathode?

(a) 0.032 g   (b) 0.037 g   (c) 0.042 g   (d) 0.048 g   (e) 0.051 g

6-14 The complete combustion of propane gas is represented by the following skeleton equation:

$$C_3H_8 + O_2 = CO_2 + H_2O$$

Which of the following most nearly equals the amount of air measured at 25°C and a pressure of 760 mmHg which would be required to burn 10,500 ft³ of propane gas measured at the same conditions of temperature and pressure? (Air is approximately 21 percent O_2 by volume and 79 percent N_2.)

(a) 250,000 ft³   (b) 260,000 ft³   (c) 275,000 ft³   (d) 282,000 ft³
(e) 288,000 ft³

6-15 A high-grade phosphate rock containing phosphate equivalent to 85 percent "bone-dry phosphate," $Ca_3(PO_4)_2$, is treated with an excess of sulfuric acid which converts 95 percent of the phosphate to phosphoric acid. Which of the following most nearly equals the amount of phosphoric acid which would be produced from 100 tons of rock? Assume that the reaction involved is given by the following skeleton equation:

$$Ca_3(PO_4)_2 + H_2SO_4 \rightarrow H_3PO_4 + CaSO_4$$

Atomic weights: Ca = 40, P = 31, O = 16, H = 1, and S = 32.

(a) 46 tons   (b) 49 tons   (c) 51 tons   (d) 53 tons   (e) 57 tons

6-16 A gas analysis by volume gives the following: CO_2, 12 percent; H_2, 4 percent; CH_4, 5 percent; CO, 23 percent; and N_2, 56 percent. Which of the following most nearly equals the amount of air which would be required for complete combustion? (Atomic weights: C = 12, O = 16, H = 1, and N = 14.)

(a) 0.60 ft³/ft³   (b) 0.75 ft³/ft³   (c) 0.80 ft³/ft³   (d) 1.00 ft³/ft³
(e) 1.10 ft³/ft³

6-17 For preparation of ferrous sulfide (FeS) 3 parts by weight of iron filings are heated with 2 parts by weight of sulfur. The product from such a process is found to contain 60 percent by weight of FeS. Which of the following most nearly equals the amount of iron in the final product from such a process, assuming that no sulfur is lost and that no other sulfides are formed?

(a) 16 wt %   (b) 17.5 wt %   (c) 19 wt %   (d) 22 wt %
(e) 24 wt %

6-18 Which of the following most nearly equals the amount of SO_2 in the gases obtained from burning FeS_2 with 50 percent excess air, assuming that all the iron goes to Fe_2O_3 and all the sulfur to SO_2?

(a) 11 vol %   (b) 14 vol %   (c) 17 vol %   (d) 18.5 vol %
(e) 19.5 vol %

6-19 In a gaseous mixture collected over water at 14°C, the partial pressures of the components are: H_2, 300 mm; C_2H_6, 100 mm; O_2, 50 mm; and C_2H_4, 189 mm. The aqueous tension at 14°C is 12 mm. Which of the following most nearly equals the amount of hydrogen in the mixture?

(a) 42 vol %   (b) 46 vol %   (c) 49 vol %   (d) 51 vol %
(e) 53 vol %

6-20 In the electric-furnace method of producing phosphorous, the raw materials are phosphate rock, silica, and coke. The products are calcium silicate, phosphorous, and carbon monoxide. The process is represented by the following skeleton equation:

$$Ca_3(PO_4)_2 + SiO_2 + C \rightarrow CaSiO_3 + P + CO$$

Which of the following most nearly equals the amount of carbon which would be required per mole of phosphate?

(a) 2 mol   (b) 3.5 mol   (c) 4 mol   (d) 4.5 mol   (e) 5 mol

6-21 One of the principal scale-forming constituents of water is calcium bicarbonate, $Ca(HCO_3)_2$. This substance may be removed by treating the water with lime, $Ca(OH)_2$, in accordance with the following reactions:

$$?Ca(HCO_3)_2 + Ca(OH)_2 \rightarrow ?CaCo_3 + ?H_2O$$

Which of the following most nearly equals the amount of lime required to remove 1 lb of calcium carbonate?

(a) 0.32 lb   (b) 0.38 lb   (c) 0.46 lb   (d) 0.52 lb   (e) 0.55 lb

6-22 Gasoline may be represented approximately by the formula C_8H_{18}. Which of the following most nearly equals the amount of oxygen required for the complete combustion of 1 lb of fuel?

(a) 2.8 lb   (b) 3.1 lb   (c) 3.5 lb   (d) 3.7 lb   (e) 3.85 lb

6-23 Chlorine may be prepared by the electrolysis of an aqueous solution of sodium chloride. If the decomposition efficiency of NaCl in the electrolyte cell is 50 percent, which of the following most nearly equals the volume of chlorine gas, at standard conditions of temperature and pressure, which would be obtained per ton of salt electrolyzed?

(a) 3100 ft^3 (b) 3300 ft^3 (c) 3450 ft^3 (d) 3700 ft^3 (e) 3900 ft^3

6-24 If 1 lb of cane sugar ($C_{12}H_{22}O_{11}$) is burned with the theoretical amount of air to give complete combustion, which of the following most nearly equals the amount of CO_2 in the combustion products?

(a) 14.2 vol %   (b) 15.3 vol %   (c) 16.1 vol %   (d) 16.8 vol %
(e) 17.6 vol %

6-25 An ultimate analysis of a Southern coal is as follows:

Carbon	0.7161
Hydrogen	0.0526
Oxygen	0.9979
Nitrogen	0.0123
Sulfur	0.0074
Ash	0.1137
	1.0000

Atomic weights: C = 12, H = 1, O = 16, N = 14, and S = 32. The flue gas weight per pound of fuel at theoretically perfect combustion would most nearly equal which of the following?

(a) 8.7 lb   (b) 10.5 lb   (c) 11.2 lb   (d) 11.4 lb   (e) 11.7 lb

6-26 The weight of copper formed by the reaction of hydrogen with 10 g of cupric oxide would most nearly equal which of the following?

(a) 6.9 g   (b) 7.3 g   (c) 7.8 g   (d) 8.0 g   (e) 8.2 g

Multiple-Choice Sample Quiz

For each question select the most nearly correct answer from the five given possibilities. For practice, try to complete the following 10 problems in 15 minutes or less.

6M-1 If 1 mol of carbon combines with 1 mol of oxygen, which of the following will form?
   (*a*) 1 mol of carbon monoxide
   (*b*) 1 mol of carbon dioxide
   (*c*) 2 mol of carbon dioxide
   (*d*) 1 mol of carbon dioxide and 1 mol of carbon monoxide
   (*e*) none of these

6M-2 Hardness in water supplies is primarily due to the solution in it of:
   (*a*) carbonates and sulfates of calcium and magnesium
   (*b*) alum
   (*c*) soda ash
   (*d*) sodium sulfate
   (*e*) sodium chloride

6M-3 The group of elements containing bromine, iodine, and chlorine is known as the:
   (*a*) reactants   (*d*) disaccharides
   (*b*) ceramics   (*e*) halogens
   (*c*) colloids

6M-4 Which of the following statements is true?
   (*a*) Equal volumes of gases at the same temperature and pressure have equal numbers of molecules.
   (*b*) The viscosity of a gas is decreased by an increase in temperature if the pressure remains constant.
   (*c*) All gases and vapors may be treated as ideal gases.
   (*d*) All gases have the same specific heat at constant pressure.
   (*e*) Gases are not soluble in water.

6M-5 Which of the following chemical formulas is incorrect? (The valence of the various elements is as follows: $Al = 3+$; $Ba = 2+$; $So_4 = 2-$; $NO_3 = 1-$; $O = 2-$.)
   (*a*) Al_2O_3 (*d*) $Al_2(SO_4)_3$
   (*b*) $Al(NO_3)_3$ (*e*) $BaNO_3$
   (*c*) $BaSO_4$

6M-6 The process in which a solid changes directly to the gaseous state is called:
   (*a*) sublimation   (*d*) vaporization
   (*b*) homogenization   (*e*) distillation
   (*c*) crystallization

6M-7 The number of molecules in 22.4 liters (under standard conditions) of a substance in its gaseous state is called:
   (*a*) Dulong's number   (*b*) Petit's number

(c) Avogadro's number   (e) Graham's number
(d) Gay-Lussac's number

6M-8 One equivalent weight of H_2SO_4 is equal to:

(a) 98.06 g
(b) 2 g
(c) 49.03 g
(d) 96.06 g
(e) 32.06 g
Atomic weights: H = 1.00   S = 32.06   O = 16.00

6M-9 H_2, Cl_2, and HCl are in equilibrium in a sealed box. What would be the effect on the concentration of the Cl_2 if more H_2 were injected into the sealed box?

(a) Increase
(b) Decrease
(c) Remain the same
(d) Equal the concentration of the HCl divided by the concentration of the H_2
(e) Equal the concentration of the H_2 minus the concentration of the HCl

6M-10 If 2 mol of sodium react with 2 mol of water, which of the following will result?

(a) 1 mol of sodium hydroxide and 1 mol of hydrogen
(b) 2 mol of sodium hydroxide and 2 mol of hydrogen
(c) 2 mol of sodium hydroxide and 1 mol of hydrogen
(d) 1 mol of sodium hydroxide and 2 mol of hydrogen
(e) none of these

7
THERMODYNAMICS

7-1 EXAMINATION COVERAGE

Thermodynamics is covered in the morning session and is one of the five optional subjects in the afternoon session.

The subdivisions of the major subject which may be covered in an examination, and which should be reviewed, include the following:

Application of basic laws
  First law—closed system
  Second law—steady flow
  Third law—open system

Chemical reactions

Thermal cycles
  Carnot cycle
  Otto cycle
  Diesel cycle
  Brayton cycle
  Rankine cycle
  Sterling cycle
  Refrigeration cycles

Air conditioning

Combustion

Gas mixtures

Flow through nozzles, turbines, and compressors

Throttling

Phase change

Heat

Work

Ideal gas

Adiabatic processes

Polytropic processes

Availability and reversibility

Simple heat transfer

7-2 GENERAL ENERGY EQUATION

Bernoulli's equation is universally used in fluid mechanics and, with the necessary modifications, is also widely applied in thermodynamics. Bernoulli's equation states that the energy contained in a flowing fluid at point A must be equal to the energy in that same fluid at a later or downstream point B plus the energy lost or minus that gained between points A and B. With incompressible fluids the energy contained in the fluid at any point is essentially equal to the sum of the kinetic energy and the potential energy due to its elevation and pressure. The difference in the internal energy of the incompressible fluid between points A and B was so slight as to be inconsequential. This is not the case, however, with a compressible fluid; account must be taken of the change in internal energy, and additional terms must be added to Bernoulli's equation to make it applicable to compressible fluids. When these terms are added, Bernoulli's equation becomes, for 1 lb of fluid:

$$\frac{P_1}{w} + \frac{v_1^2}{2g} + \text{internal energy}_1 + \text{heat added} + Z_1 = \frac{P_2}{w} + \frac{v_2^2}{2g}$$
$$+ \text{internal energy}_2 + Z_2 + W$$

where W is the work done by the gas between points 1 and 2.

In gases the change in energy due to the difference in elevation between the two points considered is seldom sufficient to affect the accuracy of the result, so Z_1 and Z_2 are usually omitted. Energy is given in British thermal units per pound, so all the terms in the equation are given in British thermal units. The specific weight w is generally replaced by its reciprocal for gases $(1/w) = V$, specific volume. The internal energy is denoted by the symbol u, and the heat added to the substance by q, both in British thermal units per pound. The work done by the substance W is also measured in British thermal units per pound of gas. The equation given then becomes

$$\frac{P_1V_1}{J} + \frac{v_1{}^2}{2gJ} + u_1 + q = \frac{P_2V_2}{J} + \frac{v_2{}^2}{2gJ} + u_2 + W$$

where $J = 778$ ft·lb/Btu, the mechanical equivalent of heat.

The internal-energy term and the PV/J term are usually combined, since $u + (PV/J) = h$, the enthalpy. The general energy equation then becomes

$$\frac{v_1{}^2}{2gJ} + h_1 + Q = \frac{v_2{}^2}{2gJ} + h_2 + W$$

where the value of h depends on the pressure and the temperature and may be obtained from a table giving the properties of that particular gas.

7-3 FIRST LAW OF THERMODYNAMICS

The general energy equation is actually an expression of the first law of thermodynamics, which can be stated: "Energy can be neither created nor destroyed, but only converted from one form to another."

■ In thermodynamics the following terms are most generally employed in connection with the steady flow of thermodynamic fluids: Q, h, v^2, Z, and W. Write the general energy equation between two points 1 and 2 for velocity in a perfect nozzle, assuming that point 1 is the entrance of the nozzle and that point 2 is the exit of the nozzle.

In a perfect nozzle, no heat is added or lost and no work is done by the fluid. The nozzle is a device for converting some of the internal energy into kinetic energy. The general energy equation then reduces to

$$\frac{v_1{}^2}{2gJ} + h_1 = \frac{v_2{}^2}{2gJ} + h_2$$

since $Z_1 = Z_2$, $Q = 0$, and $W = 0$. This is probably a converging nozzle with a relatively large entrance area, in which case v_1 will be small and may be disregarded, giving $v_2{}^2/2gJ = h_1 - h_2$, limited, however, by the sonic velocity at the throat or discharge section of a converging nozzle, which is the smallest section of the nozzle. In a converging-diverging nozzle the gas will expand in the diverging section and the velocity will be supersonic, provided the critical pressure ratio is exceeded.

▪ The pressure and temperature in a throttling calorimeter are 14.7 lb/in² and 240°F, respectively. If the pressure of the steam in the main is 150 lb/in², which of the following most nearly equals its quality?

   (*a*) 96.5%   (*b*) 98.2%   (*c*) 97.5%   (*d*) 91.4%   (*e*) 93.8%

The quality of a mixture is the percent by weight that is gas. It might be noted here that steam (i.e., "dry" steam) is a gas, just as are air, nitrogen, carbon dioxide, and all other gases. And it conforms to the same laws as do the other gases. However, steam is not a "perfect" gas (neither is any other gas). Since steam is so widely used, it has been found desirable to determine what the true properties are at different state points.

The concept of steam quality can be explained with the aid of a temperature-entropy diagram (see Fig. 7-1). At point A the water is 100 percent liquid. At point B the water is 100 percent gas. At some intermediate point, C, it is a mixture of gas and liquid. Note that the entire mixture is at the same temperature; i.e., the line AB is an isotherm.

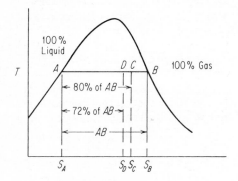

Figure 7-1

The change in entropy times temperature $T\,\Delta S$ is equal to the change in enthalpy less $V\,dP$. For this case, evaporation of water, the pressure and temperature are constant from A to B. The change in heat content over the span is then directly proportional to the change in entropy. Thus the increase in heat content of the steam from A to B is directly proportional to the percentage of the water that is in the gaseous state. If, for example, the temperature is 250°F, then, from the steam tables, the heat content of the 100 percent liquid water at point A equals 218.5 Btu/lb. The heat content of the saturated steam at point B is 1164.0 Btu/lb. The difference between the two heat contents is 945.5 Btu/lb, the heat of vaporization. Similarly, $T \, \Delta S$ between saturated liquid and saturated vapor equals $710°R \times 1.332 = 945.7$ Btu/lb.

The steam flows into the calorimeter through a throttling orifice so that no heat is added or subtracted and no work is done in the process. The elevations of the two points are the same, and the velocity at point 1 is essentially zero, as is the velocity in the calorimeter. The general energy equation then reduces to $h_1 = h_2$, and the enthalpy at point 2 can be found in a steam table. At point 2 the enthalpy of the steam (14.7 lb/in² abs and 240°F) is 1164.2 Btu/lb. This means that the enthalpy at point 1 must also equal 1164.2 Btu/lb.

The pressure in the main is 150 lb/in² abs. The enthalpy of the saturated liquid at 150 lb/in² abs is 330.51 Btu/lb, that of the saturated vapor is 1194.1 Btu/lb, and the heat of vaporization is 863.6 Btu/lb. Since the enthalpy of the steam in the main is only 1164.2 Btu/lb, it must be wet steam. If we take $1164.2 - 330.5 = 833.7$ Btu/lb, this means that only 833.7 Btu/lb is available for evaporating the saturated liquid into steam, whereas 863.6 Btu/lb is required for total evaporation of the saturated liquid.

Then, $833.7/863.6 = 96.5$ percent of the liquid has been evaporated, and the quality is 96.5 percent. The correct answer is (a).

■ A steam turbine receives 3600 lb of steam per hour at 110 ft/s velocity and 1525 Btu/lb enthalpy. The steam leaves at 810 ft/s and 1300 Btu/lb. Which of the following values most nearly equals the horsepower output?

(a) 110 hp   (b) 175 hp   (c) 240 hp   (d) 300 hp
(e) 320 hp

Assume that all the energy given up by the steam appears as work done by the turbine. Applying the general energy equation between points 1 entering the turbine and 2 leaving the turbine, we obtain ($Z_1 = Z_2$ and Q is assumed to be zero)

$$\frac{110^2}{2gJ} + 1525 = \frac{810^2}{2gJ} + 1300 + W \qquad \text{for each pound of steam}$$

$$W = (1525 - 1300) + (0.24 - 13.1) = 212 \text{ Btu/lb of steam}$$

$$\frac{3600}{60} \text{ lb/min} \times \frac{1}{33,000 \text{ ft·lb/(min)(hp)}} \times 212 \text{ Btu/lb} \times 778 \text{ ft·lb/Btu} = 300 \text{ hp}$$

The correct answer is (d).

■ In an impulse turbine, 5 lb of steam per second impinges upon the buckets at an initial velocity of 4000 ft/s and the steam leaves the buckets at a velocity of 1000 ft/s. Disregarding frictional losses, which of the following most nearly equals the efficiency of the turbine?

(a) 87.2%   (b) 91.6%   (c) 93.8%   (d) 95.2%   (e) 96.8%

The general energy equation applies, but $h_1 = h_2$, $Q = 0$, and $Z_1 = Z_2$:

$$\frac{v_1^2}{2gJ} = \frac{v_2^2}{2gJ} + W$$

$$W = \frac{v_1^2 - v_2^2}{2g} = \frac{4000^2 - 1000^2}{2g} = 233{,}000 \text{ ft·lb/lb of steam}$$

$$\text{hp} = \frac{5 \text{ lb/s} \times 233{,}000 \text{ ft·lb/lb}}{550 \text{ ft·lb/(s)(hp)}} = 2120 \text{ hp developed}$$

When the steam leaves the turbine it still has a velocity of 1000 ft/s and possesses $1000^2/2g = 15{,}500$ ft·lb/lb of mechanical energy. When the steam entered the turbine it had $4000^2/2g = 248{,}500$ ft·lb/lb of mechanical energy. The turbine then extracted 233,000 ft·lb/lb of mechanical energy from the steam, and the efficiency of the turbine is then $233{,}000/248{,}500 = 93.8$ percent. The correct answer is (c).

7-4 SECOND LAW OF THERMODYNAMICS

There are many different ways of stating the second law. One of these is the first enunciation as given by Lord Kelvin as: "It is impossible to transfer heat from a colder system to a warmer system without other simultaneous changes occurring in the two systems or in their environment." The second enunciation is: "It is impossible to take heat from a system and to convert it into work without other simultaneous changes occurring in the system or in its environment." Clausius stated the second law as: "Heat cannot, of itself, pass from a colder to a hotter system." The proof of these different statements of the second law is only that they have never been disproved.

7-5 THIRD LAW OF THERMODYNAMICS

To complete the discussion of the "laws" of thermodynamics, the third law will be given here. The third law is also known as "Nernst's postulate." It can be stated as: "The entropy of a substance becomes zero in a state at the absolute zero of temperature."

7-6 EQUATION OF STATE

The equation of state may be given, for 1 lb of a gas, as

$$PV = RT$$

where

P = pressure, lb/ft^2
V = specific volume, ft^3/lb
T = temperature, °R or °F absolute
R = specific gas constant, ft/°R or ft·lb/(lb)(°R)

The equation of state is written correctly in a number of different ways which are all equivalent. It is sometimes written $PV = WRT$, where P, R, and T are as before, W = pounds of gas, and V = total volume in cubic feet. If we let $W = M$ lb, where M is equal numerically to the molecular weight, we have $PMV = MRT$, where V = specific volume in cubic feet per pound. Rearranging terms gives

$$MR = \frac{P(MV)}{T}$$

From Avogadro's law all perfect gases have the same number of molecules in any given volume at a given temperature and pressure. It follows from Avogadro's law that a pound mole [a weight of gas equal in pounds to the molecular weight (MW) of the gas] of any perfect gas will occupy the same volume as that of any other perfect gas for the same conditions of temperature and pressure. At 32°F and atmospheric pressure this volume is called a *mole volume* and equals 359 ft^3. [In the centimeter-gram-second (cgs) system the mole volume equals 22.4 liters, which is called the "gram-molecular volume" and contains 6.024×10^{23} molecules.]

Referring to the above equation, we can see that MV (lb $\times$ ft^3/lb) will be a constant number for all perfect gases at a particular pressure and temperature. At standard conditions $MV = 359$ ft^3. We know that PV/T is a constant (equal to R) for a particular gas. Then $[P(MV)]/T$ equals a constant for any perfect gas and MR is a constant for that particular gas; MR is called the "universal gas constant" or the "molar gas constant" and is equal to (very nearly) 1544. With this we can determine R for a given gas from the relationship $R = 1544/\text{MW}$.

The equation of state will always hold for a perfect gas and will very nearly hold for all common gases in moderate pressure and temperature ranges.

For constant pressure we have that, since $PV = RT$,

$$\frac{V_1}{T_1} = \frac{V_2}{T_2} = \text{constant}$$

which is called "Charles's law."

For constant temperature we have that $P_1V_1 = RT = P_2V_2$, or

$$P_1V_1 = P_2V_2 \qquad (T = \text{constant})$$

This is called "Boyle's law."

■ When air at atmospheric pressure is pumped into a tire, its volume is compressed to one-third of its original value and the temperature of the air rises from 60 to 100°F. The gauge pressure in the tire will most nearly equal which of the following?

(*a*) 29.4 lb/in² gauge   (*b*) 47.5 lb/in² gauge   (*c*) 44.1 lb/in² gauge   (*d*) 26.2 lb/in² gauge   (*e*) 32.8 lb/in² gauge

The equation of state gives us

$$\frac{P_1V_1}{T_1} = \frac{P_2V_2}{T_2} = R$$

where R is very nearly constant over the range considered. Then

$$P_2 = P_1 \frac{T_2}{T_1} \frac{V_1}{V_2}$$

The initial pressure of the air is 14.7 lb/in² abs:

$$T_1 = 60°F = 60 + 460 = 520°R$$
$$T_2 = 100°F = 100 + 460 = 560°R$$

and the final volume is one-third the initial volume, or $V_2 = \tfrac{1}{3}V_1$:

$$P_2 = 14.7 \times \frac{560}{520} \times \frac{1}{\tfrac{1}{3}} = 47.5 \text{ lb/in}^2 \text{ abs}$$

The gauge pressure of the air in the tire would then equal 47.5 − 14.7 = 32.8 lb/in² gauge. The correct answer is (*e*).

■ A completely airtight cylinder 300 cm long is divided into two parts by a freely moving, airtight, heat-insulating piston. When the temperatures in the two compartments are equal, and equal 27°C, the piston is located 100 cm from one end of the cylinder. Which of the following most nearly equals the distance that the piston will move if the gas in the smaller part of the cylinder is heated to 74°C, if the temperature in the larger portion remains constant? Assume that the perfect gas laws hold, and disregard the thickness of the piston.

(a) 70 cm   (b) 85 cm   (c) 110 cm   (d) 125 cm
(e) 100 cm

For this problem it is best to draw the figure first (Fig. 7-2).

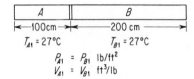

Figure 7-2

Assume that the piston is held stationary while the heat is added, and calculate the pressure increase in volume A (this would be a constant-volume process):

$$P_{A2} = P_{A1} \frac{T_2}{T_1} = P_{A1} \frac{74 + 273}{27 + 273} = 1.16 P_{A1}$$

Then let the piston move until the pressures in chambers A and B are equalized, holding the temperatures constant:

$$P_{A2} V'_{A2} = P_{A3} V'_{A3} \quad \text{and} \quad P_{B1} V'_{B1} = P_{B3} V'_{B3}$$

plus the equation

$$V'_{B1} - V'_{B3} = V'_{A3} - V'_{A2}$$

or

$$V'_{A2} + V'_{B1} = V'_{A3} + V'_{B3}$$

since the total volume remains constant. Here we are using the relationship $PV' = WRT$, where V' is measured in cubic centimeters. Then

$$V'_{A2} = 100A \qquad V'_{B1} = 200A$$

where A is the cross-sectional area.

The same relationships will hold regardless of what the cross-sectional area might be. To simplify the calculation, let us use $A = 1$:

$$P_{A2} = 1.16P_{A1} = 1.16P_{B1}$$
$$V'_{A1} = 100 = V'_{A2} \quad \text{and} \quad V'_{B1} = 200$$
$$1.16P_{A1} \times 100 = P_{A3}V'_{A3}$$
$$P_{B1}V'_{B1} = P_{A1} \times 200 = P_{B3}V'_{B3}$$
$$P_{A3} = P_{B3} \quad \text{and} \quad V'_{A3} + V'_{B3} = 300$$

Solving these equations simultaneously gives

$$V'_{A3} = 110 \quad \text{and} \quad V'_{B3} = 190$$

The piston will then move 10 cm to the right in the figure as drawn.

To determine the new volume in one step we utilize the same reasoning, regarding cross-sectional area as before, and we have

$$\frac{P_{A1}V'_{A1}}{T_{A1}} = \frac{P_{A2}V'_{A2}}{T_{A2}} = W_A R \quad \text{constant}$$

and
$$\frac{P_{B1}V'_{B1}}{T_{B1}} = \frac{P_{B2}V'_{B2}}{T_{B2}} = W_B R \quad \text{constant}$$

where $P_{A1} = P_{B1}$, $P_{A2} = P_{B2}$, $V'_{A1} = 100$, $V'_{B1} = 200$, $T_{B1} = T_{B2}$, and $V'_{A2} + V'_{B2} = 300$. Then

$$\frac{100P_{A1}}{27 + 273} = \frac{P_{A2}V'_{A2}}{74 + 273} \quad \text{or} \quad P_{A1} = 0.00864V'_{A2}P_{A2}$$
$$P_{B1} \times 200 = P_{B2}V'_{B2} \quad P_{A1} = 0.005P_{A2}V'_{B2}$$

Combining these two gives $V'_{B2} = 1.728V'_{A2}$, which, combined with the relationship $V'_{A2} + V'_{B2} = 300$, gives $V'_{A2} = 110$. The correct answer is (c).

- A tank with a volume of 10 ft^3 contains oxygen at 500 lb/in^2 abs and 85°F. Part of the oxygen was used, and later the pressure was observed to have dropped to 300 lb/in^2 abs and the temperature, to 73°F. Which of the following most nearly equals the volume that the used oxygen would occupy at standard atmospheric pressure and 70°F?

  (a) 128 ft^3 (b) 143 ft^3 (c) 132 ft^3 (d) 138 ft^3
  (e) 135 ft^3

At the beginning we have

$$P_1 V_1 = W_1 R T_1$$

$$R = \frac{1544}{MW} = \frac{1544}{32} = 48.2$$

$$P_1 = 500 \times 144 \qquad V_1 = 10 \text{ ft}^3 \qquad T_1 = 460 + 85$$

$$W_1 = \frac{(500 \times 144) \times 10}{48.2 \times 545} = 27.4 \text{ lb} \qquad \text{oxygen at start}$$

$$W_2 = \frac{P_2 V_2}{R T_2} = \frac{300 \times 144 \times 10}{48.2 \times 533} = 16.8 \text{ lb} \qquad \text{oxygen remaining}$$

The amount of oxygen used equaled $27.4 - 16.8 = 10.6$ lb. Further, we have

$$V = \frac{WRT}{P} = \frac{48.2(460 + 70)}{14.7 \times 144} \times 10.6 = 128 \text{ ft}^3$$

The correct answer is (a).

▪ A hydrocarbon has the following composition: $C = 82.66$ percent and $H = 17.3$ percent. The density of the vapor is 0.2308 g/liter at 30°C and 75 mmHg. Which of the following most nearly equals its molecular weight?

    (a) 58 (b) 62 (c) 49 (d) 55 (e) 64

The equation of state gives $PV = RT$, or $w = P/RT$, where $w = 1/V = \text{g/cm}^3$. Knowing w, P, and T, we can solve for R, where $R = P/wT$, $p = 75$ mmHg, and mercury weighs 13.6 g/cm³. Therefore,

$$P = 13.6 \times 7.5 \text{ cm} \times 980 \text{ dyn/g} = 10^5 \text{ dyn/cm}^2$$
$$w = 2.308 \times 10^{-4} \text{ g/cm}^3$$
$$T = 273 + 30 = 303 \text{ K}$$

$$R = \frac{10^5}{2.308 \times 10^{-4} \times 303} = 1.43 \times 10^6 \frac{\text{dyn·cm}}{\text{(g)(K)}} = 1.43 \times 10^6 \frac{\text{ergs}}{\text{(g)(K)}}$$

The universal gas constant in the cgs system equals 8.312×10^7 ergs/(mol)(K). So

$$\frac{8.312 \times 10^7}{1.43 \times 10^6} = 58.1 \text{ g/mol}$$

or the molecular weight of the gas is 58.1.

The equation of the gas is C_nH_m, and the molecular weight is $n \times 12 + m \times 1.008$, where 12 is the atomic weight of carbon and 1.008 is the atomic weight of hydrogen; thus

$$n \times 12 = 0.8266 \times 58.1 = 48$$

and $$m \times 1.008 = 0.1734 \times 58.1 = 10.1$$

The required answers are, then: molecular weight = 58.1 and molecular formula = C_4H_{10}. The correct answer is (a).

7-7 WORK AND THERMODYNAMIC PROCESSES

The equation of state can be used to find the pressure, volume, or temperature of a gas at any point at which two of the three properties are known. It tells nothing, however, about the way in which any of these properties varies between the two points. The work required to change a gas from one state point to another or the work which can be obtained from a gas as it expands from one state point to another depends on the path which the gas follows as it changes from one point to another. This can be easily shown by an examination of a graph showing the relation between the pressure and volume of a gas as it changes from one state point to another (Fig. 7-3).

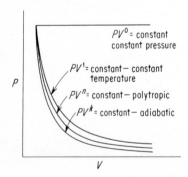

Figure 7-3

We know from our study of mechanics that work = $\int F\,ds$. For a gas, force = pressure $\times$ area, and $ds = (dV)/A$, so we have

$$F\,ds = PA\,\frac{dV}{A} = P\,dV$$

and the work done by an expanding gas equals $\int P\ dV$. We recognize this as equivalent to the area under the P-V diagram. This means that the process by which the gas changes from one state point to another is of extreme importance when it comes to determining the work required by or obtainable from a gas.

The more important processes are:

Constant temperature, isothermal	PV = constant
Constant pressure, isobaric or isopiestic	V/T = constant, P = constant
Constant volume, isometric	P/T = constant, V = constant
Adiabatic	PV^k = constant
Polytropic	PV^n = constant

where $k = c_p/c_v$, the ratio of specific heats, and n is an exponent, ordinarily somewhat less than k.

The different processes are shown in Fig. 7-3, and the differences in the work required, or developed, are apparent. The work required to go from point 1 to point 2 can be calculated from the relationship

$$\text{Work} = \int_1^2 P\ dV$$

For constant pressure we have

$$W = P \int_1^2 dV = PV_2 - PV_1$$

For the constant temperature, polytropic, and adiabatic processes, we have a similar condition in that the relationship PV^n = constant applies to all three cases, with n equaling 1, n, and k, respectively. Here we have, taking logarithms of both sides,

$$\ln P + n \ln V = \ln C$$

and differentiation gives

$$\frac{dP}{P} + n\frac{dV}{V} = 0$$

or

$$dV = -\frac{V}{n}\frac{dP}{P}$$

then

$$W = \int_1^2 -\frac{1}{n} V\ dP$$

but

$$V = \left(\frac{C}{P}\right)^{1/n} \quad \text{or} \quad PV^n = C$$

Then

$$W = -\frac{1}{n} \int_1^2 C^{1/n} P^{-1/n} \, dP = -\frac{C^{1/n}}{n} \frac{1}{1 - 1/n} P^{(1-1/n)} \Big|_1^2$$

$$= -\frac{P^{1/n}V}{n - 1} P^{(1-1/n)} \Big|_1^2$$

which gives

$$W = \frac{P_2 V_2 - P_1 V_1}{1 - n}$$

for work done by the gas; W will be negative if work is done on the gas. This equation can be used to calculate the work done by (or on) the gas for $n = k$ and $n = n$, the polytropic exponent. For an isothermal process where PV = constant and $n = 1$, this relationship gives the indeterminate fraction, $0/0$, so the relationship for work for an isothermal process must be found in another way:

$$PV = RT \text{ constant} \qquad P = \frac{RT}{V}$$

$$W = \int P \, dV = \int \frac{RT}{V} \, dV = RT \ln \frac{V_2}{V_1} = RT \ln \frac{P_1}{P_2}$$

$$\text{Work} = P_1 V_1 \ln \frac{P_1}{P_2}$$

In the case of constant volume, $dV = 0$ and no work is done by the gas. There will, however, be a change in the enthalpy of the gas for a change in the pressure when the volume remains constant. This follows from the fact that $h = u + (PV)/J$ or $h = u + (RT)/J$ and also that $du = c_v \, dT$, where u, the internal energy of the gas, is a function of temperature alone. There will, then, be a change in both the internal-energy term and also the PV/J term for a constant-volume process.

Utilizing the same relationships, it can be seen that an isothermal process is a constant-enthalpy process and also a constant-energy process for a perfect gas.

With the relationships given, the work done by a gas (positive or negative) to go from one state point to another can be calculated.

The work may also be determined by the application of the general energy equation; often this will be simpler than calculating the integral of $P \, dV$. We have

$$W = \frac{v_1^2 - v_2^2}{2gJ} + h_1 - h_2 + Z_1 - Z_2 + Q$$

In most cases this can be reduced to $W = h_1 - h_2 + Q$, and it is then necessary only to determine Δh and Q to calculate W.

The enthalpy h is equal to the internal energy u plus the quantity PV/J. For a perfect gas and a reversible adiabatic process, the following relationships will hold:

$$du + P\,dV = 0$$

$$du = c_v\,dT \quad \text{or} \quad \Delta u = c_v(T_2 - T_1)$$

$$dh = c_p\,dT \quad \text{or} \quad \Delta h = c_p\,\Delta T$$

$$R = c_p - c_v \quad \text{for the specific gas, numerically } R = (c_p - c_v)J$$

$$W = c_v(T_1 - T_2) = \text{work done by gas on a piston}$$

$$h_1 - h_2 = \frac{k}{k-1}(P_1V_1 - P_2V_2)$$

also, for an isentropic process, $dh - V\,dP = 0$. And since

$$h_2 - h_1 = \int_1^2 dh$$

we have

$$\frac{k}{k-1}(P_1V_1 - P_2V_2) = \int_1^2 V\,dP$$

For a reversible polytropic process ($PV^n = \text{constant}$), it can be shown that

$$Q = U_2 - U_1 + \int_1^2 P\,dV = \left(\frac{R}{n-1} - c_v\right)(T_1 - T_2) = \frac{c_p - nc_v}{n-1}(T_1 - T_2)$$

- A centrifugal compressor pumps 100 lb of air per minute from 14.7 lb/in^2 abs and 60°F to 50 lb/in^2 abs and 270°F by an irreversible process. The temperature rise of 50 lb/min of circulating water about the casing is 12°F. Which of the following most nearly equals the horsepower required? Disregard any changes in kinetic energy.

(a) 109 hp (b) 121 hp (c) 135 hp (d) 142 hp
(e) 147 hp

The general energy equation will always apply:

$$\frac{v_1^2}{2gJ} + h_1 + Z_1 + Q = \frac{v_2^2}{2gJ} + h_2 + Z_2 + W$$

Disregard changes in kinetic energy, and assume that the difference in elevation between the inlet and outlet of the compressor is small:

$$W = h_1 - h_2 + Q$$

The heat Q lost by the air during the compression process can easily be determined from the data given for the cooling water:

$$Q = -50 \text{ lb/min} \times 12°F \times 1 \text{ Btu/(lb)(°F)} = -600 \text{ Btu/min}$$

From the relationship $\Delta h = c_p \Delta T$, which holds for a steady-flow process,

$$h_2 - h_1 = 0.24(210) = 50.4 \text{ Btu/lb} = -(h_1 - h_2)$$
$$W = -(100 \text{ lb/min} \times 50.4 \text{ Btu/lb}) - 600 = -5640 \text{ Btu/min}$$
$$\frac{5640 \text{ Btu/min} \times 778 \text{ ft·lb/Btu}}{33,000 \text{ ft·lb/(min)(hp)}} = 133 \text{ hp}$$

which is the required answer, considering the gas as a "perfect" gas.

The process is given as an irreversible one so the work cannot be correctly calculated from the relationship

$$\text{Work} = \int P \, dv$$

This follows from the fact that the process is not reversible and, therefore, is not frictionless and that the amount of work done against friction cannot be calculated from the reversible (or frictionless) work relationships. The path of the process can, however, be described by means of the relationship $PV^n = $ constant. Let us calculate what the work requirement would have been had the compression been a reversible polytropic process. This process can be represented by a sketch in the P-V plane as shown in Fig. 7-4.

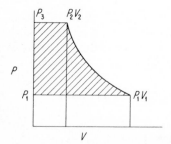

Figure 7-4

A centrifugal compressor is essentially a centrifugal pump which converts velocity head to pressure head and then pushes the compressed gas at constant pressure to wherever it goes. The curve of the process would be first a constant-pressure intake of air at pressure P_1 from zero volume to V_1. The gas would then be compressed polytropically, in accordance with the relationship PV^n = constant, to the state point represented by P_2V_2. Finally, the compressed gas would be pushed out at a constant pressure to the state point represented by $P_3(P_3 = P_2)$. We can assume a clearance of zero since it is a centrifugal compressor. The shaded area of the diagram in Fig. 7-4 indicates the work done on the gas and equals the area under that portion of the curve. The shaded area equals

$$\int_1^2 P\, dV + P_2V_2 - P_1V_1$$

where P_2V_2 is the work represented by the constant-pressure discharge and P_1V_1 is the work represented by the constant-pressure intake at atmospheric pressure:

$$\int_1^2 P\, dV = \int_1^2 CV^{-n}\, dV$$

where $PV^n = C$, giving $P = CV^{-n}$, and

$$\int_1^2 CV^{-n}\, dV = \frac{C}{1-n} V^{1-n} \Big|_1^2 = \frac{PV^n}{1-n}(V_2^{1-n} - V_1^{1-n}) = \frac{P_2V_2 - P_1V_1}{1-n}$$

which is negative, indicating, according to our convention, that work is being done on the gas. This is the work done on the gas during the polytropic compression.

The constant-pressure process work is expressed as

$$\int_2^3 P\, dV = P(V_3 - V_2) = -P_2V_2$$

This is negative, indicating work done on the gas.

The work done during the intake stroke equals $\int_0^1 P\, dV = P_1V_1$, which is positive, indicating work done by the gas. This gives

$$\text{Work} = \int_0^1 P\, dV + \int_1^2 P\, dV + \int_2^3 P\, dV = P_1V_1 + \frac{P_2V_2 - P_1V_1}{1-n} - P_2V_2$$

which combines to give $n/(1-n)(P_2V_2 - P_1V_1)$ work done by the gas.

This could also have been obtained from the relationship $W = \int_1^2 V\, dP$, which also would give the shaded area.

Since PV^n = constant, we find by differentiation that

$$nV^{n-1} P\, dV - V^n\, dP = 0$$

or $V\, dP = -nP\, dV$. This gives $\int V\, dP = -n \int P\, dV$, indicating that the work required to go from $P_1 V_1$ to $P_2 V_2$ equals

$$\frac{n}{1-n}\,(P_2 V_2 - P_1 V_1)$$

This also represents the difference between flow work and nonflow work. The nonflow work is represented by the shaded area in Fig. 7-5. In this process a given volume of gas is compressed from V_1 to V_2 (or expands from V_2 to V_1). There is no flow of gas here; the quantity of the gas is constant, no gas being added and none removed. Thus we have the term "nonflow work," with the work being represented by $\int_1^2 P\, dV$.

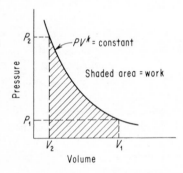

Figure 7-5

In the case of Fig. 7-4 there is actually a flow of gas. It is taken in at a pressure of P_1 to a volume of V_1, compressed to P_2 and V_2, and then discharged at a pressure P_2. There is an actual flow of gas through the compressor—thus the term "flow work" with the work equal to $\int_1^2 V\, dP$.

Since $PV^n =$ constant, $P_1 = 14.7$ lb/in^2 abs, $P_2 = 50$ lb/in^2 abs, $P_1 V_1^n = P_2 V_2^n$, and

$$\ln P_1 + n \ln V_1 = \ln P_2 + n \ln V_2$$

$$n = \frac{\ln P_1 - \ln P_2}{\ln V_2 - \ln V_1} = \frac{\ln (P_1/P_2)}{\ln (V_2/V_1)}$$

$$V_1 = RT_1/P_1 = 53.3 \times (460 + 60)/(14.7 \times 144) = 13.10 \text{ ft}^3/\text{lb}$$

$$V_2 = 5.40 \text{ ft}^3/\text{lb} \qquad n = \frac{\ln (14.7/50)}{\ln (5.42/13.1)} = \frac{-1.225}{-0.884} = 1.386$$

which gives $W = -40,035$ ft·lb/lb, or $40,035$ ft·lb/lb work done on the gas for a reversible (frictionless) polytropic cycle. This amount of work would require 121.3 hp for 100 lb/min of air. The frictional work would equal the difference, or 11.5 hp. The correct answer is (c).

- A satellite contains a gas generator which produces gas at a pressure of 100 lb/in² gauge and 150°F. The gas has an equivalent molecular weight of 22 and flows out of an ideal nozzle into space. The specific heat of the gas at constant pressure equals 0.26. The design atmospheric pressure is 4 lb/in² abs. The throat of the nozzle is 1 in in diameter.

These data can be used as the core for a number of subproblems.

- Which of the following most nearly equals the weight rate of flow of gas through the nozzle?
  (a) 4.8 lb/s   (b) 4.2 lb/s   (c) 3.6 lb/s   (d) 2.9 lb/s
  (e) 2.6 lb/s

The problem gives the initial gas pressure as 100 lb/in² gauge and the local atmospheric pressure as 4 lb/in² abs. The chamber pressure then equals 104 lb/in² abs. Similarly, the initial temperature is given as 150°F, which equals 610°R. The gas constant equals 1544/MW, giving $R = \frac{1544}{22} = 70.18$ ft/°R.

The velocity in the throat of the nozzle would be sonic because of the large expansion ratio, $\frac{104}{4} = 26$. Then throat velocity is

$$\sqrt{kgRT} = \sqrt{1.3 \times 33.2 \times 70.2 \times 610} = 1339 \text{ ft/s}$$

The ratio of specific heats is assumed to equal 1.3 since it is assumed that the generated gas is formed of molecules made up of three or more atoms. To a good approximation the ratio of specific heats will equal 1.66 for monatomic gases, 1.40 for diatomic gases, and 1.30 for gases which consist of molecules containing three or more atoms. Steam, for example, has a k value of 1.28.

The throat area equals 0.7854 in², or 0.00545 ft².

The specific volume of the gas at throat conditions would equal

$$V_1 = \frac{RT_1}{P_1} = 70.18 \times \frac{610}{104 \times 144} = 2.86 \text{ ft}^3/\text{lb}$$

The corresponding specific weight $w = 0.350$ lb/ft³.

The weight rate of flow would equal $v \times A \times w$, or $W = 1339 \times 0.00545 \times 0.350 = 2.55$ lb/s. The correct answer is (e).

■ Which of the following most nearly equals the exit diameter of the nozzle?

(a) 1.8 in   (b) 2.3 in   (c) 2.6 in   (d) 2.9 in   (e) 3.1 in

To determine the exit diameter of the nozzle, it is necessary to calculate the specific weight of the gas at the exit conditions and the velocity of the gas at exit. Then, knowing the weight flow rate, we can determine the required area:

$$A = \frac{W}{vw} = \frac{WV}{v}$$

The gas does no work, and no heat is transferred as it flows out of the nozzle, so the expansion would be adiabatic, or

$$PV^{1.3} = \text{constant}$$

$$V_2 = V_1 \left(\frac{P_1}{P_2}\right)^{1/k} = 2.86 \left(\frac{104}{4}\right)^{0.769} = 35.06 \text{ ft}^3/\text{lb}$$

$$T_2 = T_1 \left(\frac{P_2}{P_1}\right)^{(k-1)/k} = 610 \left(\frac{4}{104}\right)^{0.231} = 288°\text{R, or } -172°\text{F}$$

Using the universal gas law at the exit state point, we have $4 \times 144 \times 35.06 = 70.18 \times T$, giving $T = 288°\text{R}$.

The velocity of the gas would increase as it expands and the temperature lowers. The decrease in heat energy would be matched by an increase in kinetic energy. From the general energy equation

$$\frac{v_1^2}{2gJ} + h_1 = \frac{v_2^2}{2gJ} + h_2$$

$$h_1 - h_2 = c_p(T_1 - T_2) = 0.26 \times 322 = 83.72 \text{ Btu/lb}$$

Then the change in the kinetic energy of the gas would equal

$$83.72 \times 778 = 65,134 \text{ ft·lb/lb}$$

The kinetic energy at the throat $\text{KE}_1 = \frac{1}{2}mv_1^2$

$$\text{KE}_1 = \frac{1}{2g} \times 1339^2 = 27,840 \text{ ft·lb/lb}$$

$$\text{KE}_2 = 27,840 + 65,134 = 92,974 \text{ ft·lb/lb}$$

$$v_2 = \sqrt{2g \times 92,974} = 2447 \text{ ft/s}$$

$$A = \frac{W}{vw} = \frac{WV}{v} = \frac{2.55 \times 35.06}{2447} = 0.0365 \text{ ft}^2, \text{ or } 5.26 \text{ in}^2$$

The exit diameter would equal 2.59 in. The correct answer is (c).

■ Which of the following most nearly equals the thrust developed by the gas generator and the nozzle?

(a) 240 lb   (b) 226 lb   (c) 194 lb   (d) 187 lb   (e) 175 lb

The thrust would equal the force due to the change in momentum or

$$F = \frac{m}{t} \, \Delta v = (2.55/g) \text{ slugs/s} \times 2447 \text{ ft/s} = 194 \text{ lb}$$

The correct answer is (c).

If the discharge cone had not been added to the nozzle, the discharge velocity would have equaled only the throat velocity of 1339 ft/s. The thrust would have equaled $(2.55/32.2) \times 1339 = 106$ lb. By adding the discharge cone to the nozzle, the thrust was increased by 83 percent.

7-8 CYCLES

If we put together a series of processes so that the beginning and ending points are identical, the sequence is termed a "cycle." One important cycle is the Carnot cycle, which consists of a constant-temperature process, a constant-entropy process, a second constant-temperature process, and a second constant-entropy process, as shown in Fig. 7-6.

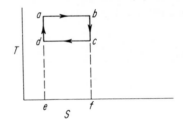

Figure 7-6

7-9 ENTROPY

This brings up the question which is so bothersome to many engineers: "What is entropy?"

Entropy cannot be pictured; it has no physical shape or substance. It is actually a mathematical concept and is the name given to the quantity $\int dQ/T$, which appears so frequently in thermodynamic analyses that it seemed desirable to give it a name. There is no absolute value of entropy; we deal with differences of entropy between state points. We can equate differences of entropy only for reversible processes, and for a reversible process

$$S_2 - S_1 = \int_1^2 \frac{dQ}{T}$$

from which can be obtained other expressions which can be used to solve for differences of entropy for a perfect gas. Thus

$$ds = \frac{dQ}{T} = \frac{du + P\,dV}{T} = c_v \frac{dT}{T} + R \frac{dV}{V}$$

which can be integrated to give

$$s_2 - s_1 = c_v \ln \frac{T_2}{T_1} + R \ln \frac{V_2}{V_1}$$

$$s_2 - s_1 = c_p \ln \frac{V_2}{V_1} + c_v \ln \frac{P_2}{P_1} = c_p \ln \frac{T_2}{T_1} - R \ln \frac{P_2}{P_1}$$

Also, the increase of entropy during a polytropic process can be shown to equal

$$s_2 - s_1 = \frac{c_p - nC_v}{n - 1} \ln \frac{T_1}{T_2} = c_v \frac{k - n}{1 - n} \ln \frac{T_2}{T_1}$$

- A volume of gas having an initial entropy of 3000 Btu/°R is heated at a constant temperature of 1000°F until the entropy equals 4500 Btu/°R. Which of the following most nearly equals the heat that was added?

    (a) 219×10^6 Btu   (b) 1.85×10^6 Btu   (c) 1.67×10^6 Btu
    (d) 1.78×10^6 Btu   (e) 2.77×10^6 Btu

We know that $S_2 - S_1 = \int_1^2 dQ/T$, and if T is a constant we can integrate easily, obtaining $S_2 - S_1 = (1/T)(Q_2 - Q_1)$. Then we have $(S_2 - S_1)T = \Delta Q$:

$$\Delta Q = (4500 - 3000)(1000 + 460) = 2,190,000 \text{ Btu}$$

The correct answer is (a).

- Given: 1 lb of saturated steam at 400°F expands isothermally to 60 lb/in² abs. First, which of the following most nearly equals the change of entropy?

    (a) 0.12 Btu/(lb)(°F)   (b) 0.16 Btu/(lb)(°F)   (c) 0.19 Btu/(lb)(°F)   (d) 0.21 Btu/(lb)(°F)   (e) 0.24 Btu/(lb)(°F)

From a steam table the pressure corresponding to a saturated-steam temperature of 400°F is 247.3 lb/in² abs. Also h = 1201 Btu/lb for saturated vapor, s = 1.5272 Btu/(lb)(°F), and V = 1.863 ft³/lb.

Since the expansion is isothermal, the final product will be superheated steam at 400°F and 60 lb/in² abs. The steam table gives h = 1233.6 Btu/lb, s = 1.7135 Btu/(°F)(lb), and V = 8.357 for this condition.

The change of entropy will then be 1.7135 − 1.5272 = 0.1863 Btu/(lb)(°F). The correct answer is (c).

■ For the same 1 lb of steam, which of the following most nearly equals the work done by the steam as it expands?

(a) 115 Btu   (b) 120 Btu   (c) 125 Btu   (d) 130 Btu
(e) 135 Btu

For a perfect gas, $\Delta h = c_p(T_2 - T_1)$, which is zero for an isothermal process. Also, $\Delta u = c_v(T_2 - T_1) = 0$ for an isothermal process for a perfect gas. Steam is not, however, a perfect gas, and the change of enthalpy is 1233.6 − 1201.0 = 32.6 Btu/lb. From the general energy equation:

$$\frac{v_1^2}{2gJ} + h_1 + Z_1 + Q = \frac{v_2^2}{2gJ} + h_2 + Z_2 + W$$

We see that

$$Q = W + h_2 - h_1$$

since $v_1 = v_2 = 0$ and $Z_1 = Z_2$.

The work $= \int_1^2 P \, dV$, which equals $PV \ln (P_1/P_2)$, as determined previously. Unfortunately,

$$P_1 V_1 = 247.3 \times 144 \times 1.863 = 66,340 \text{ ft}$$

and $P_2 V_2 = 60 \times 144 \times 8.357 = 72,200$ ft, so PV is not a constant and the work calculated by this method will not be quite correct. Averaging the two values, this relationship gives

$$W = 69,200 \ln \frac{247.3}{60} = 98,110 \text{ ft·lb/lb} \qquad \text{work done by the gas}$$

For a reversible isothermal process, $Q = T \, \Delta s$. This is not a reversible process, but applying this relationship gives $Q = (400 + 460) \times 0.1863 = 160.2$ Btu/lb heat added to the gas.

Checking, we have $Q - (h_2 - h_1) = 160.2 - 32.6 = 127.6$ Btu/lb, or 99,300 ft·lb/lb, which is close to the average value for W of 98,110 ft·lb/lb.

We recall that the work done by a gas depends on the path it follows, and since PV is not a constant, we do not know the equation of the path accurately. Also, the relationship $Q = T \, \Delta s$ Btu/lb is for a reversible process, which this is not. For an ideal gas $\Delta Q = \Delta u + \int P \, dV$. Since $h = u + (PV/J)$, Δh is the same, in the case of an isothermal process, as Δu. In this actual case it is not. In this case

$$u_1 = h_1 - \frac{PV_1}{J} = 1201.0 - \frac{66,200}{778} = 1115.7 \text{ Btu/lb}$$

and $$u_2 = 1233.6 - \frac{72,200}{778} = 1140.8 \text{ Btu/lb}$$

The work done will equal the heat added, $T \, \Delta S = 160.2$ Btu/lb, less the increase in internal energy of 25.1 Btu/lb, or 135.1 Btu/lb.

The value of 135.1 Btu/lb of work can also be obtained by iteration—calculating the values of $P \, \Delta V$ over small increments of pressure from 247.3 to 60 lb/in² abs and adding the increments to obtain the total over the complete range of pressure. By the iteration process the actual process can be closely approximated even though the actual P-V path cannot be expressed mathematically.

7-10 CARNOT CYCLE

The ideal Carnot cycle consists of four reversible processes, two isothermals and two adiabatics. Referring to Fig. 7-6:

1. Points a to b, heat enters the engine at a constant temperature, reversible isothermal expansion
2. Points b to c, reversible adiabatic expansion until temperature drops
3. Points c to d, heat removed from the engine at constant temperature, reversible isothermal compression
4. Points d to a, reversible adiabatic compression until temperature rises to the initial temperature

This cycle is a work cycle, with the engine doing work during the adiabatic expansion. A Carnot cycle can also be run in reverse to act as a refrigerator. Consider the following example.

- If a Carnot refrigerator which extracts heat from a cold-storage plant at $-20°$F and gives it to the atmosphere at $100°$F has a 100 ton/day capacity, which of the following values most nearly equals the horsepower needed to drive the refrigerator?

(*a*) 87 hp   (*b*) 92 hp   (*c*) 117 hp   (*d*) 129 hp
(*e*) 132 hp

A "ton" of refrigeration capacity is the amount needed to freeze 2000 lb of water in 24 h. Since the heat of fusion of water is 144 Btu/lb, this means 288,000 Btu/24-h day, 12,000 Btu/h, or 200 Btu/min.

When the Carnot heat engine cycle is reversed, a refrigeration cycle is obtained. The work supplied to the refrigerator is represented by the area *abcd* in Fig. 7-6. The heat extracted is represented by the area *dcfe*. This means that work in the amount of $(T_a - T_d) \Delta S$ is required for the operation of the refrigerator and that heat in the amount of $T_d \Delta S$ is extracted from the cold body. From this we can see that for the normal operating temperature range, more heat is extracted from the cold body than is required to operate the Carnot refrigerating machine. In this particular case we have $(460 - 20) \Delta S = 440 \Delta S$ Btu extracted from the cold source, while only $(560 - 440) \Delta S$ Btu is required to operate the engine. This means that for every Btu of energy used to run the refrigerator,

$$\frac{440 \, \Delta S}{(560 - 440) \, \Delta S} = 3.67 \text{ Btu}$$

is removed from the cold source. This value is called the "coefficient of performance" (c.o.p.), which is defined as the ratio of the amount of refrigeration effected to the work supplied. The c.o.p. may be either greater or less than 1. Usually it is greater than 1. It might be noted in passing that the thermodynamic advantage as evidenced by a c.o.p. is the reason for the development of the heat pump. In the case of the heat pump all the energy will be transferred to the hot body, and the c.o.p. of a heat pump is equal to the c.o.p. of the same machine used as a refrigerator plus 1. In the illustration the c.o.p. for this machine used as a heat pump would be

$$\frac{560 \, \Delta S}{(560 - 440) \, \Delta S} = 4.67 \quad \text{or} \quad 3.67 + 1.00$$

Since 1 Btu of refrigeration in the example would require only 1/3.67 Btu of work, 100 tons of refrigeration would require

$$\frac{100 \times 200}{3.67} = 5450 \text{ Btu/min} \qquad \frac{5450 \times 778}{33,000} = 128.5 \text{ hp}$$

The horsepower required is 128.5 The correct answer is (*d*).

It might be noted here that the Carnot efficiency, which equals the work done divided by the heat supplied, is the maximum efficiency possible for any thermal engine operating between two temperatures. No other cycle may exceed this efficiency, and it is not necessary for a different cycle efficiency to equal it. Other reversible cycles are as efficient as a Carnot cycle; irreversible cycles are not so efficient as a Carnot cycle.

▪ A constant-temperature heat source at $t = 1000°F$ supplies heat to a Carnot engine which receives it at 600°F. The engine rejects heat at 100°F to a cold body (or sink) which is at a constant temperature of 60°F. Ten percent of the energy available to the Carnot engine is converted into friction and lost, and 1000 Btu is supplied by the hot body. Which of the following most nearly equals the efficiency of the cycle?

(a) 64%   (b) 57%   (c) 53%   (d) 47%   (e) 37%

The maximum cycle efficiency possible between the two reservoir temperatures would be $(1460 - 520)/1460 = 64.4$ percent. If the engine receives the heat at 600°F and discharges it at 100°F, the maximum possible engine efficiency would be

$$\frac{(1060 - 560)}{1060} = 47.2 \text{ percent}$$

The actual efficiency of the cycle will be somewhat less than 47.2 percent, however, because 10 percent of the heat supplied (10 percent × 1000, or 100 Btu) is lost in overcoming friction. This will give us a modified type of Carnot cycle as shown in Fig. 7-7. The actual engine efficiency will equal the work done by the engine divided by the heat supplied to the engine. Since the heat is supplied to the engine at a constant temperature of 600°F (1060°R), the difference in entropy at the start would equal heat

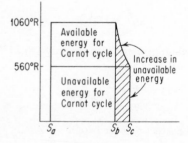

Figure 7-7

divided by the temperature, or $\Delta S = Q/T = \frac{1000}{1060} = 0.944$ Btu/°F. The heat discharged at the cold temperature by an ideal Carnot engine would be $Q = T \Delta S = 560 \times 0.944 = 528$ Btu. The engine loses another 100 Btu in overcoming friction, so the total heat lost would equal 628 Btu. The heat utilized in doing useful work would then be $1000 - 628 = 372$ Btu. The cycle efficiency would be 372/1000, or 37.2 percent. The correct answer is (e).

■ A mechanical refrigerator has a capacity of 200 tons and operates on a vapor-compression cycle. The refrigerant is R-12 (formerly known as Freon 12). The cooling coil is maintained at 10°F, and the vapor leaves the compressor at 200°F and 180 lb/in² abs. During compression 3 Btu/lb of refrigerant are transferred to the cylinder walls of the compressor. The condensed liquid enters the control valve at a temperature of 90°F. Which of the following most nearly equals the coefficient of performance?

(a) 2.1   (b) 1.8   (c) 1.6   (d) 1.4   (e) 2.7

The refrigerator will operate on a modified Carnot cycle (see Fig. 7-8). From a to b the refrigerant throttles at constant enthalpy by expanding to a lower pressure. Part of the liquid flashes into vapor and the mixture cools to the evaporator temperature. In this case that is 10°F. From b to c the remaining cold liquid boils at constant pressure and temperature and absorbs heat. From c to d is the compression portion of the cycle in which the gas which has boiled off is compressed to some point in the superheat region. For this case that is 200°F and 180 lb/in² abs. The superheated vapor enters the condenser at these conditions and

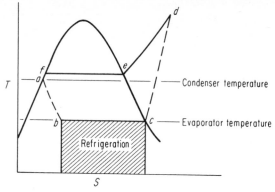

Figure 7-8

heat is discharged to the atmosphere. The refrigerant cools to the condensation temperature at point e. Heat is extracted at constant pressure and temperature from e to f, at which point the refrigerant is a saturated liquid. The liquid is cooled further to point a, still at the pressure of 180 lb/in² abs. The cooled liquid then enters the expansion valve.

First look up the properties of R-12 at state points a, c, and d:

Point a (saturated liquid at 90°F): enthalpy $h_f = 28.713$ Btu/lb

Point c (saturated vapor at 10°F): enthalpy $h = 78.335$ Btu/lb; entropy $s_g = 0.16798$ Btu/(lb)(°F)

Point d (superheated vapor 200°F): enthalpy $h = 103.291$ Btu/lb; entropy $s = 0.18556$ Btu/(lb)(°F)

The compression of the vapor from point c to point d is not isentropic. This is apparent from the fact that the entropy of the vapor is higher at point d than it is at point c. In addition, it is stated that heat in the amount of 3 Btu/lb is transferred to the cylinder walls during the compression of the gas.

The work done by the compressor from point c to point d would equal the increase in the enthalpy of the fluid plus the heat lost during the compression. The engine must supply both the heat retained and the heat lost:

$$h_d - h_c = 103.291 - 78.335 = 24.956 \text{ Btu/lb}$$

To this must be added the heat lost, so the work done would equal $24.956 + 3 = 27.956$ Btu/lb.

The heat extracted from the cold box by the refrigerant would equal the difference in enthalpies between points b and c. The expansion from a to b is a constant-enthalpy process, so

$$h_a = h_b$$

Heat extracted $= h_c - h_a = 78.335 - 28.713 = 49.622$ Btu/lb

The coefficient of performance equals the ratio of the heat extracted divided by the work done.

$$\text{c.o.p.} = \frac{\text{refrigeration}}{\text{net work}} = \frac{49.622}{24.956} = 1.775$$

The correct answer is (b).

Utilizing the same data as above, we can solve the following problem.

- Which of the following most nearly equals the amount of refrigerant that would be recirculated?

(a) 786 lb/min   (b) 806 lb/min   (c) 823 lb/min
(d) 843 lb/min   (e) 851 lb/min

The ideal coefficient of performance of a refrigeration machine operating between a high temperature of T_1 and a low temperature of T_2 equals

$$\text{c.o.p.} = \frac{T_2}{T_1 - T_2}$$

For this case the ideal c.o.p. would equal

$$\frac{470}{660 - 470} = 2.774$$

For a refrigeration capacity of 200 tons the refrigeration system would have to extract heat in the amount of $200 \times 200 = 40{,}000$ Btu/min. Since 1 lb of refrigerant extracts 49.622 Btu, the required rate of flow of refrigerant would equal

$$\frac{40{,}000}{49.622} = 806 \text{ lb/min}$$

The correct answer is (b).

Again using the same set of data, consider this example.

■ Which of the following most nearly equals the power that would be required to operate the refrigerator?

(a) 530 hp   (b) 560 hp   (c) 575 hp   (d) 580 hp
(e) 595 hp

The work required to compress 1 lb of refrigerant vapor equals 27.956 Btu/lb. The required input power would then equal $806 \times 27.956 = 22{,}532.5$ Btu/min.

Since 1 hp equals 33,000 ft·lb/min, or

$$\frac{33{,}000}{778} = 42.42 \text{ Btu/min}$$

the power required to operate the refrigeration unit would equal

$$\frac{22{,}532.5}{42.42} = 531 \text{ hp}$$

The horsepower per ton of refrigeration equals 4.71/c.o.p.:

$$\frac{4.71}{1.775} = 2.65 \text{ per ton, or } 531 \text{ hp}$$

The correct answer is (a).

Three other cycles briefly covered in previous examinations are the Rankine, Diesel, and Otto cycles.

7-11   RANKINE CYCLE

The Rankine cycle is a more practical cycle for a vapor-liquid system than is the Carnot cycle. It has a lower thermal efficiency than the Carnot cycle but a higher work ratio. This is the theoretical cycle for a steam power plant.

- Draw a Rankine vapor cycle on the pressure-volume diagram shown in Fig. 7-9. Indicate the thermodynamic process involved in each of the following parts of the diagram:

  a. Expansion through turbine
  b. Condensation in condenser
  c. Pumping of condensate into boiler.
  d. Change of liquid to saturated vapor in boiler

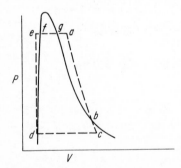

Figure 7-9

The Rankine vapor cycle has been sketched over the saturated-steam line in Fig. 7-9. The expansion through the turbine is represented by the line abc. This is a reversible adiabatic expansion of superheated steam, denoted by a, into wet steam at point c. The transition point from dry to wet steam occurs at point b, where the cycle line crosses the saturated-steam line.

From c to d the wet steam condenses into water at a constant temperature and a constant pressure, reversibly.

The condensate is then pumped into the boiler along a reversible adiabatic line de. This is practically a constant-volume process.

The liquid is then changed to superheated steam along the constant-pressure line $efga$. From e to f the water is heated at a constant pressure to the boiling temperature; this is a reversible process. From f to g the water is vaporized into steam reversibly and isobarically. From g to a there is a reversible isobaric heating of the steam into superheated steam at a higher temperature.

7-12   DIESEL CYCLE

The Diesel cycle is the basis of the operation of the diesel engine.

- Sketch an ideal Diesel cycle diagram (pressure vs. volume). Label the suction stroke, compression stroke, expansion stroke, and exhaust stroke.

The Diesel cycle is shown in Fig. 7-10. The intake stroke is shown by the line ab, a constant-pressure process. The air is next compressed adiabatically from b to c. Heat is then added (fuel injected) at constant pressure from c to d, and the gas expands adiabatically from d to e. The exhaust valve opens at e and exhaust takes place from e to b and during the exhaust stroke, ba. This is a four-cycle engine as shown, with suction stroke ab, compression bc, expansion cde, and exhaust stroke eba.

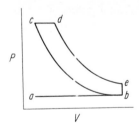

Figure 7-10

7-13   OTTO CYCLE

The Otto cycle, shown in Fig. 7-11, is the cycle on which the operation of the ordinary gasoline engine is based. The intake stroke is from a to b. The mixture is compressed adiabatically from b to c, at which point

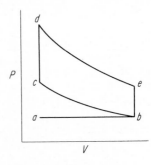

Figure 7-11

ignition takes place and heat is added, raising the temperature and pressure to d. An adiabatic expansion takes place from d to e, at which point the exhaust valve opens and combustion gases flow out. The path eb represents a constant-volume drop in temperature and an extraction of heat. The exhaust stroke is represented by the line ba.

The cycle efficiency of an Otto cycle can be shown to equal $1 - (1/r^{k-1})$, where r is the compression ratio.

7-14 INDICATOR DIAGRAMS

Indicator cards are P-V diagrams obtained during engine operation. The area under a P-V diagram equals the work done during a cycle, so the net area (work is required to draw air into the cylinder during the intake stroke) is a measure of the work done on a piston during one cycle. The area of the P-V diagram is frequently divided by the length of the stroke to give average pressure; this is known as the "mean effective pressure" (MEP). The indicated horsepower is then

$$\frac{\text{MEP (lb/ft}^2) \times \text{volume (ft}^3) \times \text{power (strokes/min)}}{33,000 \text{ ft·lb(min)(hp)}}$$

Sometimes the mean effective pressure is converted to the output of the engine, making allowances for frictional losses in the engine and power transmission system. It is then termed the "brake mean effective pressure" (BMEP). This is usually calculated from dynamometer readings taken at the output shaft.

7-15 SPECIFIC HEAT AND LATENT HEAT

Problems concerning specific heat and latent heat merely require a balance of the heat added, lost, and retained.

■ A tank holds 1000 lb of water at a temperature of 90°F. A piece of ice at 32°F was added and completely melted in the water, reducing its temperature to 40°F. Assuming no heat losses, which of the following most nearly equals the pounds of ice added? (Latent heat of ice = 80 cal/g.)

(a) 310 lb   (b) 330 lb   (c) 350 lb   (d) 365 lb   (e) 372 lb

First, convert the value for the latent heat of ice to British thermal units per pound. Since there are 252 cal/Btu, and 454 g of mass corresponds to 1 lb of weight, the latent heat of fusion of ice will equal

$$\frac{80 \text{ cal}}{252 \text{ cal/Btu}} \times 1/\text{g} \times 454 \text{ g/lb} = 144 \text{ Btu/lb}$$

The ice was added at 32°F; it melted and rose to 40°F, so that each pound of ice absorbed $(144 + 8) = 152$ Btu. The 1000 lb of water would give up $1000 \times (90 - 40) = 50{,}000$ Btu in cooling. The total amount of ice required would thus equal $\frac{50{,}000}{152} = 329$ lb. The correct answer is (b).

■ Which of the following most nearly equals the amount of steam which must be added to the tank contents resulting from the above operation to bring the contents to 140°F? (Latent heat of steam at 212°F = 540 cal/g.)

(a) 95 lb   (b) 101 lb   (c) 112 lb   (d) 117 lb   (e) 127 lb

Convert the latent heat of steam at 212°F to British thermal units per pound: $\frac{540}{252} \times 454 = 973$ Btu/lb.

The total amount of water in the tank would equal $1000 + 329 = 1329$ lb. The heat required to raise the temperature to 140°F would equal $1329 \times (140 - 40) = 132{,}900$ Btu. Each pound of steam would give up $973 + (212 - 140) = 1045$ Btu. Steam required would equal $132{,}900/1045 = 127$ lb. The correct answer is (e).

7-16 PSYCHROMETRY

The principal points to remember here are that psychrometry is an application of Dalton's law of partial pressures and the Gibbs-Dalton law, and that the water vapor in the air is actually steam whose properties are the same as those listed in the steam tables. Many of the principles of psychrometry can be illustrated by the following multiple-part problem:

■ Atmospheric air has a dry-bulb temperature of 85°F, and a wet-bulb temperature of 64°F. The barometric pressure is 14.10 lb/in² abs. Calculate (without the use of the psychrometric chart):

a. Specific humidity in grains per pound of dry air
b. Relative humidity
c. Dew-point temperature
d. Enthalpy of mixture in British thermal units per pound of dry air

As the air passes over the wet bulb, it absorbs moisture, thus lowering the temperature of the air and of the water, and increasing the amount of the moisture in the air. The final temperature of the air, when it has absorbed all the moisture it can hold, is the wet-bulb temperature. It is also the dew-point temperature of the air after it has absorbed the additional moisture, which is higher than the dew point of the air originally. This is also called the "temperature of adiabatic saturation." The derivation of the relationship used to calculate the specific humidity is somewhat complex and will not be discussed here. The process can be described by means of a temperature-entropy diagram for steam, as shown in Fig. 7-12. Here the vapor in the air is initially at temperature T_c, and the vapor is in the superheat region, as denoted by point 1. If the air, and the vapor in the air, were cooled along the constant-pressure line 1-2, condensation would start to occur at T_a, where the constant-pressure line crosses the saturated-steam line. This is the dew point of the original air-vapor mixture. As the air absorbs moisture, the saturation temperature will rise, and the wet-bulb temperature will be between T_a and T_c at point 3 on the saturated-steam line. From this we can see that the lowest air temperature which could be obtained through evaporation of water, evaporative cooling, would be the wet-bulb temperature.

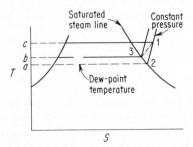

Figure 7-12

A relationship derived by Carrier often used to find humidity is

$$P_{wv} = P_{swb} - \frac{P - P_{swb}}{2830 - 1.44t_{wb}} (t - t_{wb})$$

where

P_{wv} = pressure of water vapor
P_{swb} = saturation pressure at wet-bulb temperature
P = atmospheric pressure
t_{wb} = wet-bulb temperature, °F
t = dry-bulb temperature, °F

For the example problem, P_{swb} = 0.2951 lb/in² abs; P = 14.10 lb/in² abs; saturation pressure at 85°F = 0.5959 lb/in² abs:

$$P_{wv} = 0.2951 - \frac{14.10 - 0.2951}{2830 - 1.44(64)} (85 - 64) = 0.1891 \text{ lb/in}^2 \text{ abs}$$

From Dalton's law of partial pressures we know that the pressure of the air will then be $14.10 - 0.189 = 13.91$ lb/in² abs:

$$w_a = \frac{13.91 \times 144}{53.3 \times 545} = 0.0692 \text{ lb/ft}^3$$

$$w_{wv} = \frac{0.1891 \times 144}{85.8 \times 545} = 0.000584 \text{ lb/ft}^3$$

Since there are 7000 grains/lb, 1 ft³ of water vapor would weigh 4.09 grains. One pound of dry air would occupy

$$\frac{1}{0.0692} = 14.45 \text{ ft}^3$$

so there would be $4.09 \times 14.45 = 59.1$ grains of moisture per pound of dry air.

The relative humidity will equal the ratio of the actual water vapor pressure to the water vapor pressure at saturation:

$$\text{Relative humidity} = \frac{0.1891}{0.5959} = 31.7 \text{ percent}$$

The dew-point temperature would be the saturation temperature for the amount of moisture in the air; following down the constant-pressure line in Fig. 7-12 from point 1 to point 2 gives us that the dew point (saturation temperature for $p = 0.1891$ lb/in² abs) equals 51.8°F.

The enthalpy of the mixture will equal the sum of the enthalpies of the gases making up the mixture. The steam at 85°F is superheated steam which has a saturation temperature equal to the dew point. The enthalpy of saturated steam at 51.8°F is 1084.4 Btu/lb. The steam is then superheated 33.2°F. The specific heat of steam at low temperature and pressure may be taken as 0.446 Btu/(lb)(°F), giving an enthalpy for the vapor of

$$1084.4 + 33.2 \times 0.446 = 1099.2 \text{ Btu/lb}$$

The enthalpy of saturated steam at 85°F is 1098.8 Btu/lb, which would have served just as well for this purpose.

The enthalpy of the air at low pressures is $c_p(t - t_0)$, and since the point of zero enthalpy for air is 0°F, the enthalpy of the air is 0.240(85) = 20.4 Btu/lb.

Since we have 59.1/7000 = 0.00845 lb of steam per pound of air, the total enthalpy per pound of dry air is

$$20.4 + 0.00845 \times 1099.2 = 29.68 \text{ Btu/lb dry air}$$

7-17 HEAT TRANSFER

Heat transfer has received some attention in past examinations. The emphasis has been primarily on conduction. In Fig. 7-13 we see that as heat flows from t_1 to t_2 it is hindered first by the film factor on the outside of the first layer of the wall, then by the first layer, the second layer, the third layer, and finally by the film factor on the inside of the inner layer. These may all be combined to give one equivalent heat-transfer factor for the assembly. The relationship for the combined heat-transfer factor is

$$\frac{1}{U} = \frac{1}{h_1} + \frac{L_1}{k_1} + \frac{L_2}{k_2} + \frac{L_3}{k_3} + \frac{1}{h_2}$$

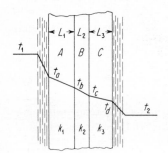

Figure 7-13

where
U = overall heat-transfer coefficient
h_1 = first film factor

k_1 = thermal conductivity of the material of first layer
L_1 = thickness of first layer
k_2 = thermal conductivity of second layer
L_2 = thickness of second layer
k_3 = thermal conductivity of third layer
L_3 = thickness of third layer
h_2 = film factor at the other face of the wall

Since k is ordinarily given in Btu/(h)(ft^2)(°F)/(ft), the units of U will be Btu/(h)(ft^2)(°F), which must be multiplied by area (square feet) times temperature differential (degrees Fahrenheit). The rate of heat transfer through the wall is then

$$Q = UA(t_1 - t_2) \qquad \text{Btu/h}$$

This relationship can be easily derived with the aid of Fig. 7-13. Here the flow of the heat is from t_1 to t_2. The temperature drop through the first film is from t_1 to t_a. The temperature drops from t_a to t_b through layer A, from t_b to t_c through layer B, and so on. Since the system has reached a steady state, the same amount of heat will flow through each film and through each layer. We shall call the rate of heat flow Q Btu/(h)(ft^2); that is, Q Btu passes through each square foot of the wall per hour. We can set up relationships for the rate of heat flow through each layer of the resistance:

$$Q = (t_1 - t_a)h_1 \qquad \text{giving } t_1 - t_a = \frac{Q}{h_1}$$

$$Q = \frac{(t_a - t_b)k_1}{L_1} \qquad \text{giving } t_a - t_b = \frac{QL_1}{k_1}$$

$$Q = \frac{(t_b - t_c)k_2}{L_2} \qquad \text{giving } t_b - t_c = \frac{QL_2}{k_2}$$

$$Q = \frac{(t_c - t_d)k_3}{L_3} \qquad \text{giving } t_c - t_d = \frac{QL_3}{k_3}$$

$$Q = (t_d - t_2)h_1 \qquad \text{giving } t_d - t_2 = \frac{Q}{h_2}$$

Adding together the relationships on the right we obtain

$$t_1 - t_2 = Q\left(\frac{1}{h_1} + \frac{L_1}{k_1} + \frac{L_2}{k_2} + \frac{L_3}{k_3} + \frac{1}{h_2}\right)$$

which reduces to $Q = U(t_1 - t_2)$ per square foot or

$$Q \text{ Btu/h} = UA(t_1 - t_2)$$

■ Given the air temperature in a room is 75°F and outside temperature 30°F. The inside temperature of the window pane is 45°F and the outside temperature 40°F. The window pane is $\frac{3}{16}$ in thick and has a conductivity $k = 1.0 \text{ Btu/(h)}(°F)(ft^2 \text{ surface})(ft \text{ thick})$. The resistances to heat flow consist of the air film on the inside surface, the outside surface air film, and the window pane itself. Which of the following most nearly equals the rate of flow of heat through 1 ft² of window pane?

(a) 320 Btu/h   (b) 335 Btu/h   (c) 352 Btu/h
(d) 360 Btu/h   (e) 368 Btu/h

We must calculate this problem in steps, remembering that the heat which flows through the inner film equals the heat which flows through the pane and the heat which flows through the outer film. This gives us three equations. A sketch of the temperature variation is shown in Fig. 7-14.

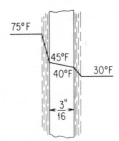

Figure 7-14

The heat which flows out is Q. The heat-transfer coefficient through this film equals h_1, which gives us the relationship $Q = h_1(75 - 45)$ Btu/h for 1 ft² of area.

This same amount of heat must flow through the pane, and the inverse of the coefficient of heat transfer through the pane will equal

$$\frac{1}{U} = \frac{L}{k} = \frac{\frac{3}{16} \times \frac{1}{12}}{1.0} = 0.01563$$

giving $U = 64 \text{ Btu/(h)}(ft^2)(°F)$. Then

$$Q = U(t_1 - t_2) = 64(45 - 40) = 320 \text{ Btu/(h)}(ft^2)$$

The correct answer is (a).

This example can be expanded a bit to illustrate another aspect of heat transfer, i.e., the determination of the two film factors. Since the flow of heat through both films and the pane must be the same, the flow through the outer film will be $Q = 10h_2$. This gives

$$Q = 320 \text{ Btu/h} \quad \text{for each square foot of window pane}$$
$$Q = 30h_1 \quad h_1 = 10.7 \text{ Btu/(h)(ft}^2)(°F)$$
$$Q = 10h_2 \quad h_2 = 32 \text{ Btu/(h)(ft}^2)(°F)$$

Flow of heat through pipe insulation is a bit different because the area of the insulation and of heat flow increases as the radius increases. This gives rise to a different relationship for the heat transfer, namely

$$Q = \frac{2\pi r_m l(t_1 - t_2)}{r_m/(r_o h_o) + (r_o - r_i)/k_m + r_m/(r_i h_i)} \quad \text{Btu/h}$$

where

r_m = the logarithmic mean radius = $(r_o - r_i)/(\ln r_o/r_i)$
r_i = inner radius (Fig. 7-15)
r_o = outer radius
l = length of pipe
k_m = mean thermal conductivity

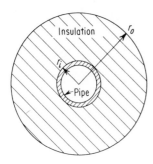

Figure 7-15

As can be seen, this equation ignores the insulating effect of the metal pipe which can safely be disregarded in most cases.

For illustrative purposes, consider a multifaceted problem.

- A $1\frac{1}{2}$-in pipe is lagged with 85 percent magnesia insulation $1\frac{1}{4}$ in thick. The pipe temperature is 230°F, and the outside temperature of the lagging is 85°F.

  a. Calculate the heat loss per hour for 200 ft of pipe.

  b. What is the average area of heat transfer per foot of pipe?
  c. Calculate the coefficient of heat transfer based on the (1) inside area, (2) outside area, and (3) average area.

Consideration of the film coefficients has been eliminated through the specification of the temperatures at the inner and outer edges of the insulation.

If we assume that the outer diameter (OD) of the pipe is 1.5 in, then $r_i = 0.75$ in and $r_o = 2.00$ in:

$$r_m = \frac{1.25}{\ln 2.67} = 1.27$$

$$Q = \frac{2\pi \times (1.27/12) \times 200(230 - 85)}{1.25/(12 \times 0.041)} = 7580 \text{ Btu/h}$$

The average area would be $2\pi(r_o + r_i)/2 \times \frac{1}{12} = 0.72$ ft^2/ft. Similarly, the inside area would be 0.393 ft^2/ft, and the outside area would be 1.05 ft^2/ft.

The coefficient of heat transfer based on the inside area would be 7580/ $(200 \times 0.393) = 96.5$ Btu/(h)(ft^2); on the outside area, 36.1 Btu/(ft^2)(h); and on the average area, 52.7 Btu/(ft^2)(h).

The principles of conductive heat transfer also govern the rate of heat transfer which takes place in a heat exchanger. Heat exchangers are made in three general types—parallel flow, counterflow, and cross flow. They are also made in single-pass and multiple-pass configurations. Examination coverage can be expected to be limited to simple parallel-flow and counterflow heat exchangers.

The variation of temperatures of the cooling fluid and the heating fluid in parallel-flow and counterflow heat exchangers is as shown in Fig. 7-16. In both cases, the temperature difference between the hotter and cooler fluids changes from inlet to outlet. To calculate an overall heat-transfer factor for the heat exchanger, it is necessary to use some mean temperature difference. The rate of heat transfer can then be calculated, using the basic relationship of

$$Q = UA \, \Delta t_m$$

where Δt_m is the mean temperature difference. However, the difference between the two fluids changes across the heat exchanger, so it is not feasible to use the arithmetic mean. It has been determined that more accurate results can be obtained by using the logarithmic mean temperature difference. The log-mean-temperature difference is denoted by

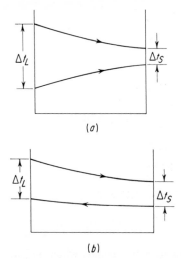

(a)

(b)

Figure 7-16 (a) Parallel-flow heat exchanger; (b) counterflow heat exchanger.

Δt_m and equals

$$\Delta t_m = \frac{\Delta t_L - \Delta t_S}{\ln (\Delta t_L / \Delta t_S)}$$

where Δt_L is the larger temperature difference and Δt_S is the smaller temperature difference (see Fig. 7-16).

▪ It is necessary to cool water from 160 to 80°F at a flow rate of 60 gal/min. Water is available from a stream at 55°F, and it is proposed to heat it 10°F and then return it to the stream. Which of the following most nearly equals the ratio of the heat-transfer area which would be required for a parallel-flow heat exchanger vs. the area required for a counterflow heat exchanger?

(a) 0.95   (b) 1.00   (c) 1.13   (d) 1.21   (e) 0.87

Given: 60 gal/min × 8.33 lb/gal × 1.0 × (160 − 80) = 39,984 Btu/min required. Flow rate of stream water = 39,984/10°F = 3998 lb/min or 480 gal/min.

For parallel flow:

$$\Delta t_L = 160 - 55 = 105°F$$

$$\Delta t_S = 80 - 15 = 15°F$$

$$t_m = \frac{105 - 15}{\ln (105/15)} = 46.25°F$$

For counterflow:

$$\Delta t_L = 160 - 65 = 95°F$$

$$\Delta t_S = 80 - 55 = 25°F$$

$$t_m = \frac{95 - 25}{\ln (95/25)} = 52.43°F$$

Assuming the same heat-transfer coefficient U, we obtain

$$Q \text{ (parallel flow)} = UA_P \times 46.25$$

$$Q \text{ (counterflow)} = UA_C \times 52.43$$

$$A_P = \frac{52.43}{46.25} \times A_C = 1.13 \times A_C$$

The correct answer is (c).

Radiant heat transfer takes place between two bodies which are at different temperatures which can "see" each other. The rate of heat transfer is proportional to the difference between the fourth powers of the two temperatures. It is also a function of the coefficient of emissivity of the hotter body and of the coefficient of absorptivity of the colder surface. The quantity of heat transferred is also a function of the areas, and there also may be a shape factor. The general equation for radiant heat transfer may be written as

$$Q = 0.173F_E F_A A \left[\left(\frac{T_1}{100}\right)^4 - \left(\frac{T_2}{100}\right)^4 \right]$$

where

Q = heat radiated (and absorbed), Btu/hr
F_E = a factor which depends on the coefficients of emissivity (coefficient of emissivity for a given surface equals the coefficient of absorptivity for that surface)
F_A = a factor depending on the geometry of the surfaces
A = area, ft^2
T_1, T_2 = temperatures of the two surfaces, °R

A body with an absorptivity, or emissivity, of one is termed a *blackbody*. The radiated heat from such a body would equal

$$Q = 0.173A \frac{T}{100^4} \text{ Btu/h}$$

The constant 0.173×10^{-8} Btu/(h)(ft^2)(deg^4) is called the "Stefan-Boltzmann constant of proportionality." In metric units, the Stefan-Boltzmann constant equals 5.672×10^{-5} erg/(s)(cm^2)(deg^4).

Sample Problems

7-1 An engine operates on the Rankine cycle with complete expansion (Fig. 7P-1). Steam, the working medium, is admitted to the engine at a pressure of 200 lb/in^2 abs and a temperature of 600°F; it is exhausted at a pressure of 4 in Hg. Which of the following most nearly equals the ideal thermal efficiency of the cycle?
    (a) 45%   (b) 37%   (c) 28%   (d) 22%   (e) 19%

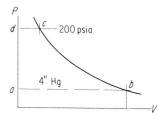

Figure 7P-1

7-2 Assume that 5 lb of a gas mixture is 30 percent CO_2 and 70 percent O_2 by weight. It is compressed from (1) a pressure of 15 lb/in^2 abs to (2) a pressure of 150 lb/in^2 abs by a process characterized by the equation $P_1 V_1{}^n = P_2 V_2{}^n$, where the value of n is 1.0. The temperature of the mixture is 60°F at the beginning of the process. Which of the following most nearly equals the work needed to compress the 5 lb of mixture?
    (a) 301,000 ft·lb   (b) 265,000 ft·lb   (c) 257,000 ft·lb
    (d) 246,000 ft·lb   (e) 239,000 ft·lb

7-3 Saturated steam is used in a mixing tank to heat and mix 200 gal of solution. The solution is to be heated from 62 to 150°F by introducing steam directly into the solution. The specific heat of the solution is about 0.85 Btu/(lb)(°F) at 62°F and remains essentially constant over the temperature range stated. The specific weight of the solution is 8.40 lb/gal. Which of the following most nearly equals the amount of saturated steam at a gauge pressure of 20 lb/in^2 which will be required for heating if all the steam is condensed in the solution and the heat loss to the surroundings is assumed to be negligible? (See Fig. 7P-3.)

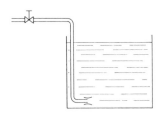

Figure 7P-3

(a) 78 lb   (b) 91 lb   (c) 98 lb   (d) 117 lb   (e) 121 lb

7-4   Given 10.0 ft³ of a gas at 25 in Hg. The gas has a specific heat at constant pressure of 0.2025 Btu/(lb)(°F) and a specific heat at constant volume of 0.1575 Btu/(lb)(°F). Which of the following most nearly equals the volume the gas would occupy at a final pressure of 5 atm if the process is adiabatic?
(a) 2.48 ft³   (b) 2.62 ft³   (c) 2.78 ft³   (d) 2.87 ft³   (e) 2.92 ft³

7-5   A system consisting of 1.0 lb of water initially at 70°F is heated at a constant pressure of 30 lb/in² abs while confined inside a cylinder equipped with a piston. If the addition of 1200 Btu of heat raises the temperature to 400°F, which of the following most nearly equals the change in internal energy of the system?
(a) 875 Btu   (b) 947 Btu   (c) 1075 Btu   (d) 1106 Btu
(e) 1189 Btu

7-6   A masonry wall has a 4-in-thick brick facing wall bonded to an 8-in-thick concrete backing. On a day when the room temperature is 68°F and the outside temperature is 12°F, the inner-surface temperature of the concrete is 57°F and the outer-surface temperature of the brick is 19°F. The thermal conductivity of the brick is 0.36 and of the concrete, 0.68 Btu/(h)(ft)(°F). Which of the following most nearly equals the overall heat-transfer coeficient?
(a) 0.271 Btu/(h)(ft²)(°F)   (b) 0.298 Btu/(h)(ft²)(°F)
(c) 0.337 Btu/(h)(ft²)(°F)   (d) 0.356 Btu/(h)(ft²)(°F)
(e) 0.401 Btu/(h)(ft²)(°F)

7-7   A small steam generating plant burns oil as a fuel. The oil has a specific gravity of 1.008 and costs $2.50 per barrel. The heating value of the oil is 18,250 Btu/lb. The boiler operates at 75 percent thermal efficiency when taking in feedwater at 180°F and delivering dry saturated steam at 100 lb/in² gauge. Which of the following most nearly equals the cost of fuel per 1000 lb of steam on this basis (1 bbl = 42 gal)?
(a) 54 cents   (b) 62 cents   (c) 73 cents   (d) 81 cents
(e) 97 cents

7-8   The internal energy of a gas is given by the relationship $u = 0.08T + 0.0027T^2$. If 1 lb of this gas expands without heat flow, but with a temperature drop from 300 to 100°F, which of the following most nearly equals the amount of work done?
(a) 217,000 ft·lb   (b) 381,000 ft·lb   (c) 423,000 ft·lb
(d) 507,000 ft·lb   (e) 572,000 ft·lb

7-9   The volumetric analysis of a gas mixture shows that it consists of 70 percent nitrogen, 20 percent carbon dioxide, and 10 percent carbon monoxide. If the pressure and temperature of the mixture are 20 lb/in² abs and 100°F, respectively, which of the following most nearly equals the weight percent of the carbon monoxide?
(a) 10 wt %   (b) 12 wt %   (c) 8 wt %   (d) 7 wt %   (e) 9 wt%

7-10 The relation between mass and energy as derived by Einstein is $E = MC^2$, where E = energy, M = mass, and C = speed of light. If 1 lb of uranium is caused to fission and there is a loss of one-tenth of 1 percent of the mass in the process, which of the following most nearly equals the amount of energy that would be liberated?

(a) 0.9×10^8 kW·h   (b) 1.1×10^7 kW·h   (c) 1.2×10^6 kW·h
(d) 2.1×10^7 kW·h   (e) 3.4×10^{11} kW·h

7-11 A steam turbine receives 3600 lb of steam per hour at 110 ft/s velocity and 1525 Btu/lb. The steam leaves at 810 ft/s and 1300 Btu/lb. Which of the following most nearly equals the horsepower output?

(a) 300 hp   (b) 325 hp   (c) 357 hp   (d) 382 hp   (e) 401 hp

7-12 Atmospheric air has a dry-bulb temperature of 85°F and a wet-bulb temperature of 64°F. The barometric pressure is 14.0 lb/in² abs. Which of the following most nearly equals the relative humidity?

(a) 51%   (b) 48%   (c) 42%   (d) 37%   (e) 32%

7-13 Air in the cylinder of a diesel engine is at 86°F and 20 lb/in² abs in compression. If it is further compressed to one-eighteenth of its original volume, which of the following most nearly equals the work done in compression if the displacement volume of the cylinder is 0.5 ft³?

(a) 4200 ft·lb   (b) 5700 ft·lb   (c) 6300 ft·lb   (d) 7800 ft·lb
(e) 8100 ft·lb

7-14 An indicator card of a four-cycle automobile engine taken with a 420-lb spring is 3 in long and has an area of 1.25 in². If the engine has 8 cylinders, 3.60-in diameter by 3.00-in stroke, and travels at 2800 r/min, which of the following most nearly equals the developed horsepower?

(a) 127 hp   (b) 142 hp   (c) 151 hp   (d) 167 hp   (e) 173 hp

7-15 A wall of 0.8-ft thickness is to be constructed from material which has an average thermal conductivity of 0.75 Btu/(h)(ft)(°F). The wall is to be insulated with material having an average thermal conductivity of 0.2 Btu/(h)(ft)(°F) so that the heat loss per square foot will not exceed 580 Btu/h. Which of the following most nearly equals the thickness of insulation required if the surface temperatures of the walls are 2400 and 80°F, respectively?

(a) 6.8 in   (b) 7.0 in   (c) 7.3 in   (d) 7.8 in   (e) 8.1 in

7-16 There was 3 in of ice on a pond when the atmospheric temperature was 2°F. Which of the following most nearly equals the rate at which the ice was increasing in thickness? [Thermal conductivity of ice $= 12$ Btu/(h)(ft²)(°F)/in.]

(a) 0.18 in/h   (b) 0.16 in/h   (c) 0.14 in/h   (d) 0.12 in/h
(e) 0.10 in/h

7-17 A copper bar 15 cm long and 6.0 cm² in cross section is placed with one end in a steam bath and the other in a mixture of ice and water at atmospheric pressure. The sides of the bar are thermally insulated. Which of the following most nearly equals the amount of ice which melts in 2 min? The thermal conductivity of copper is 0.90 cal/(cm)(s)(°C).

(a) 45 g   (b) 51 g   (c) 54 g   (d) 58 g   (e) 61 g

7-18 A steam turbine receives 3600 lb of steam per hour at 1100 ft/s velocity and 1525 Btu/lb enthalpy. The steam leaves at 810 ft/s and 1300 Btu/lb. Which of the following most nearly equals the horsepower output?

(*a*) 278 hp   (*b*) 297 hp   (*c*) 309 hp   (*d*) 328 hp   (*e*) 334 hp

7-19 If 1 lb of nitrogen with an initial volume of 6 ft^3 and pressure of 300 lb/in^2 abs expands according to the law PV^n = constant to a final volume of 30 ft^3 and a pressure of 40 lb/in^2 abs, which of the following most nearly equals the work done by the gas?
(*a*) 55 Btu   (*b*) 87 Btu   (*c*) 112 Btu   (*d*) 143 Btu   (*e*) 166 Btu

7-20 Assume that 3 mol of nitrogen, $\gamma = 1.4$, $C_v = 4.6$ (g·cal/mol), are at atmospheric pressure and 20°C. The gas is then heated at constant volume to 40°C. It then undergoes adiabatic expansion and returns to 20°C. Finally it is compressed isothermally to its original state. Which of the following most nearly equals the joules of work done in adiabatic expansion?
(*a*) 1245 J   (*b*) 1372 J   (*c*) 1485 J   (*d*) 1591 J   (*e*) 1607 J

7-21 An automobile cooling system of 26 qt of water and antifreeze has a freezing point of -12°F. Which of the following most nearly equals the amount of the above mixture which must be drained out in order to add sufficient antifreeze to protect the engine at -15°F? The table printed on the antifreeze container indicates that $2\frac{1}{2}$ qt will protect a 13-qt cooling system to -12°F and $4\frac{1}{2}$ qt to -15°F.
(*a*) 4.95 qt   (*b*) 5.07 qt   (*c*) 5.62 qt   (*d*) 5.75 qt   (*e*) 5.82 qt

7-22 A steam turbine operating with steady-flow conditions produces 1350 hp. All losses except exhaust amount to 144,000 Btu/h. The steam flow is 150 lb/min. Inlet condition is 450 lb/in^2 abs and 740°F, and exhaust occurs at a pressure of 0.8 lb/in^2 abs. Which of the following most nearly equals the quality of the exhaust steam?
(*a*) 100%   (*b*) 92%   (*c*) 89%   (*d*) 85%   (*e*) 78%

7-23 A gas has a constant-pressure specific heat of $C_p = 6.6 - 7.2 \times 10^{-4}T$ Btu/(mol)(°R). Which of the following most nearly equals the entropy of this gas at 2000°R and 1.0 atm of pressure if the base of zero entropy is taken as 520°R and 1.0 atm?
(*a*) 8.9 Btu/(mol)(°R)   (*b*) 7.8 Btu/(mol)(°R)   (*c*) 7.3 Btu/(mol)(°R)
(*d*) 6.7 Btu/(mol)(°R)   (*e*) 6.4 Btu/(mol)(°R)

7-24 A balloon weighs 425 lb and has a volume of 35,000 ft^3 when filled with hydrogen gas weighing 0.0056 lb/ft^3. Which of the following most nearly equals the load that the balloon will support in air weighing 0.081 lb/ft^3?
(*a*) 1400 lb   (*b*) 1800 lb   (*c*) 2000 lb   (*d*) 2200 lb   (*e*) 2400 lb

7-25 A gas company buys gas at 90 lb/in^2 gauge and 75°F and sells it at 3.80 in of water pressure and 28°F. Disregarding the losses in distribution, which of the following most nearly equals the number of cubic feet sold for each cubic foot purchased?
(*a*) 5.7 ft^3/ft^3 (*b*) 6.4 ft^3/ft^3 (*c*) 6.8 ft^3/ft^3 (*d*) 7.2 ft^3/ft^3
(*e*) 7.5 ft^3/ft^3

7-26 An indicator card of a four-cycle automobile engine taken with a 425-lb spring is 2.80 in long and has an area of 1.20 in^2. If the engine has six cylinders, 3.5 in in diameter, with a stroke of 3 in, and is operating at 3000 r/min, which of the following most nearly equals the developed horsepower of the engine?

(a) 240 hp   (b) 215 hp   (c) 190 hp   (d) 165 hp   (e) 120 hp

7-27 The number of British thermal units required to heat 1 lb of a certain substance is given by the equation $Q = 0.32t - 0.00008t^2$, where Q is expressed in British thermal units per pound and t is in degrees Fahrenheit. Which of the following most nearly equals the instantaneous specific heat of the substance at 150°F?

(a) 0.30 Btu(lb)(°F)   (b) 0.32 Btu(lb)(°F)   (c) 0.35 Btu(lb)(°F)
(d) 0.27 Btu(lb)(°F)   (e) 0.25 Btu(lb)(°F)

7-28 If a 100-hp heat engine operates on the ideal Carnot cycle between the temperatures of 540 and 140°F, which of the following most nearly equals the amount of heat rejected?

(a) 153,000 Btu/h   (b) 217,000 Btu/h   (c) 382,000 Btu/h
(d) 407,000 Btu/h   (e) 429,000 Btu/h

7-29 An air-speed indicator of a plane is operated on the same principle as the pitot tube. It is calibrated to read knots. If the indicator reads 200 kn at an altitude where the density of air is one-half the density at which it was calibrated, which of the following most nearly equals the true speed of the plane?

(a) 141 kn   (b) 157 kn   (c) 217 kn   (d) 283 kn   (e) 297 kn

7-30 A barometer tube 88 cm long contains some air above the mercury. It gives a reading of 68 cm when upright. When the tube is tilted to an angle of 45°, the length of the mercury column becomes 80 cm. Which of the following most nearly equals the reading of an accurate barometer?

(a) 67.8 cm   (b) 72.3 cm   (c) 75.6 cm   (d) 76.9 cm   (e) 78.2 cm

7-31 Assume that 1 mol of a monatomic ideal gas at 300 K is subjected to three consecutive changes: (1) the gas is heated at constant volume until its temperature is 900 K; (2) the gas is then allowed to expand isothermally until its pressure drops to its initial value; (3) the gas is then cooled at constant pressure until it returns to its original state. Which of the following most nearly equals the total work done by the gas? (Express the answer in calories.)

$$C_v = 3 \text{ cal/(mol)(K)} \qquad R = 2 \text{ cal/(mol)(K)}$$

(a) 690 cal   (b) 780 cal   (c) 810 cal   (d) 975 cal   (e) 1200 cal

7-32 Which of the following most nearly equals the resulting quality of the steam if 10 lb of water is injected into an evacuated vessel which is maintained at 350°F, if the volume of the vessel is 10 ft³?

(a) 30%   (b) 45%   (c) 60%   (d) 75%   (e) 100%

7-33 Steam expands behind a piston doing 50,000 ft·lb of work. If 12 Btu of heat is radiated to the surrounding atmosphere during the expansion, which of the following most nearly equals the change of internal energy?

(a) 64 Btu   (b) 67 Btu   (c) 71 Btu   (d) 73 Btu   (e) 76 Btu

7-34 Suppose that 2 lb of a dry and saturated steam is confined in a tank. A gauge on the tank indicates a pressure of 145 lb/in². Barometric pressure is 736.6 mmHg. Which of the following most nearly equals the enthalpy of the steam?

(*a*) 1120 Btu/lb   (*b*) 1123 Btu/lb   (*c*) 1125 Btu/lb
(*d*) 1127 Btu/lb   (*e*) 1130 Btu/lb

7-35 If 1400 ft³ of air at 85°F dry-bulb temperature and 50 percent relative humidity is mixed with 200 ft³ of air at 54°F dry-bulb temperature and 40 percent relative humidity, which of the following most nearly equals the relative humidity of the resulting mixture?
(*a*) 48%   (*b*) 52%   (*c*) 55%   (*d*) 57%   (*e*) 59%

7-36 Using the data from problem 7-35, which of the following most nearly equals the final wet-bulb temperature?
(*a*) 75°F   (*b*) 71°F   (*c*) 69°F   (*d*) 67°F   (*e*) 64°F

7-37 Tests of a six-cylinder 4-in bore and $3\frac{1}{8}$-in stroke aircraft engine at full throttle show a brake horsepower of 79.5 at 3400 rpm. The compression ratio is 8:1, the specific fuel consumption is 0.56 lb/(bhp)(h). The higher heating value of the fuel is 19,800 Btu/lb. Which of the following most nearly equals the brake thermal efficiency?
(*a*) 18%   (*b*) 21%   (*c*) 23%   (*d*) 27%   (*e*) 32%

7-38 A four-cycle, 225-hp, nine-cylinder diesel aircraft engine has a rated speed of 1950 r/min. The cylinder bore is $4\frac{13}{16}$ in, and the stroke is 6 in. Which of the following most nearly equals the rated brake mean effective pressure?
(*a*) 87 lb/in²   (*b*) 93 lb/in²   (*c*) 97 lb/in²   (*d*) 102 lb/in²
(*e*) 108 lb/in²

7-39 Air is compressed in a diesel engine from an initial pressure of 13 lb/in² abs and a temperature of 120°F to one-twelfth its initial volume. Which of the following, assuming adiabatic compression, most nearly equals the final temperature?
(*a*) 947°F   (*b*) 982°F   (*c*) 1052°F   (*d*) 1097°F   (*e*) 1107°F

7-40 A centrifugal compressor is to operate at 3600 rpm and handle 20,000 ft³/min of free air. Compression is according to the law $PV^{1.35}$ = constant. Intake is atmospheric (14.4 lb/in² abs at 70°F) and compression is to 25 lb/in² gauge. Which of the following most nearly equals the final delivery at 70°F?
(*a*) 7310 ft³/min   (*b*) 7820 ft³/min   (*c*) 8210 ft³/min
(*d*) 8540 ft³/min   (*e*) 9490 ft³/min

7-41 A 100-hp engine is being tested by loading it with a water-cooled Prony brake. When the engine delivers the full-rated 100 hp to the shaft, the Prony brake being cooled with tap water absorbs and transfers to the cooling water 95 percent of the 100 hp. Which of the following most nearly equals the rate at which tap water passes through the Prony brake shell, if the water enters at 65°F and leaves at 131°F?
(*a*) 6.4 gal/min   (*b*) 6.9 gal/min   (*c*) 7.3 gal/min
(*d*) 7.7 gal/min   (*e*) 8.2 gal/min

7-42 The internal energy of a gas is given by the equation $u = 0.08t + 0.002t^2$, where t = °F. If 1 lb of this gas expands without heat flow, but with a temperature drop from 300 to 100°F, which of the following most nearly equals the amount of work done?
(*a*) 167 Btu   (*b*) 176 Btu   (*c*) 182 Btu   (*d*) 191 Btu
(*e*) 198 Btu

7-43 A mixture of gases at 100°F consists by weight of 80 percent nitrogen, 12 percent carbon dioxide, 7 percent oxygen, and 1 percent water vapor. The total pressure is 14.696 lb/in² abs. Which of the following most nearly equals the amount of heat which would be removed in reducing the temperature of 1.0 lb of the mixture to the dew point at constant total pressure?
  (a) 9.8 Btu   (b) 10.1 Btu   (c) 10.3 Btu   (d) 10.6 Btu
  (e) 11.1 Btu

7-44 If 5 ft³ of a gas with $C_p = 0.55$ and $R = 96.2$ expands polytropically from 140 lb/in² abs and 120°F to a pressure of 50 lb/in² abs and a volume of 10 ft³, which of the following most nearly equals the amount of heat added or extracted?
  (a) +47 Btu   (b) +24 Btu   (c) −14 Btu   (d) −36 Btu
  (e) −51 Btu

7-45 Using the data given in problem 7-44, which of the following most nearly equals the change in entropy?
  (a) −0.103 Btu/°F   (b) −0.214 Btu/°F   (c) −0.221 Btu/°F
  (d) −0.184 Btu/°F   (e) −0.127 Btu/°F

7-46 An Otto cycle engine has a compression ratio of 6:1. On a test it uses 1 gal of gasoline in 15 min while developing 150 lb·ft of torque at 1500 r/min. The gasoline has a specific gravity of 0.70 and a higher heating value of 19,100 Btu/lb. Which of the following most nearly equals the thermal efficiency of the engine?
  (a) 21%   (b) 23%   (c) 25%   (d) 27%   (e) 31%

7-47 A gas made up of 5 lb of methane and 10 lb of ethane is placed in a tank at 60°F. If the volume of the tank is 19 ft³, which of the following most nearly equals the pressure in the tank?
  (a) 142 lb/in² gauge   (b) 153 lb/in² gauge   (c) 159 lb/in² gauge
  (d) 167 lb/in² gauge   (e) 175 lb/in² gauge

7-48 A throttling steam calorimeter has a pressure of 150 lb/in² abs on the high-pressure side of its orifice and atmospheric pressure on the discharge side. The thermometer on the discharge side reads 280°F. Which of the following most nearly equals the quality of the steam in the line?
  (a) 97.4%   (b) 98.7%   (c) 99.1%   (d) 98.4%   (e) 97.9%

7-49 Which of the following most nearly equals the heat-transfer coefficient of a wall consisting of 13 in of brick and 1.0 in of plaster? The specific conductivity for brickwork and for plaster is 5.00. The surface resistances of brick and plaster are 4.02 and 1.40, respectively.
  (a) 0.95 Btu/(h)(ft²)(°F)   (b) 0.91 Btu/(h)(ft²)(°F)
  (c) 0.87 Btu/(h)(ft²)(°F)   (d) 0.84 Btu/(h)(ft²)(°F)
  (e) 0.81 Btu/(h)(ft²)(°F)

7-50 If atmospheric air has a dry-bulb temperature of 87°F and a wet-bulb temperature of 65°F, which of the following most nearly equals the specific humidity of dry air?
  (a) 47 grains/lb   (b) 51 grains/lb   (c) 56 grains/lb
  (d) 62 grains/lb   (e) 67 grains/lb

7-51 A steel tube 1.315 in OD and 1.049 in ID transmits 38,000 Btu/(h)(ft²) of outside-wall area to the surroundings. Which of the following most nearly equals the outside-wall temperature if the inner wall is at 212°F? [Given: $k = 19.1$ Btu/(h)(°F)(ft) for tube metal.]

(a) 187°F   (b) 185°F   (c) 183°F   (d) 182°F   (e) 179°F

7-52 A jet-propelled plane has a fuel flow of 0.9 lb/s with a heating value of 20,000 Btu/lb. Air inlet flow is 40 lb/s. The plane is flying at a speed of 460 mi/h, and the exit velocity of the jet is 2100 ft/s. Which of the following most nearly equals the overall thermal efficiency?

(a) 13%   (b) 32%   (c) 29%   (d) 23%   (e) 17%

7-53 Water at 200 lb/in² abs and 300°F is introduced into an insulated chamber maintained at 50 lb/in² abs. Which of the following most nearly equals the percentage of the water so injected that flashes into steam?

(a) 1.8%   (b) 2.5%   (c) 2.9%   (d) 3.2%   (e) 3.4%

7-54 A tank contains air at 800 lb/in² abs and 80°F. An amount of air measuring 700 ft³ at 14.7 lb/in² abs and 60°F is removed. The pressure in the tank is now found to equal 300 lb/in² abs when the temperature is 80°F. Which of the following most nearly equals the volume of the tank?

(a) 17 ft³   (b) 19 ft³   (c) 21 ft³   (d) 23 ft³   (e) 25 ft³

7-55 Steam is supplied to a nozzle at 100 lb/in² abs and 400°F. The nozzle exhaust pressure is 20 lb/in² abs. Which of the following most nearly equals the ideal exit velocity?

(a) 1500 ft/s   (b) 1700 ft/s   (c) 2100 ft/s   (d) 2500 ft/s
(e) 2800 ft/s

7-56 A furnace fired with a hydrocarbon fuel oil has a dry-stack analysis of CO_2, 12 percent; O_2, 7 percent; and N_2, 81%. Which of the following most nearly equals the weight percent of hydrogen in the fuel oil?

(a) 8.2%   (b) 7.9%   (c) 7.7%   (d) 7.2%   (e) 6.6%

7-57 A hollow steel cylinder having an internal volume of 4 ft³ contains air at an atmospheric pressure of 15 lb/in² abs. Which of the following most nearly equals the number of cubic feet of outside air which must be pumped into the cylinder to raise the pressure to 45 lb/in² gauge? Assume that the temperature remains constant.

(a) 10 ft³   (b) 12 ft³   (c) 14 ft³   (d) 16 ft³   (e) 15 ft³

7-58 In an air cycle (Fig. 7P-58), process 1 to 2 is constant volume, 2 to 3 is adiabatic isentropic, and 3 to 1 is constant pressure. Assume that V_1 is 1.0

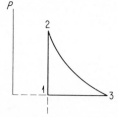

Figure 7P-58

ft³, V_3 is 2.0 ft³, p_1 and p_3 are each 1 atm, and t_1 is 60°F. Which of the following most nearly equals the efficiency of the cycle?
(a) 10%   (b) 12%   (c) 15%   (d) 17%   (e) 19%

7-59   At point 1 (Fig. 7P-59), water is supplied at the rate of 500 lb/min at 3 lb/in² gauge and 60°F. At point 2, steam is supplied at 3 lb/in² gauge and 95 percent dry. At point 3, the mixture leaves at 212°F as a saturated liquid. Assuming no heat loss from the chamber, which of the following most nearly equals the amount of steam per minute required at point 2?
(a) 84 lb/min   (b) 81 lb/min   (c) 79 lb/min   (d) 77 lb/min
(e) 75 lb/min

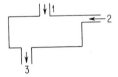

Figure 7P-59

7-60   An ammonia refrigeration plant is to operate between a saturated liquid at 120 lb/in² abs at the condenser outlet and a saturated vapor at 15 lb/in² abs at the evaporator outlet. If 30-ton capacity is required, computed on the basis of an ideal cycle, which of the following most nearly equals the amount of cooling water which would be required, assuming a 15°F temperature rise?
(a) 375 lb/min   (b) 392 lb/min   (c) 468 lb/min   (d) 510 lb/min
(e) 525 lb/min

7-61   An ideally insulated tank of helium, pressure 3000 lb/in² gauge, temperature 60°F, is connected to an empty weather balloon through a valve. The system is mounted on a weighing platform. The whole has a tare of 200 lb, including valve and weather balloon. The weather balloon is inflated to 4 ft in diameter, pressure 16 lb/in² abs, average temperature 60°F in air at 14.7 lb/in² abs, 60°F. If the tank pressure at the end of inflation is 200 lb/in² gauge, which of the following most nearly equals the volume of the tank?
(a) 0.18 ft³   (b) 0.22 ft³   (c) 0.25 ft³   (d) 0.28 ft³   (e) 0.31 ft³

Multiple-Choice Sample Quiz

For each question select the correct answer from the five given possibilities. For practice, try to complete the following 10 problems in 15 minutes or less.

7M-1   In an office building containing 100 employees, the design allowed for 10 ft³/min per person of outside air for ventilation purposes. If the supply fan delivers 6000 ft³/min, what is the temperature of the air entering the cooling coil if the outside air is 105°F and the recirculated air is 80°F?
(a) 75.3°F   (b) 80.6°F

(c) 84.2°F   (e) 95.0°F
(d) 93.7°F

7M-2 To determine the relative humidity of a sample of air, the easiest way to proceed would be as follows:
(a) Measure the wet-bulb and dry-bulb temperatures and use a psychrometric chart
(b) Measure the wet-bulb and dry-bulb temperatures and divide the smaller value by the larger
(c) Measure the wet-bulb and dry-bulb temperatures and use a M. llier chart
(d) Measure the barometric pressure and the dry-bulb temperature and divide the smaller value by the larger
(e) Measure the barometric pressure and use a psychrometric chart

7M-3 A steam turbine uses steam having an initial pressure of 140 lb/in² abs and an enthalpy of 1192 Btu/lb. The turbine exhausts steam at 23 lb/in² abs and 1158 Btu/lb enthalpy. The energy used by the turbine is:
(a) 2350 Btu/lb   (d) 2342 Btu/lb
(b) 34 Btu/lb   (e) none of these
(c) 42 Btu/lb

7M-4 If an ideal gas is compressed from a lower pressure to a higher pressure at constant temperature, which of the following is true?
(a) The work required will be zero.
(b) The volume remains constant.
(c) The volume will vary inversely as the absolute pressure.
(d) The volume will vary directly as the gauge pressure.
(e) Heat is being absorbed by the ideal gas during the compression.

7M-5 An adiabatic throttling process is one in which:
(a) the entropy is constant
(b) the enthalpy is constant
(c) the specific volume is constant
(d) the available energy is constant
(e) none of these

7M-6 An adiabatic process is one in which:
(a) the pressure is constant
(b) internal energy is constant
(c) no work is done
(d) no heat is transferred
(e) friction is not considered

7M-7 How much heat is required to raise 10 g of water from 0 to 1°C?
(a) 10 Btu   (d) 5 J
(b) 1 Btu   (e) 10 cal
(c) 1 cal

7M-8 If a gas is cooled at constant volume, which of the following will be true?
(a) The pressure will remain constant.
(b) The absolute pressure will increase.

(c) The gauge pressure will decrease by the same ratio as the temperature decrease in degrees Fahrenheit.

(d) The absolute pressure will decrease by the same ratio as the absolute temperature decrease.

(e) Cooling the gas has no effect on the pressure.

7M-9 In the ordinary commercial steam turbine, the change in enthalpy of the steam as it passes through the reaction section:

(a) occurs almost entirely in the nozzles of the turbine

(b) occurs almost entirely in the turbine blading

(c) is due to the condensation of the steam

(d) is negligible

(e) occurs across the throttling valve of the turbine

7M-10 The maximum thermal efficiency that can be obtained in an ideal reversible heat engine operating between 1540 and 340°F is closest to:

(a) 100 percent   (d) 40 percent

(b) 60 percent   (e) 22 percent

(c) 78 percent

8

ELECTRICITY AND ELECTRONICS

Some of the problems contained in the more recent examinations have covered fairly specialized topics such as single-stage, resistance-coupled amplifiers; two-stage, triode amplifiers; cathode followers; vacuum-tube and pentode amplifiers; ac signal generators; and transistor circuitry and amplifiers. These subjects fit well into an electrical engineer's background but are somewhat foreign to other branches of engineering. The material covered in this chapter, therefore, is limited to those topics which are of more general importance—those subjects which have been frequently treated in past examinations and have the greatest probability of being treated in future examinations.

8-1 EXAMINATION COVERAGE

Electricity is covered in the morning session and is also one of the required subjects in the afternoon session. The subdivisions of the major subject which may be covered, and which should be reviewed, include the following:

DC circuits
AC circuits
Three-phase circuits
Steady-state and transient circuits
Electric fields
Magnetic fields

Electronics

Power and machines

8-2 ELECTRON THEORY

Many of the various electrical phenomena can be explained by means of the electron theory. Current is the flow of electrons. Voltage differential between two points is the potential difference between those two points and is equal to the work done per unit charge against electrical forces when a charge is transported from one point to the other. Also remember that like charges repel and unlike attract. The force of repulsion or attraction is proportional to the product of the two charges divided by the square of the distance between them.

8-3 OHM'S LAW AND RESISTANCE

Whenever there is a flow of electrons through a conductor, there is a resistance to that flow. The resistance is described by Ohm's law, $V = IR$. The electrical resistance between two points is proportional to the distance and inversely proportional to the cross-sectional area of the conductor.

■ A copper wire $\frac{1}{2}$ in in diameter and 1000 ft long has a resistance of 0.05 Ω at 20°C. Which of the following most nearly equals the resistance of 500 ft of $\frac{1}{2}$-in square wire of similar material?

(a) 0.014 Ω (b) 0.012 Ω (c) 0.142 Ω (d) 0.020 Ω
(e) 0.091 Ω

$$\frac{R_{\text{square}}}{R_{\text{round}}} = \frac{A_r}{A_s}\frac{L_s}{L_r}$$

$$R_s = 0.05 \times \frac{0.196}{0.250} \times \frac{500}{1000} = 0.0196 \ \Omega$$

The correct answer is (d).

It should also be remembered that the equivalent resistance for a group of resistances in series is equal to the sum of the individual resistances: $R = R_1 + R_2 + R_3 + \cdots$. The equivalent resistance for a group of resistances in parallel can be determined from the relationship $1/R = 1/R_1 + 1/R_2 + 1/R_3 + \cdots$.

■ Direct current flows through the circuit sketched in Fig. 8-1a. If the

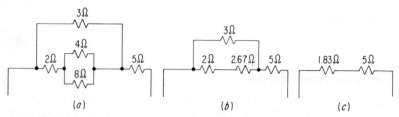

Figure 8-1

impressed voltage on terminals A and B is 25 V, which of the following most nearly equals the line current?

(a) 3.7 A   (b) 4.2 A   (c) 5.1 A   (d) 5.6 A   (e) 5.9 A

We have a series-parallel circuit here, and the easiest way to determine the current is to determine the equivalent circuit resistance and apply Ohm's law. Do this in steps. The first step is to calculate the equivalent resistance for the parallel circuit of 4 and 8 Ω. Then $1/R = \frac{1}{4} + \frac{1}{8}$; $R = 2.67$, which gives the equivalent circuit shown in Fig. 8-1b. Adding 2 plus 2.67 gives 4.67 Ω, series resistance in the branch. This gives another parallel circuit, for which the equivalent resistance is

$$\frac{1}{R_2} = \frac{1}{3} + \frac{1}{4.67}$$

Here, $R_2 = 1.83$ Ω, and we have the equivalent circuit shown in Fig. 8-1c. The line current is $I = V/R_T = 25/6.83 = 3.66$ A. The correct answer is (a).

An important application of resistance and Ohm's law is the Wheatstone bridge.

▪ In the circuit shown in Fig. 8-2, which of the following most nearly equals the required resistance of X so that $V_1 = 0$?

(a) 700 Ω (b) 800 Ω (c) 900 Ω (d) 1000 Ω (e) 850 Ω

First add the letters A, B, C, and D to the figure to aid in identification of the points. For a general discussion also label the resistances R_1, R_2, R_3, and R_4 and the currents I_1 and I_2.

For voltage V_1 to be zero, the voltage drops I_1R_1 and I_2R_2 must be equal. This means that I_1R_3 and I_2R_4 must also be equal, giving $I_1R_1 = I_2R_2$ and $I_1R_3 = I_2R_4$. Dividing one equation by the other gives

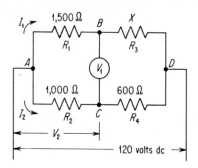

Figure 8-2

$(R_1/R_3) = (R_2/R_4)$:

$$R_3 = R_1 \frac{R_4}{R_2} = 1500 \frac{600}{1000} = 900 \ \Omega$$

The correct answer is (c).

- With the value of X found above, and 120 V applied, which of the following most nearly equals the voltage V_2?

  (a) 60 V (b) 65 V (c) 70 V (d) 75 V (e) 80 V

To determine the voltage differential V_2, calculate either I_1 or I_2. Since the voltage difference between B and C is zero ($V_1 = 0$), there will be a drop of 120 V across the branch ABD and across the branch ACD. Since $V = IR$,

$$I_1 = \frac{120}{1500 + 900} = 0.05 \ \text{A}$$

and the voltage drop from A to B equals $0.05 \times 1500 = 75$ V.

To check, $I_2 = 120/(1000 + 600) = 0.075$ A, giving $V_2 = 0.075 \times 1000 = 75$ V. The correct answer is (d).

Closely associated with the general subject of circuit resistance is the internal resistance of a battery. Any battery will have some internal resistance which will serve to reduce the effective potential as the current increases. A battery can be considered as consisting of a source of electromotive force (emf) with a resistance, the internal resistance of the battery, in series. This equivalent circuit is shown in Fig. 8-3.

- Two batteries, identical in every way, are connected in series across a 5-Ω resistor, and the current is found to be 0.2 A. The same cells

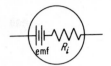

Figure 8-3

are then connected in parallel with the same 5-Ω resistor, and the current is found to be 0.16 A across the 5-Ω resistor. Which of the following most nearly equals the internal resistance of the batteries?

(a) 2.2 Ω (b) 2.5 Ω (c) 2.8 Ω (d) 3.1 Ω (e) 3.3 Ω

First, draw the equivalent circuits of the two hookups (Fig. 8-4). For the series circuit we have

$$2V - 2I_1R_i - 5I_1 = 0$$
$$2V = 0.4R_i + 1.00$$

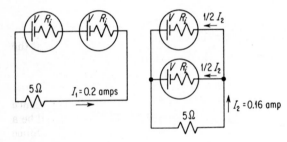

Figure 8-4

For the parallel circuit

$$V - \tfrac{1}{2}I_2R_i - 5I_2 = 0$$
$$V = 0.08R_i + 0.80$$

Combining equations gives

$$0.2R_i + 0.50 = 0.08R_i + 0.80$$
$$R_i = 2.50 \ \Omega$$

The correct answer is (b).

■ Which of the following most nearly equals the open-circuit voltage?

(a) 0.5 V   (b) 0.6 V   (c) 0.7 V   (d) 0.9 V   (e) 1.0 V

From the preceding equation: $2V = 0.4R_i + 1.00$ and $V = (0.4 \times 2.50 + 1.00)/2 = 1.0$. The correct answer is (e).

8-4 METER INTERNAL RESISTANCE

- An electrical milliammeter, having an internal resistance of 500 Ω, reads full scale when 1 mA of current is flowing. If the meter is to be used as an ammeter reading 0.1 A full scale, which of the following most nearly equals the size (ohms resistance) of resistor which should be used in parallel with it?

  (a) 4.85 Ω (b) 4.90 Ω (c) 4.95 Ω (d) 5.00 Ω
  (e) 5.05 Ω

An ammeter is essentially a voltmeter which measures the voltage drop across a resistance or shunt. The milliammeter in the problem reads full scale when measuring 0.5 V differential: $V = 0.001 \times 500 = 0.5$ V. To be used as an ammeter reading 0.1 A full scale, it would be connected as shown in Fig. 8-5a. Since a current of 0.001 A will flow through the meter when it reads full scale, 0.099 A must flow through the shunt resistance. The voltage drop across the two parallel resistances must equal 0.5 V. Then $R_1 = 0.5/0.099 = 5.05$ Ω. The correct answer is (e).

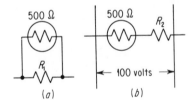

(a) (b) **Figure 8-5**

- If the milliammeter described above were to be used as a voltmeter reading 100 V full scale, which of the following would most nearly equal the size of the resistor which should be used in series with the milliammeter?

  (a) 100,000 Ω (b) 99,500 Ω (c) 99,000 Ω (d) 89,500 Ω
  (e) 89,000 Ω

To use the milliammeter as a voltmeter reading 100 V full scale, it should be connected as shown in Fig. 8-5b. The total resistance in the meter circuit is $(500 + R_2)$ Ω. The current through the meter must be

0.001 A for full-scale reading at 100 V. Then $100 = 0.001R; R = 100,000$ Ω. The total resistance $R = 500 + R_2; R_2 = 99,500 \; \Omega$ required. The correct answer is (b).

8-5 KIRCHHOFF'S LAWS

Two principles which hold in the analysis of electric circuits are known as "Kirchhoff's laws." These may be stated as follows:

1. The algebraic sum of all currents flowing into (and out of) any junction of an electric circuit equals zero.
2. The algebraic sum of all voltage drops around a closed circuit equals zero.

■ In the circuit shown in Fig. 8-6, which of the following most nearly equals the current which would flow through the 30-Ω resistance?

(a) 0.88 A   (b) 0.77 A   (c) 0.66 A   (d) 0.55 A
(e) 0.44 A

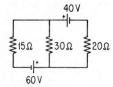

Figure 8-6

Redraw the circuit as in Fig. 8-7 and assume the directions of the branch currents as shown. In this case it has been assumed that the positive direction of current flow is out of the positive terminal of each battery. If one or more of the assumed directions is incorrect, the value obtained for that current when the circuit equations are solved will be negative.

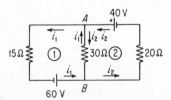

Figure 8-7

Currents i_1 and i_2 meet at junctions A and B. From the first law the sum of all currents meeting at a junction must equal zero. That is, i_1 enters junction B, so current in the amount of i_1 must leave. Similarly, i_2 leaves junction B, so current in the amount of i_2 must enter junction B. Thus the current in the 30-Ω resistance, going in the positive direction around loop 1, i.e., in the direction of the assumed current flow, will equal $i_1 - i_2$, since that is the net current that flows through that branch. Similarly, as loop 2 is traversed in the direction of assumed current flow, the current through the 30-Ω resistance will equal $i_2 - i_1$. This is the same current in amperes, but in the opposite direction.

Summing the voltages and voltage drops around loop 1 gives

$$60 - (i_1 - i_2) \times 30 - i_1 \times 15 = 0$$

which reduces to $4 = 3i_1 - 2i_2$. Note that the voltage drop through the 30-Ω resistance equals 30 times the net current flowing through that resistance, $i_1 - i_2$. This will also be true when the voltage drops through loop 2 are calculated.

Summing the voltages and voltage drops around loop 2 gives

$$40 - (i_2 - i_1) \times 30 - i_2 \times 20 = 0$$

which reduces to $4 = 5i_2 - 3i$. Solving the two equations simultaneously gives

$$i_1 = 3.11 \text{ A} \qquad i_2 = 2.67 \text{ A}$$

The current through the 30-Ω resistor would equal $i_1 - i_2 = 3.11 - 2.67 = 0.44$ A. The correct answer is (e).

8-6 INDUCTANCE AND CAPACITANCE

In addition to resistance, ac circuits usually contain inductance and/or capacitance.

- A single-phase load takes 855 W and 9.75 A from a source when a 60-cycle voltage of 120 V is applied. It is known that the current lags the voltage. Determine the:
  - a. Resistance of the load
  - b. Impedance of the load
  - c. Reactance of the load
  - d. Power factor of the load
  - e. Power-factor angle
  - f. Inductance or capacitance, whichever predominates

8-7 POWER FACTOR

The power taken by an electric load is measured in watts, where watts = volts × amperes × power factor. The power factor equals the in-phase component of the current divided by the total current, or watts divided by volt-amperes. In this case the power factor pf = $855/(120 \times 9.75)$ = 73 percent. The impedance of the load $Z = V/I$ = $120/9.75$ = $12.3 \ \Omega$. The impedance also equals $\sqrt{(X_L - X_C)^2 + R^2}$, where X_L is the inductive reactance and X_C is the capacitive reactance. The power factor is equal to R/Z. The resistance of the load is then 0.73×12.3 = $8.98 \ \Omega$, and the impedance is $12.3 \ \Omega$. The reactance $X = X_L - X_C$ is equal to $X = \sqrt{Z^2 - R^2}$ = $\sqrt{151.3 - 80.8}$ = $8.40 \ \Omega$. The power factor, already determined, equals 0.73, or 73 percent. The power-factor angle, more commonly called "phase angle" φ, equals the arc cosine of the power factor. In this case φ = $\cos^{-1} 0.73$ = $43.1°$ lagging.

8-8 PHASE ANGLES— LEADING OR LAGGING

It is given that the current lags the voltage. From this we know that the inductance exceeds the capacitance. There are different methods of remembering this. One is by remembering the phrase, "*ELI* the *ICE*man." The *ELI* indicates that *E*, or voltage (now expressed by *V*), comes before, or leads, *I* the current for an inductive load *L*; *ICE* indicates that current *I* comes before, or leads, the voltage *E* for a capacitive load *C*. The current lags; therefore the voltage leads and the inductance predominates. Then $X_L = 2\pi f L$, or $L = X_L/(2\pi f)$, where *L* is in henrys:

$$L = \frac{8.40}{2\pi \times 60} = 0.0223 \text{ H}$$

or 22.3 mH. Capacitive impedance $X_c = 1/(2\pi f C)$, where *C* is in farads.

8-9 VECTOR DIAGRAMS

Figure 8-8a shows the vector diagram of the current and the phase angle. Note that the in-phase component of the current will equals watts/volts = $\frac{855}{120}$ = 7.12 A:

$$I \cos \phi = 9.75 \times 0.73 = 7.12 \text{ A}$$

The time diagram is shown in Fig. 8-8b. Note that the current starts, or passes through zero, at a later time, $\omega t = 0$. This indicates that the current lags, since it arrives at similar points (maximum, zero, minimum) after the voltage, or later than the voltage.

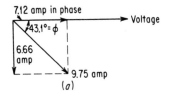

7.12 amp in phase

$43.1° = \phi$

6.66 amp

9.75 amp

(a)

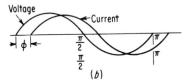

Voltage

Current

ϕ

$\dfrac{\pi}{2}$

$\dfrac{\pi}{2}$

π

π

(b)

Figure 8-8

8-10 RESONANT CIRCUITS

Resonance occurs when the inductive impedance of a circuit equals the capacitive impedance. In the case of a series circuit the resulting impedance will equal the resistance for resonance (series resonance) and the current will be a maximum:

$$Z = \sqrt{R^2 + (X_L - X_C)^2} = R \qquad \text{for resonance}$$

In the case of resonance the current will be in phase with the voltage and the power factor will equal 1. For series resonance the inductive reactance equals the capacitive reactance. For the case just considered, if the current had been limited only by the resistance, it would have been $120/8.98 = 13.36$ A, and the power would have equaled $W = I^2R = 13.36^2 \times 8.98 = 1605$ W, instead of 855 W as before.

- Which of the following most nearly equals the resonant frequency of a series circuit consisting of a $0.005\text{-}\mu F$ capacitor and a 10-mH inductance?

  (a) 32,000 Hz (b) 29,700 Hz (c) 27,400 Hz
  (d) 24,800 Hz (e) 22,500 Hz

Given: $X_L = 2\pi fL$, and $X_c = 1/(2\pi fC)$. For resonance, $X_L = X_C$, and

$$f = \frac{1}{2\pi\sqrt{LC}} = \frac{1}{2\pi\sqrt{0.005 \times 10^{-6} \times 10 \times 10^{-3}}} = 22,500 \text{ Hz}$$

The correct answer is (e).

Parallel resonance (i.e., resonance in a parallel circuit) also occurs when the inductive impedance equals the reactive impedance and the phase angle equals zero. However, in general, in the case of a parallel circuit, this condition will not occur when the inductive and capacitive reactances are equal. Take, for example, the circuit shown in Fig. 8-9. The inductive and capacitive reactances are equal, but the current is not in phase with the voltage. In order for the parallel circuit shown to be resonant, holding the inductive branch constant, the capacitive reactance must be increased until the leading branch current is reduced to 19.2 A, the lagging component of the lagging branch current:

$$X_c = \frac{120}{19.2} = 6.25$$

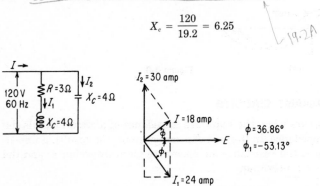

Figure 8-9

The resulting resonant circuit with the corresponding phasor diagram is shown in Fig. 8-10.

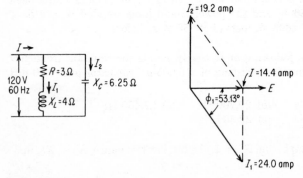

Figure 8-10

Alternatively, if the capacitive reactance were held constant, the inductive reactance would have to be reduced to 2.01 Ω. The new circuit, phasor diagram, and calculations are shown in Fig. 8-11. Note how construction of the phasor diagram simplifies the analysis of the circuit.

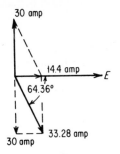

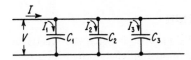

Figure 8-11

8-11 CAPACITORS IN SERIES AND PARALLEL

An equivalent capacitor to replace three capacitors in parallel is equal to the sum of the three capacitors. Looking at Fig. 8-12, we see that the voltage across the three capacitors is the same and equals V. The current is made up of the three branch currents: $I = I_1 + I_2 + I_3$. From $V = IX_C$, we get $I = V/X_C$; since $X_C = 1/(2\pi fC)$, $I = V \times 2\pi fC$. Similarly $I_1 = V \times 2\pi fC_1$, $I_2 = V \times 2\pi fC_2$, and $I_3 = V \times 2\pi fC_3$, which combine to give $C = C_1 + C_2 + C_3$.

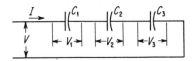

Figure 8-12

The case of the series circuit is shown in Fig. 8-13. Here

$$V = V_1 + V_2 + V_3 = IX_{C1} + IX_{C2} + IX_{C3} = IX_C$$

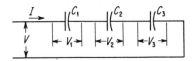

Figure 8-13

Replacing the X_C values by their equivalents and canceling like terms gives

$$V = \frac{I}{C} = \frac{I}{C_1} + \frac{I}{C_2} + \frac{I}{C_3} \qquad \frac{1}{C} = \frac{1}{C_1} + \frac{1}{C_2} + \frac{1}{C_3}$$

8-12 INDUCTANCES IN SERIES AND PARALLEL

A circuit of inductances in series is shown in Fig. 8-14. Here $V = V_1 + V_2 + V_3$, or $IX_L = IX_{L_1} + IX_{L_2} + IX_{L_3}$; since $X_L = 2\pi fL$, this gives $L = L_1 + L_2 + L_3$ for inductances in series.

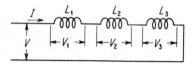

Figure 8-14

Inductances in parallel (Fig. 8-15) give us $I = I_1 + I_2 + I_3$ and $V/X_L = V/X_{L_1} + V/X_{L_2} + V/X_{L_3}$, which reduces to $\dfrac{1}{L} = \dfrac{1}{L_1} + \dfrac{1}{L_2} + \dfrac{1}{L_3}$ for inductances in parallel.

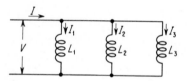

Figure 8-15

8-13 COMPLEX NUMBER NOTATION

Complex number notation is commonly used to describe the electrical vector quantities. Electrical vectors are now frequently called "phasors" to differentiate them from force vectors. Since a phasor has either a leading component or a lagging component (Fig. 8-16), it is convenient to use the complex number notation to describe it. The out-of-phase component corresponds to the imaginary part, and the in-phase component corresponds to the real part of the complex number. Thus in Fig. 8-16 $I_1 = 5 + j2$, $I_2 = 3 - j3$, $I_3 = 4$. $I = I_1 + I_2 + I_3 = 12 - j$.

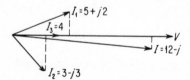

Figure 8-16

■ If the admittance of an a-c circuit is given by the relationship $Y =$

$4 - j3$, which of the following most nearly equals the magnitude of the impedance?

(a) $0.18 \ \Omega$ (b) $0.20 \ \Omega$ (c) $0.22 \ \Omega$ (d) $0.24 \ \Omega$
(e) $0.26 \ \Omega$

The admittance of a circuit is defined as the reciprocal of the impedance, or $Y = 1/Z$. In this problem, then,

$$Z = \frac{1}{Y} = \frac{1}{4 - j3}$$

Simplification of this fraction, which is an example of division by a complex number, is effected by multiplying both numerator and denominator by the conjugate of the denominator. The conjugate of a complex number is formed by changing the sign of the complex term. For this case the conjugate of the denominator would be $4 + j3$. Simplification is achieved as follows:

$$Z = \frac{1}{4 - j3} \times \frac{4 + j3}{4 + j3} = \frac{4 + j3}{16 + 9} = 0.16 + j0.12$$

and $\sqrt{0.12^3 + 0.16^2} = 0.200 \ \Omega$. The correct answer is (b).

Figure 8-17 shows the reactive component above the axis, which is in accordance with our convention. The voltage will equal current times impedance. This gives an identically shaped diagram for the voltage, shown also in Fig. 8-17, where the voltage leads the current, or the current lags the voltage.

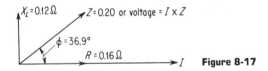

$X_L = 0.12 \ \Omega$ $Z = 0.20$ or voltage $= I \times Z$

$\phi = 36.9°$

$R = 0.16 \ \Omega$ $\rightarrow I$ **Figure 8-17**

The complex number notation serves to simplify many of the ac circuit calculations. For a series circuit, the total impedance $Z = Z_1 + Z_2 + Z_3$, which follows directly from the fact that $Z = \sqrt{R^2 + (X_C - X_L)^2}$. Parallel-circuit analysis is much more troublesome, however; it can be handled more simply if one remembers that

$$\frac{1}{Z} = \frac{1}{Z_1} + \frac{1}{Z_2} + \frac{1}{Z_3} + \cdots$$

where all the Z values are in the complex number notation.

Phasors can also be written in polar form where the length of the phasor and the angle it makes with the abscissa are specified. Referring to Fig. 8-16, the current $I_1 = 5 + j2$. This can be written as $I_1 = 5.385\underline{/21.80°}$ where $5.385 = \sqrt{5^2 + 2^2}$ and is the length of the phasor. The phase angle $\theta = \tan^{-1}\frac{2}{5} = 21.80°$.

A phasor in polar form is written as $P = M\underline{/\theta}$, where M equals the magnitude of the phasor and θ is the angle through which the vector is rotated from zero, the right-hand horizontal axis. A positive angle denotes counterclockwise rotation; a negative angle denotes clockwise rotation. Again referring to Fig. 8-16, the current $I_2 = 3 - j3$ can also be written as $I_2 = 4.24\underline{/-45°}$.

The product of two phasors P_1 and P_2 equals

$$P_1P_2 = M_1M_2\underline{/\theta_1 + \theta_2}$$

Division of two phasors $\dfrac{P_1}{P_2} = \dfrac{M_1}{M_2}\underline{/\theta_1 - \theta_2}$

- Two parallel branches of an electrical circuit, one consisting of 5 Ω resistance and 0.023 H inductance and the other, 8.66 Ω resistance and 530 μF capacitance, take 30 A from a 60-Hz power supply. Which of the following most nearly equals the current in the branch with the capacitor?

  (a) 21.2 A   (b) 22.4 A   (c) 23.1 A   (d) 23.3 A
  (e) 24.1 A

First sketch the circuit (Fig. 8-18) and calculate the individual impedances:

$$X_L = 2\pi fL = 2\pi \times 60 \times 0.023 = 8.67 \text{ } \Omega$$

$$X_C = \frac{1}{2\pi fC} = \frac{1}{2\pi \times 60 \times 530 \times 10^{-6}} = 5.00 \text{ } \Omega$$

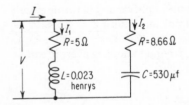

Figure 8-18

These give $Z_1 = 5 + j8.67$ and $Z_2 = 8.66 - j5.00$:

$$\frac{1}{Z} = \frac{1}{5 + j8.67} + \frac{1}{8.66 - j5.00} = 0.1366 - j0.0366$$

$$Z = \frac{1}{0.1366 - j0.0366} = \frac{0.1366 + j0.0366}{0.02} = 6.83 + j1.83$$

or
$$Z = \sqrt{6.83^2 + 1.83^2} = 7.07\ \Omega$$

The applied voltage $V = IZ = 30 \times 7.07 = 212$ V. The same voltage acts across each branch, so

$$I_1 = \frac{V}{Z_1} = \frac{212}{5 + j8.67} = 10.6 - j18.36 = 21.2\ \text{A}$$

$$I_2 = \frac{V}{Z_2} = \frac{210}{8.66 - j5.00} = 18.36 + j10.6 = 21.2\ \text{A}$$

The correct answer is (a).

These could also have been calculated using the polar form for the phasors:

$$I_1 = \frac{212}{5 + j8.67} = \frac{212\underline{/0^\circ}}{10\underline{/+60^\circ}} = 21.2\underline{/-60^\circ} = 10.6 - j18.27$$

$$I_2 = \frac{210}{8.66 - j5.00} = \frac{210\underline{/0^\circ}}{10\underline{/-30^\circ}} = 21.0\underline{/30^\circ} = 18.19 + j10.5$$

- Which of the following most nearly equals the instantaneous current which would occur in branch 2 of Fig. 8-18 when the instantaneous current in the supply main had a value of 36.7 A?

  (a) 20 A (b) 23 A (c) 26 A (d) 29 A (e) 31 A

The current in the main is given as 30 A. This is the effective, or root mean square, current. The maximum current would equal

$$I_{\text{eff}} \times \sqrt{2} = 30 \times 1.414 = 42.4\ \text{A}$$

The line current would equal the vector sum of the two branch currents $I = I_1 + I_2 = 28.96 - j7.76$. The phase angle would be $\tan^{-1} 0.268 = 15^\circ$ lagging. Similarly for the two branch currents, $\phi_1 = 60^\circ$ lagging and $\phi_2 = 30^\circ$ leading. These three currents can then be plotted, since they are all of the form $I = I_{\text{max}} \sin(\omega t + \phi)$. The main current will have an instantaneous value of 36.7 A when $\sin(\omega t - 15) = 36.7/42.4 = 0.866$, or $\omega t - 15 = 60^\circ$ and $\omega t = 75^\circ$.

The instantaneous value of branch current

$$I_1 = 21.2\sqrt{2} \sin (\omega t - 60°)$$

giving $I_1 = 30 \sin 15° = 7.76$ A.

The instantaneous value of branch current

$$I_2 = 21.2\sqrt{2} \sin (\omega t + 30°)$$

giving $I_2 = 30 \sin 105° = 29$ A. The correct answer is (d).

Taking the voltage as the reference phasor, plot the effective currents as in Fig. 8-19.

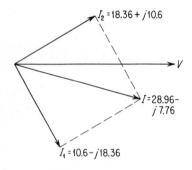

$I_2 = 18.36 + j10.6$

V

$I = 28.96 - j7.76$

$I_1 = 10.6 - j18.36$

Figure 8-19

One type of problem which appears rather frequently in fundamentals examinations is one where the examinee is asked to determine what capacitance must be added to an inductive circuit to raise the power factor to a particular value.

- A single-phase, 10-hp, 230-V, 85 percent efficient induction motor operates at full load with a 75 percent power factor. The power factor drops to 65 percent when the motor is partially loaded to 4 hp, where the efficiency drops to 68 percent. The frequency of the supply current is 60 Hz. Which of the following most nearly equals the size of capacitor which would have to be added to the circuit to raise the power factor to 95 percent for either condition?

  (a) 165 μF (b) 179 μF (c) 193 μF (d) 217 μF
  (e) 243 μF

The circuit can be sketched as shown in Fig. 8-20. The equivalent circuit of the motor will be a resistance and an inductance in series. The capacitor will be connected in parallel with the motor load.

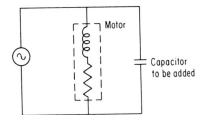

Figure 8-20

At full load the in-phase component of the current would equal

$$\frac{10 \times 746}{230 \times 0.85} = 38.16 \text{ A}$$

The phase angle $\phi = \cos^{-1} 0.75 = 41.41°$ lagging. The out-of-phase component of the current is

$$38.16 \times \tan 41.41° = 33.65 \text{ A}$$

It is desired to raise the power factor to 95 percent which would mean a phase angle of $\phi = \cos^{-1} 0.95 = 18.19°$ and an out-of-phase component equal to $38.16 \times \tan 18.19° = 12.54$ A, so the current through the capacitor would have to equal $33.65 - 12.54 = 21.11$ A. Then

$$X_c = \frac{230}{21.11} = 10.90 \text{ }\Omega$$

This can be shown graphically (see Fig. 8-21). For the unloaded condition,

$$\text{In-phase current} = \frac{4 \times 746}{230 \times 0.68} = 19.08 \text{ A}$$

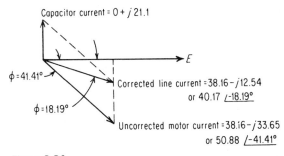

Figure 8-21

The phase angle $\phi = \cos^{-1} 0.65 = 49.46°$. The out-of-phase component would equal $19.08 \times \tan 49.46° = 22.31$ A. The out-of-phase component must be corrected to $19.08 \times \tan 18.19° = 6.27$ A, which requires a current equal to $22.31 - 6.27 = 16.04$ A through the capacitor. The required impedance

$$X_c = \frac{230}{16.04} = 14.34 \ \Omega$$

This condition is shown graphically in Fig. 8-22.

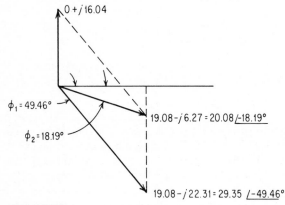

Figure 8-22

The capacitive component lowers the total current through the system and thus reduces the line losses (I^2R losses in the transmission line). It also reduces the voltage drop from the source to the load.

The product of reactive current and voltage is termed *reactive power* or *wattless power* and is measured in reactive volt-amperes (VA reactance, volt-amperes reactance). The reactive power for the fully loaded motor equals

$VI \sin \phi = 230 \times 50.88 \times \sin 41.41°$

$= 7740$ VA reactance, or 7.74 kVA reactance

This is termed "positive reactive power." Capacitive reactance is termed "negative reactive power." The capacitive or negative reactive power required for the fully loaded motor to correct the system power factor to 95 percent would equal

$230 \times 21.11 = 4855$ VA reactance, or 4.86 kVA reactance

Similarly, the reactive power for the unloaded motor would equal

$230 \times 29.35 \times \sin 49.46° = 230 \times 22.30$

$\qquad\qquad\qquad\qquad = 5130$ VA reactance, or 5.13 kVA reactance

and the corrective factor required equals

$\qquad 230 \times 16.04 = 3689$ VA reactance, or 3.69 kVA reactance

For the case of the fully loaded motor, $X_c = 10.90 \ \Omega$ and

$$C = \frac{1}{2\pi f X_c} = \frac{1}{2\pi \times 60 \times 20.70} = 2.43 \times 10^{-4}\text{F, or } 243 \ \mu\text{F}$$

For the case of the unloaded motor,

$$C = \frac{1}{2\pi \times 60 \times 16.04} = 1.65 \times 10^{-4} \text{ F, or } 165 \ \mu\text{F}$$

The situation which occurs with the loaded motor controls; the size of the capacitor required to correct the power factor to 95 percent for the case equals 243 μF. The correct answer is (e).

8-14 TRANSFORMERS

A transformer is a device for increasing or decreasing an ac voltage. The losses in a transformer are usually divided into two groups: iron losses, or core losses, and the I^2R losses due to the resistance of the windings. The iron losses are made up of the hysteresis loss and the eddy-current loss. The hysteresis loss is proportional to the frequency of the flux reversals and a power of the flux density. It can be expressed by the equation

$$P_h = k_1 f B^{1.6} \qquad \text{watts}$$

where

the 1.6 power = a representative value (The power of the flux density, called "the Steinmetz exponent," may vary from approximately 1.5 to over 2.0, with 1.6 being the value usually taken.)

f = frequency, Hz

k_1 = a constant depending on the characteristics of the core

B = flux density, lines or kilolines/in^2

The eddy-current loss in the core may be represented by the equation

$$P_e = k_2 f^2 B^2 \qquad \text{watts}$$

where the symbols have the same meaning as before except that k_2 is a different constant.

The copper loss equals $I_1^2 R_1 + I_2^2 R_2$, where I_1, R_1, I_2, and R_2 are the current and resistance in the primary and secondary windings. The resistances are usually small and can be approximated by a dc test. The ac resistance will, however, be several percent higher, because of eddy-current losses in the copper.

The efficiency of a transformer is equal to

$$\frac{\text{output}}{\text{input}} \qquad \text{or} \qquad \frac{\text{input} - \text{losses}}{\text{input}}$$

The voltage regulation of a transformer in percent equals (no-load secondary voltage minus full-load secondary voltage) divided by (full-load secondary voltage):

$$\text{Regulation} = \frac{V_2' - V_2}{V_2} \qquad \text{percent}$$

Single-phase transformers are made with two windings and also with just one winding. When made with a single winding, they are termed "autotransformers." Schematics of a single-winding transformer are shown in Fig. 8-23 connected for voltage step-down and step-up. In each case $V_2 = V_1(N_2/N_1)$.

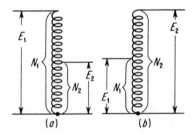

(a)   (b)   **Figure 8-23**

The common part of the winding carries both currents I_1 and I_2, but these currents buck each other, so this part of the winding is more lightly loaded than the other part. The current diagrams for the two cases are shown in Fig. 8-24 for an ideal autotransformer.

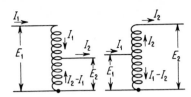

Figure 8-24

■ A 2400/240-V, single-phase, additive transformer is connected as shown in Fig. 8-25. Which of the following most nearly equals the current in the supply circuit at the instant that the load current is 40 A, assuming no transformer losses?

(a) 40 A   (b) 42 A   (c) 44 A   (d) 46 A   (e) 48 A

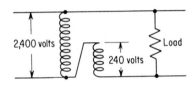

Figure 8-25

First redraw the diagram in the form of an autotransformer, as shown in Fig. 8-26. It is then immediately apparent that it is acting as a step-up transformer and is increasing the voltage. When the load current is 40 A, the supply current will be $\frac{2640}{2400} \times 40 = 44$ A. It is interesting that in this case the current in the primary would be only $44 - 40 = 4$ A. The correct answer is (c).

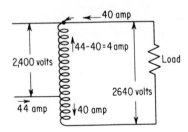

Figure 8-26

8-15 GENERATORS

The generated voltage of the armature of a generator is

$$V = \frac{N}{\text{paths}} \times \phi \times \text{poles} \times \frac{\text{rpm}}{60} \times 10^{-8}$$

where N is the total number of inductors and ϕ is the magnetic flux per pole.

■ It is desired to design a dc generator which will generate an average emf of 230 V. The armature used has 180 series inductors between brushes. Which of the following most nearly equals the number of poles which must be used if the flux per pole is 1.5×10^6 maxwells (Mx), and the desired speed is 520 r/min?

(a) 4 poles (b) 6 poles (c) 8 poles (d) 10 poles
(e) 12 poles

A line of induction is called a maxwell, so $\phi = 1.5 \times 10^6$:

$$V = 180 \times 1.5 \times 10^6 \times \text{poles} \times \tfrac{520}{60} \times 10^{-8} = 230 \text{ V}$$

giving $P = 9.82$, or 10 poles. The speed could be reduced slightly or, more probably, the flux lowered a small amount to give the voltage desired. The correct answer is (d).

8-16 DC MOTORS

The purpose of an electric motor is to convert electric power to mechanical power. It does this by utilizing the force exerted on a charge which lies in a magnetic field. The field is supplied by the field coils of the motor and the charge lies in the conductors wound on the armature. The force on a charge $F = qvB$, where v is the velocity of the charge. The charge q that is of interest here is due to the number of electrons affected by the magnetic field. The number of electrons in the conductor, which are also in the magnetic field, is proportional to the current in the conductor divided by the charge velocity. The equations describing this are

$$i = nq_evA$$
$$N = nlA$$
$$f = q_evB$$
$$F = N \times f = nlAq_eBv = ilB$$

where
$\quad i$ = current
$\quad n$ = number of moving charges per unit volume (number of electrons)
$\quad q_e$ = charge on an electron

v = velocity with which the electrons are moving
f = force on one electron
N = number of electrons
l = length of the conductor in the magnetic field
A = cross-sectional area of the conductor
B = magnetic-field flux density

Since the force on a conductor times the distance to the axis gives torque, it follows that motor torque is proportional to the armature current times the field. For constant field, then, the torque will be proportional to the armature current.

The static armature current in a dc machine will equal the armature voltage divided by the resistance of the armature winding. Since the armature resistance is ordinarily low, this can mean a dangerously high current. The armature voltage during motor operation equals the applied voltage minus the back emf or the counter emf. The back emf is the voltage generated in the armature because the conductors in the armature winding cut the lines of force in the field as the armature rotates. This generates a voltage in the armature winding which opposes the applied voltage and gives a net armature voltage equal to the difference of these two. This limits the armature current. The back emf equals speed times flux times a constant: $V = kNB$.

Applied voltage = back emf + (armature current × armature resistance)

This explains why reduced voltages are necessary to start large motors. When the armature is not rotating, there is no back emf and the current is limited by the applied voltage. The positioning of starting resistors for dc motors is shown in Fig. 8-27.

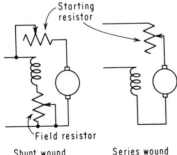

Starting resistor

Field resistor

Shunt wound   Series wound   **Figure 8-27**

8-17 SHUNT MOTORS

The field current of the shunt-wound motor is essentially constant. If the load is reduced, the motor will tend to speed up a little until the back emf has increased enough to balance the loss in I_aR_a:

$$V = \text{back emf} + I_aR_a$$

As the load reduces, the required torque will reduce, thus reducing the armature current, and the speed must increase to increase the back emf enough to make up the difference. This also explains what happens when the field current of a shunt motor is reduced. A reduction in field current will reduce the flux density and the rotor must travel faster to generate an equivalent back emf. If the field current should be reduced to zero the speed would, theoretically, become infinite.

8-18 SERIES MOTORS

In the case of the series motor a reduction in load will cause a reduction in torque and thus in armature current. Since the armature current is also the field current, this would also cause a reduction in the flux density, and the increase in speed required to maintain a balance I_aR_a and back emf would rise as a power function. The curves for series and shunt motors are shown in Fig. 8-28. Compound-wound motors would have characteristics in between these two; a characteristic curve for this type is also shown in Fig. 8-28.

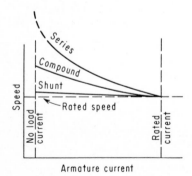

Figure 8-28

This explains why the load should never be removed from a series motor and why such a motor should be directly connected to its load or through gears and not through belts which might slip, come off, or break.

It also explains why a shunt-wound motor should not be started or run with zero field current.

- A dc series-wound motor has a field resistance of $R_f = 0.80\ \Omega$ and an armature resistance of $R_a = 0.50\ \Omega$. At 1200 r/min the motor delivers 5 hp with a line voltage and line current of 100 V and 46 A, respectively.

  a. What is the efficiency of the motor?
  b. What is the counter emf generated by the armature at 1200 r/min?
  c. If the load is decreased so that the speed rises to 1500 r/min, what is the line current?
  d. What is the available torque for condition c, assuming constant efficiency?

  Given the following relations: (1) torque varies directly as the armature current squared; (2) speed varies directly as the armature voltage; (3) flux varies directly as the armature current.

First, we have the output of the motor, 5 hp, and the input, 100 V × 46 A. The efficiency is

$$\frac{5.00 \times 746}{46 \times 100} = 81 \text{ percent}$$

since 746 W equals 1 hp, which is the answer to part a. This is a series-wound motor, so there is a drop through the field equal to 46 × 0.80 = 36.8 V. The voltage across the armature is then 100 − 36.8 = 63.2 V. The drop across the armature due to the armature resistance $I_a R_a = 23$ V. The counter emf must then equal 63.2 − 23 = 40.2 V, which is the answer to part b.

If the speed increases to 1500 r/min, the counter emf must rise, and taking the second given relationship, armature voltage (counter emf) would rise to $\frac{1500}{1200}$ × 40.2 = 50.2 V. As before, we have

$$iR_f + iR_a + \text{back emf} = 100 \text{ V}$$

so $i = 49.8/1.3 = 38.3$ A, line current (c).

Since for this case it is given that torque varies as the armature current squared, the torque would equal $(38.3/46)^2 = 0.695$ percent of the original torque. The initial torque from hp $= 2\pi N$ Tq/33,000 equals 21.9 ft·lb; so the torque for condition c is 0.695 × 21.9 = 15.2 ft·lb, answer to d.

The relationship for horsepower is derived from the work done per revolution times the number of revolutions per minute N. Torque Tq is measured in foot-pounds or pounds force at a distance of 1 ft. The work done per revolution is then force acting times distance through which it acts 2π Tq ft·lb. This multiplied by revolutions per minute gives foot-pounds per minute, which, divided by 33,000 ft·lb/(min)(hp), gives horsepower.

We might study this problem a little further, since the second of the given relationships does not seem to be quite in accord with the theory. It states that speed varies directly as the armature voltage. This would imply that the armature voltage varies directly as the speed; but while this is correct for a shunt motor with its constant field, it is not correct for a series motor where the flux is not constant but varies with the current. Since back emf is proportional to $N \times B$ and the field flux B is proportional to the field current (which is also the armature current), we have that back emf = KNi, and we can determine the constant K from the data given for the first case when $N = 1200$ r/min, $i = 46$ A, and the back emf was determined to be 40.2 V.

This gives $K = 40.2/(1200 \times 46) = 7.29 \times 10^{-4}$. When the speed increases to 1500 r/min, the back emf will equal

$$7.29 \times 10^{-4} \times 1500 \times I_2 = 1.093 \times I_2$$

The sum of the back emf and the IR drops must equal the line voltage of 100 V. The IR drops equal $I_3(0.80 + 0.50)$, giving $1.30I_2 + 1.093I_2 = 100$, or $I_2 = 41.7$ A (part c), instead of the 38.3 A calculated previously. This is the correct current based on the theory of series-wound motors. The 38.3 A would be correct for a shunt-wound motor. The back emf would then equal $1.093 \times 41.7 = 45.6$ V, instead of 50.2 V, as calculated previously.

The answer for part d would then be $(41.7/46)^2 \times 21.9 = 18$ ft·lb.

8-19 INDUCTION MOTORS

The commonest type of motor used on alternating current is the induction motor. Induction motors are made in two types, squirrel cage and wound rotor; they differ only in the construction of the rotor. The principle of operation of these two types is the same. The wound rotor is used to supply a higher starting torque and a limited amount of speed control. There is no electrical connection between the rotor circuits of an induction motor and the stator circuits of the supply line. The currents in the rotor are induced currents. The windings in the rotor are short-circuited and thus act like the secondary of a transformer, with the stator windings acting as the primary. When the rotor is stationary, the frequency of the induced current is the same as the line frequency. As the rotor gains speed, the frequency of the induced current will reduce, until the point at which, if the rotor were to be turned at the synchronous speed, the frequency and the induced current would fall to zero, since no lines of flux would be cut. We recall that induced voltage is proportional to the rate at which the lines of flux are cut. If the rotor were to rotate at the synchronous speed, the conductors would be in phase with the alternating field and no lines of flux would be cut. Therefore, an induction motor cannot operate at synchronous speed and there must be a certain amount of slip. This slip usually runs between 3 and 6 percent, where slip equals 1 minus the ratio of actual speed to synchronous speed, or

$$\text{Slip} = 1 - \frac{\text{actual speed}}{\text{synchronous speed}}$$

The synchronous speed in revolutions per minute equals $120f/p$, where f is the frequency and p is the number of poles. This formula is primarily for single-phase current but can be generalized by dividing the number of poles by the number of phases, so that p should actually be poles per phase.

We have here the same sort of starting problem that we had with dc motors, viz., excessively high starting current. When the rotor is not moving, its conductors are being cut by the greatest number of lines of flux per second; thus, when the voltage is a maximum, the current is also a maximum. The torque should also be a maximum, but in a single-phase motor, for instance, two equal and opposite forces act at the positive and negative poles; thus there is no operating torque. Various methods are used to start single-phase induction motors.

The basic torque equation $\text{Tq} = KBI$ must be modified for alternating current to include the effects of phase angle. Since we are working with an inductive circuit, the current lags behind the voltage. The voltage is in phase with the flux, so there is a phase angle between the space distribution of the flux and the current that is equal to the phase angle between the voltage and the current. The torque is then proportional not only to the product of the flux and the current but to the cosine of the phase angle as well. In other words, torque is proportional to the product of voltage times the component of the current which is in phase with the voltage: $\text{Tq} = KBI \cos \phi$.

Under operating conditions, the slip is large enough to provide a frequency of the rotor voltage large enough to produce a reactance which cannot be neglected.

Starting torque of an induction motor is given by the equation

$$\mathrm{Tq} = \frac{KBV_2R_2}{R_2{}^2 - X_2{}^2}$$

where

$R_2{}^2 - X_2{}^2 =$ the square of the rotor impedance, which is constant for zero rotor speed

$K =$ a constant depending on the physical characteristics of the motor

$R_2 =$ resistance of the rotor

$V_2 =$ rotor voltage

$B =$ flux of the rotating field

This means that starting torque is proportional to BV_2. The flux is proportional to the stator voltage, and, of course, the stationary rotor voltage is also proportional to the stator voltage, so the starting torque is proportional to the square of the stator or applied voltage.

8-20 SYNCHRONOUS MOTORS

Synchronous motors operate on a different basis than induction motors. The rotor may be considered as a permanent magnet, and it follows the rotating field at synchronous speed. If the load is increased to the point where the torque required to maintain the speed exceeds the torque supplied by the motor, the rotor will slow down and stop. This is termed the "breakdown," or "pull-out," torque. If the load is maintained constant and the rotor field current is increased sufficiently, the current will lead the voltage and the synchronous motor will operate as a synchronous condenser.

8-21 THREE-PHASE CIRCUITS

- A three-phase, 60-Hz, 220-V-between-wires, three-wire line supplies current to three single-phase heaters. Each heater has a resistance of 3 Ω and a reactance of 4 Ω. Determine:
  a. The power supplied to the three heaters connected in three-phase Δ
  b. The power which would be supplied by the same line if the heaters are connected in three-phase Y
  c. The power factor of three-heater load in a and b

Start by listing some of the important factors in balanced three-phase network analysis:

1. For a Y connection, line currents and phase currents are equal.
2. For a Y connection, the line voltages equal $\sqrt{3}$ times the phase voltages.
3. For a $\varDelta$ connection, the line voltages and the phase voltages are equal.
4. For a $\varDelta$ connection, the line currents equal $\sqrt{3}$ times the phase currents.
5. The instantaneous power for a three-phase system is constant and is equal to three times the average power per phase.
6. A balanced $\varDelta$ connection may be replaced by a balanced Y connection if the circuit constants per phase obey the relation: Y impedance = $\frac{1}{3}\varDelta$ impedance.

Solve part *b* first. Draw an equivalent four-wire circuit (Fig. 8-29).

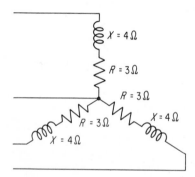

Figure 8-29

Since this current is balanced, the neutral wire will carry no current and may be removed without affecting the circuit. We may simplify this circuit by sketching one branch (Fig. 8-30). The voltage from line to neutral, phase voltage, will equal $220/\sqrt{3} = 127$ V. The current will

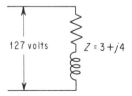

Figure 8-30

equal

$$\frac{V}{Z} = \frac{127}{3 + j4} = \frac{127}{3 + j4} \times \frac{3 - j4}{3 - j4} = \frac{381 - j508}{25} = 15.2 - j20.3$$

The line current would then equal $\sqrt{15.2^2 + 20.3^2} = 25.4$ A. The phase angle would equal $\tan^{-1}(20.3/15.2) = 53.1°$. Since we do not know whether the impedance is reactive or inductive, we do not know whether the phase angle will be leading or lagging. The power per phase would equal $127 \times 25.4 \times \cos 53.1° = 1940$ W, or since $Z = \sqrt{3^2 + 4^2} = 5$, line current equals $\frac{127}{5} = 25.4$ A; phase angle $= \tan^{-1}(X/R) = \tan^{-1}(\frac{4}{3}) = 53.1°$:

$$\text{Power} = I^2R = (25.4)^2 \times 3 = 1940 \text{ W/phase}$$

The total power supplied would be three times the power per phase, or $3 \times 1940 = 5820$ W.

Total power can also be calculated from the relationship line-to-line voltage times line current times power factor times $\sqrt{3}$, which would be $220 \times 25.4 \times 0.60 \times 1.732 = 5820$ W, which is the answer to part *b*.

Part *a* may be determined by taking a Y network with one-third the impedance per phase and calculating as above, or it may be determined directly. The Δ network is shown in Fig. 8-31. First solve the Δ network as shown. The phase voltage equals the line voltage, or 220 V.

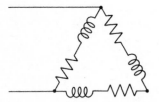

Figure 8-31

The phase impedance is again 5 Ω, so the phase current is $\frac{220}{5} = 44$ A. The phase angle is $\tan^{-1}\frac{4}{3} = 53.1°$, and the power per phase equals $220 \times 44 \times 0.60 = 5820$ W. The total power supplied is then $3 \times 5820 = 17,460$ W—the answer to part *a*.

We can determine this also from the line current and line-to-line voltage. The line current for a Δ connection equals $\sqrt{3}$ times the phase current, or $44\sqrt{3} = 76.2$ A. Total power would be $220 \times 76.2 \times 0.60 \times \sqrt{3} = 17,460$ W.

The answer to part *c* is the cosine of the phase angle, which equals 0.60.

Sample Problems

8-1 A 60-Hz 13,200-V transformer has a core loss of 525 W. When it is operated at the same maximum flux density of 50 kilolines/in² on 25 cycles, the core loss is 140 W. Which of the following most nearly equals the normal hysteresis at 60 Hz?

$$P_h = k_1 f \beta_m^{1.6} \qquad P_e = k_2 f^2 \beta_m^2$$

(a) 197 W   (b) 201 W   (c) 208 W   (d) 217 W   (e) 224 W

8-2 A single-phase ac motor delivers 5 hp and is 80 percent efficient. It operates at 220 V at a power factor of 0.707 lagging. Which of the following most nearly equals the reactance of a capacitor in parallel to bring the power factor to unity?
(a) 8.6 Ω   (b) 9.2 Ω   (c) 9.8 Ω   (d) 10.4 Ω   (e) 10.7 Ω

8-3 The power drawn by a three-phase 220-V motor is measured by two watt-meters. One reads 2000 W, and the other reads −400 W. An ammeter in one of the lines reads 12 A. Which of the following most nearly equals the power factor?
(a) 0.29   (b) 0.31   (c) 0.35   (d) 0.48   (e) 0.71

8-4 Which of the following most nearly equals the current in the 25-Ω resistor shown in Fig. 8P-4?
(a) 0.81 A   (b) 0.92 A   (c) 0.98 A   (d) 1.03 A   (e) 1.14 A

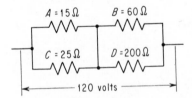

$A = 15\,\Omega$ $B = 60\,\Omega$

$C = 25\,\Omega$ $D = 200\,\Omega$

|← ——— 120 volts ———— →| **Figure 8P-4**

8-5 A dc shunt-connected motor is being supplied 220 V and 54 A. The armature has a resistance, at 75°C, of 0.08 Ω. The shunt field has a resistance of 55 Ω at 75°C. The stray power loss for this motor has been measured and equals 180 W. For this input, assume the motor to be operating continuously at a constant temperature of 75°C and determine which of the following most nearly equals the efficiency of the motor at this load?
(a) 97%   (b) 93%   (c) 89%   (d) 85%   (e) 82%

8-6 Two capacitors of 6 and 12 μF, respectively, are charged by a 120-V battery. The capacitors are in series. After being charged, the capacitors are connected by joining the two positive terminals to each other and doing the same with the two negative terminals. Which of the following most nearly equals the charge which will be on the 6-μF capacitor?
(a) 720 μC   (b) 670 μC   (c) 540 μC   (d) 470 μC   (e) 320 μC

8-7 A 110-V 60-Hz supply is connected to an unidentified circuit. The circuit draws a current of 7.5 A and uses 600 W of power. The oscilloscope picture of current and voltage is shown in Fig. 8P-7. Which of the following most nearly equals the reactance of the circuit?

(*a*) 4 Ω (*b*) 6 Ω (*c*) 8 Ω (*d*) 10 Ω (*e*) 12 Ω

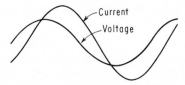

Figure 8P-7

8-8   A circuit consists of a 200-Ω resistor connected in series with the parallel combination of a coil and a 100-μF capacitor. The coil has a resistance of 10 Ω and an inductance of 1 H. Which of the following most nearly equals the voltage across the capacitor when this circuit is connected across a source of power delivering 125 V at a frequency of $100/2\pi$ Hz?

(*a*) 125 V   (*b*) 117 V   (*c*) 104 V   (*d*) 98 V   (*e*) 91 V

8-9   A group of small induction motors on a 230-V 60-Hz feeder in an industrial plant requires a total power of 25 kW at 0.707 power factor lagging. Which of the following most nearly equals the size of the capacitor to be connected in parallel with this load to correct the power factor to 0.90 lagging?

(*a*) 645 μF   (*b*) 638 μF   (*c*) 627 μF   (*d*) 617 μF   (*e*) 604 μF

8-10 A load of 110 kVA of induction motors on the three-phase, 208-V, 60-Hz power supply to a shop is balanced and operates at a power factor of 80 percent. It is proposed to bring the power factor to 90 percent lagging by means of a delta-connected bank of capacitors. Which of the following most nearly equals the total load of capacitors required in this bank?

(*a*) 89 kVA reactance   (*b*) 57 kVA reactance   (*c*) 32 kVA reactance   (*d*) 21 kVA reactance   (*e*) 10 kVA reactance

8-11 A 1000-hp, three-phase, 2200-V induction motor is loaded to rated capacity. The efficiency is 92 percent and the power factor is 87 percent lagging. Which of the following most nearly equals the total load of capacitors required to be connected in parallel with the motor to bring the power factor to 100 percent?

(*a*) 510 kVA reactance   (*b*) 460 kVA reactance   (*c*) 435 kVA reactance   (*d*) 412 kVA reactance   (*e*) 398 kVA reactance

8-12 A "trouble" lamp designed for plugging into the cigarette lighter receptacle in an automobile has a 50-ft cord of no. 18 copper wires. The lamp is rated 60 W at 10 V. Assuming that 12 V is maintained at the receptacle and the resistance of the lamp does not change with temperature, which of the following most nearly equals the wattage input to the lamp?

(*a*) 60 W   (*b*) 53 W   (*c*) 45 W   (*d*) 38 W   (*e*) 32 W

8-13 Given the circuit shown in Fig. 8P-13, which of the following most nearly equals V_{R1} when switch A is closed?

(a) 115 V   (b) 107 V   (c) 92 V   (d) 78 V   (e) 49 V

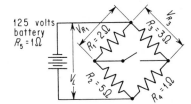

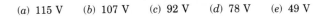

Figure 8P-13

8-14 A coil has a reactance of 37.7 Ω and a resistance of 12 Ω. It is connected to a 110-V 60-Hz line. Which of the following most nearly equals the reading of a wattmeter connected to the circuit?
(a) 93 W   (b) 102 W   (c) 117 W   (d) 121 W   (e) 127 W

8-15 A single-phase transmission line 12 mi long from source to load has a conductor with a resistance of 0.27 Ω/mi and a reactance of 0.62 Ω/mi. The voltage at the load end is 13,200 V. Which of the following most nearly equals the voltage required at the powerhouse when the load is 1500 kVA with a power factor of 0.8 leading?
(a) 12,750 V   (b) 12,900 V   (c) 13,200 V   (d) 13,350 V
(e) 13,510 V

8-16 Which of the following most nearly equals the current which would flow through a series circuit consisting of a 10-mH inductor, a 10-μF capacitor, and a 10-Ω resistor when 120 V, 60 Hz is applied across the combination?
(a) 0.23 A   (b) 0.32 A   (c) 0.38 A   (d) 0.46 A   (e) 0.51 A

8-17 A 2-μF capacitor has an air dielectric between plates separated by 0.5 cm. If the capacitor plates have a potential difference of 1000 V, which of the following most nearly equals the energy stored in the capacitor?
(a) 0.010 C   (b) 0.008 C   (c) 0.006 C   (d) 0.004 C   (e) 0.002 C

8-18 Which of the following most nearly equals the line current in L_1 in Fig. 8P-18? The loads are as given.

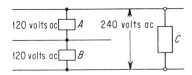

Figure 8P-18

Load	
A	1200 watts, 0.8 power factor
B	1200 watts, 1.0 power factor
C	$^1/_2$-hp motor, 80 percent efficient at 0.6 power factor

(a) 14.2 A (b) 15.1 A (c) 15.6 A (d) 16.0 A (e) 16.6 A

8-19 A storage battery for farm lighting produces 64 V across its terminals on open circuit. The battery has 0.035 Ω internal resistance and is connected through a circuit of copper wire, which has a resistance of 1.835 Ω per 1000 ft, to a load 200 ft distant. The load draws 15 A under these conditions. Which of the following most nearly equals the voltage at the load 200 ft from the battery?
(a) 52.5 V (b) 55.1 V (c) 58.3 V (d) 61.2 V (e) 63.5 V

8-20 Two batteries, which are identical in every way, are connected in series across a 5-Ω resistor, and the current is found to be 0.2 A. The same cells are then connected in parallel with the same 5-Ω resistor, and the current is found to be 0.16 A through the 5-Ω resistor. Which of the following most nearly equals the open-circuit voltage of the batteries?
(a) 0.94 V (b) 1.00 V (c) 1.03 V (d) 1.06 V (e) 1.08 V

8-21 A 60-Hz single-phase motor draws 8.5 A at 120 V and has an inductive power factor of 87 percent at this load. If a 150-μF capacitor is connected in parallel with the motor, which of the following most nearly equals what the new power factor will be?
(a) 100% (b) 95% lagging (c) 92% lagging (d) 94% leading
(e) 97% leading

8-22 A dc generator has two poles, with each pole face having an area of 12 in². The flux density of the air gap is 40,000 lines/in². Which of the following most nearly equals the average emf for one turn on the armature when the machine is running at 1200 r/min? Assume that there is no stray flux outside the pole face area.
(a) 0.143 V (b) 0.167 V (c) 0.192 V (d) 0.204 V (e) 0.212 V

8-23 The impedance of a series ac circuit is $Z = 4 + j3$. The circuit consists of a resistor, a capacitor, and an inductor. If a 200-V 60-Hz source is connected to the circuit, which of the following would most nearly equal the magnitude of the current flowing?
(a) 25 A (b) 30 A (c) 33 A (d) 37 A (e) 40 A

8-24 The resistance of the armature of a 50-hp 550-V shunt-wound dc motor is 0.35 Ω. The full-load armature current of this motor is 76 A. The field current under full load is 3.0 A. Which of the following most nearly equals the stray power losses of the motor when it is delivering 50 hp?
(a) 2480 W (b) 2280 W (c) 2190 W (d) 2020 W (e) 2007 W

8-25 It is necessary to measure the voltage across a circuit known to be about 220 V. Two voltmeters are available: (a) one with a 150-V scale and internal resistance of 15,000 Ω and (b) another with a 100-V scale and internal resistance of 12,000 Ω. If the line voltage were actually 225 V, which of the following most nearly equals the reading of the 15,000-Ω meter if the two are connected in series across the line?
(a) 120 V (b) 125 V (c) 130 V (d) 135 V (e) 138 V

8-26 A slide-wire bridge is set up to measure an unknown resistance R_x (Fig. 8P-26). The voltage V impressed on the circuit is 2.5 V, and the resistance R_s

is 50.5 Ω. The slide wire has a resistance of 12.92 Ω/ft at 70°F. When the slide is adjusted so that the galvanometer reads zero, $L_1 = 15.0$ in and $L_2 = 25$ in. Which of the following most nearly equals the magnitude of the resistor R_x?

(a) 22.4 Ω (b) 25.2 Ω (c) 28.4 Ω (d) 30.3 Ω (e) 31.7 Ω

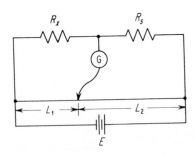

Figure 8P-26

8-27 Which of the following most nearly equals the sizes of the transformers required for a 500-hp, 2300-V, three-phase motor of 90 percent power factor and 93 percent efficiency supplied from a 6600-V three-phase power line?
(a) 157 kVA reactance   (b) 207 kVA reactance   (c) 257 kVA reactance   (d) 263 kVA reactance   (e) 275 kVA reactance

8-28 The current in a certain circuit varies with time according to the relationship: $i = 4 + 2t^2$, where i is in amperes and t is in seconds. Which of the following most nearly equals the amount of charge which passes a point in the circuit in the time interval between $t = 5$ s and $t = 10$ s?
(a) 167 C   (b) 248 C   (c) 359 C   (d) 537 C   (e) 600 C

8-29 Battery A has a no-load terminal voltage of 9 V and an internal resistance of 2 Ω. Battery B has a no-load terminal voltage of 6.5 V with an internal resistance of 1 Ω. When the positive terminals of the two batteries are connected together, the negative terminals are connected together and a 3-Ω resistor is connected between the positive and negative terminals, which of the following most nearly equals the current which will flow through the resistor?
(a) 2.0 A   (b) 2.2 A   (c) 2.4 A   (d) 2.6 A   (e) 2.8 A

8-30 An industrial plant has the following loads: 25 kVA at 0.85 power factor inductive; 50 kW at 1.0 power factor; and 250 hp in motors at 0.90 power factor inductive and 86 percent motor efficiency. Which of the following most nearly equals the magnitude of the reactance necessary to correct the plant power factor to 1?
(a) 109 kVA reactance   (b) 114 kVA reactance   (c) 118 kVA reactance   (d) 123 kVA reactance   (e) 129 kVA reactance

8-31 Three single-phase electric furnaces are each rated at 2300 V, 60 Hz, 2500 kVA, and 80 percent power factor lagging current. If these three furnaces were connected in wye to a three-phase 2300-V 60-Hz supply, which of the

following most nearly equals the total kilowatts they would take from the circuit?
(a) 1500 kW   (b) 2000 kW   (c) 3200 kW   (d) 4700 kW
(e) 6000 kW

8-32   For the circuit shown in Fig. 8P-32, which of the following most nearly equals the equivalent resistance from A to B?
(a) 1.6 Ω   (b) 2.1 Ω   (c) 2.8 Ω   (d) 3.3 Ω   (e) 3.6 Ω

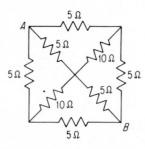

Figure 8P-32

8-33   Three dry cells, each of 1.5 V emf and 0.10 Ω internal resistance, are connected in series through two external resistances of 0.5 Ω each. A resistance of 1.0 Ω is connected from a point between the two external resistances to a point between the first and second dry cells. Which of the following most nearly equals the current I_3 as shown in Fig. 8P-33?
(a) 1.2 A   (b) 0.98 A   (c) 0.84 A   (d) 0.62 A   (e) 0.44 A

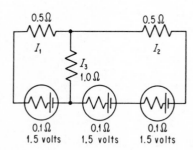

Figure 8P-33

8-34   An induction motor draws 25 A from a 125-V source and has a power factor of 0.80. Which of the following most nearly equals the line current when a capacitor of 300 μF is connected across the terminals of the motor? The frequency is 60 Hz.
(a) 18.5 A   (b) 20.0 A   (c) 22.6 A   (d) 23.1 A   (e) 24.2 A

8-35   A circuit containing 0.1·H inductance and 20-Ω resistance in series is connected across 100-V 25-Hz mains. Which of the following most nearly equals the power input?

(*a*) 187 W   (*b*) 217 W   (*c*) 268 W   (*d*) 310 W   (*e*) 327 W

8-36 A series circuit consisting of a resistance of 50 Ω, a capacitance of 25 μF, and an inductance of 0.15 H is connected across 120-V 60-Hz mains. Which of the following most nearly equals the line current?
(*a*) 1.7 A   (*b*) 1.9 A   (*c*) 2.1 A   (*d*) 2.3 A   (*e*) 2.5 A

8-37 A 30-hp, 220-volt, three-phase, squirrel-cage induction motor with normal voltage applied to the stator has a starting torque of 1.25 times the running torque. Which of the following most nearly equals the voltage which should be applied to give full-load starting torque?
(*a*) 220 V   (*b*) 209 V   (*c*) 197 V   (*d*) 192 V   (*e*) 187 V

8-38 A 0.60 power factor inductive load takes 10 A at 115 V. Which of the following most nearly equals the value of pure resistance which may be placed in series with this load to make it operate normally from a 230-V source?
(*a*) 8.7 Ω (*b*) 9.8 Ω (*c*) 11.9 Ω (*d*) 14.2 Ω (*e*) 15.1 Ω

8-39 A three-phase induction motor under test shows an input of 25 A in each line. The voltage is 223 on each phase. The manufacturer's data show 85 percent power factor and 90 percent efficiency at this load. Which of the following most nearly equals the motor output?
(*a*) 15 hp   (*b*) 17 hp   (*c*) 19 hp   (*d*) 22 hp   (*e*) 27 hp

8-40 For the circuit shown in Fig. 8P-40, which of the following most nearly equals the current A_1 if 120 V 60-Hz alternating current is applied to the circuit?
(*a*) 10.2 A   (*b*) 11.7 A   (*c*) 12.3 A   (*d*) 13.1 A   (*e*) 13.5 A

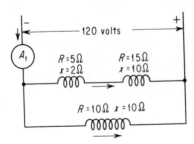

Figure 8P-40

8-41 When a certain coil is connected to a 60-V battery, 2 A of current flows. When the same coil is connected to a 50-V 60-Hz ac source, only 1 A flows. Which of the following most nearly equals the inductance of the coil?
(*a*) 0.07 H   (*b*) 0.09 H   (*c*) 0.11 H   (*d*) 0.13 H   (*e*) 0.15 H

8-42 Three resistors, $R_1 = 0.2\ \Omega$, $R_2 = 0.3\ \Omega$, and $R_3 = 0.6\ \Omega$, are connected in parallel across the terminals of a storage battery having an internal resistance of 0.02 Ω and an emf of 6.0 V on open circuit. With this circuit closed, which of the following most nearly equals the current which would flow through the 0.2-Ω resistor?

    (*a*) 25 A   (*b*) 22 A   (*c*) 19 A   (*d*) 17 A   (*e*) 14 A

8-43   A 200-hp, three-phase, four-pole, 60-Hz, 440-V, squirrel-cage induction motor operates at full load with an efficiency of 85 percent, a power factor of 91 percent, and a slip of 3 percent. For this full-load condition, which of the following most nearly equals the line current fed to the motor?
    (*a*) 187 A   (*b*) 214 A   (*c*) 235 A   (*d*) 253 A   (*e*) 271 A

8-44   Three batteries have terminal voltages of 2.00, 1.95, and 1.84 V, with internal resistances of 0.050, 0.075, and 0.064 Ω, respectively. The three batteries are connected in series with a 0.110-Ω resistance. (Assume that the voltages and internal resistances remain constant.) Which of the following most nearly equals the voltage across the resistance?
    (*a*) 1.8 V   (*b*) 2.1 V   (*c*) 2.3 V   (*d*) 2.5 V   (*e*) 2.7 V

8-45   The wiring diagram for a dc voltmeter is shown in Fig. 8P-45. When 0.588 V is applied across *AB*, the meter registers a full-scale deflection. The scale is graduated from 0 to 15. Which of the following most nearly equals the reading on the meter when 20 V is applied across *OP*?
    (*a*) 5.0   (*b*) 6.7   (*c*) 7.9   (*d*) 9.1   (*e*) 10.0

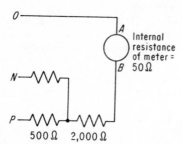

Figure 8P-45

8-46   A power company delivers 500 kW to factory A, 250 kVA at 0.85 power factor lagging to factory B, and 750 kVA at 0.97 power factor leading to factory C. Which of the following most nearly equals the total power load in kilowatts for the three companies?
    (*a*) 1390 kW   (*b*) 1420 kW   (*c*) 1440 kW   (*d*) 1460 kW
    (*e*) 1480 kW

8-47   A dc shunt-wound motor has the following constants: full-load speed, 1200 r/min; rated voltage, 230 V; full-load current, 58 A; armature resistance, 0.15 Ω. Which of the following most nearly equals the speed of the motor when it is delivering one-half the rated torque? Assume that the field remains constant.
    (*a*) 1215 r/min   (*b*) 1224 r/min   (*c*) 1232 r/min   (*d*) 1239 r/min
    (*e*) 1247 r/min

8-48   A 5-kVA 2300/230-V transformer is connected to give a 10 percent boost on a 2300-V power line. Which of the following most nearly equals the load which can be safely supplied with this booster transformer?
    (*a*) 5 kVA   (*b*) 32 kVA   (*c*) 47 kVA   (*d*) 55 kVA   (*e*) 61 kVA

8-49 In the series circuit (Fig. 8P-49), consisting of three elements, the current is I and the voltages are as follows: $V_1 = 120$ V leading 80°, $V_2 = 240$ V lagging 70°, and $V_3 = 150$ V leading 30°. Which of the following most nearly equals the voltage V which is imposed on the circuit?
(*a*) 120 V   (*b*) 155 V   (*c*) 187 V   (*d*) 204 V   (*e*) 235 V

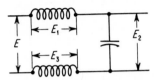

Figure 8P-49

8-50 An electrical series circuit has a resistance of 4.5 Ω and an inductive impedance of 15 Ω. Which of the following most nearly equals the real power if 240 V alternating current is impressed across the system?
(*a*) 0.97 kW   (*b*) 0.99 kW   (*c*) 1.10 kW   (*d*) 1.13 kW
(*e*) 1.15 kW

8-51 An alternating current has a sine waveform which, in turn, has a maximum value of 200 V (Fig. 8P-51), and $e = V_m \sin \theta$. Which of the following most nearly equals the effective voltage V_{eff}?
(*a*) 141 V   (*b*) 152 V   (*c*) 167 V   (*d*) 189 V   (*e*) 195 V

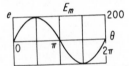

Figure 8P-51

8-52 It is desired to supply a 250-W 120-V lamp 500 ft from the source of power with full 120 V over a circuit of no. 14 copper wire. Which of the following most nearly equals the voltage which is required at the source? Disregard reactance.
(*a*) 132 V   (*b*) 130 V   (*c*) 129 V   (*d*) 127 V   (*e*) 125 V

8-53 Which of the following most nearly equals the current in the 50-Ω resistor shown in Fig. 8P-53?
(*a*) 2.0 A   (*b*) 1.98 A   (*c*) 1.96 A   (*d*) 1.94 A   (*e*) 1.91 A

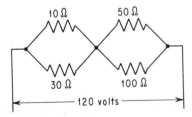

Figure 8P-53

8-54 A load consists of three-phase motors and totals 750 kW at 80 percent power factor lagging. A three-phase 300-kVA reactance static capacitor is connected in parallel with the load. Which of the following most nearly equals the resulting power factor?

(a) 0.87   (b) 0.89   (c) 0.94   (d) 0.96   (e) 0.98

8-55 Referring to the sketch in Fig. 8P-55, there is a difference in potential of 110 volts between A and D. Which of the following most nearly equals the current flowing through the 200-Ω resistance between A and B?

(a) 0.034 A   (b) 0.057 A   (c) 0.063 A   (d) 0.075 A   (e) 0.091 A

Figure 8P-55

Multiple-Choice Sample Quiz

For each question select the most nearly correct answer from the five given possibilities. For practice, try to complete the following 10 problems in 15 minutes or less.

8M-1 A direct-current electric heater has a resistance of 36.5 ± 0.5 Ω. The current flowing in the heater is 12.8 ± 0.2 A. What is the maximum power (in watts) consumed under these conditions?

(a) 5980   (d) 467
(b) 481   (e) 5715
(c) 6253

8M-2 The ratio of the actual power input (kilowatts) divided by the apparent power input (kilovolt-amperes) in a single-phase alternating current is:

(a) the power factor
(b) the phase angle in degrees
(c) the impedance
(d) the reactive power
(e) the line current

8M-3 The magnitude of the generated voltage (induced in the armature) in a shunt-wound generator is:

(a) directly proportional to the armature speed
(b) inversely proportional to the field current
(c) directly proportional to the power generated
(d) inversely proportional to the field flux
(e) none of these

8M-4 A series ac circuit consisting of a capacitor, an inductor, and a resistor is in resonance when:
  (a) the in-phase current equals the out-of-phase current
  (b) maximum current passes through the circuit
  (c) minimum current passes through the circuit
  (d) no current flows through the circuit
  (e) the voltage across the resistor is equal to the combined voltage across the capacitor and the inductor

8M-5 An electric motor has a nameplate rating of the following: 220 V, 23.0 A, 5 hp, 3 phase, 60 Hz, 1750 r/min. Which of the following statements is true?
  (a) The motor uses 220 V, 23.0 A at no load.
  (b) The power input to the motor is 5 hp.
  (c) The 1750 r/min is the no-load speed.
  (d) The motor will deliver 5 hp at full load.
  (e) None of these is true.

8M-6 The power consumed by a single-phase alternating-current circuit may be calculated if the
  (a) resistance and impedance are known
  (b) line current and resistance are known
  (c) inductive and capacitive reactance are known
  (d) line current and line voltage are known
  (e) none of these

8M-7 An electric circuit consists of a resistance of 127 Ω in series with a 150-mH inductance and a capacitor. What must be the capacitance of the capacitor for the circuit to be resonant for a 100-Hz applied current?
  (a) 150 μF
  (b) 17 μF
  (c) 5 μF
  (d) 28 μF
  (e) 98 μF

8M-8 The phase angle of a single-phase circuit can be found by:
  (a) measuring the voltage and the current flowing through the circuit
  (b) measuring the watts consumed by the circuit
  (c) measuring the capacitance and inductance in the circuit at a known frequency
  (d) measuring the pure resistance, the current, and the voltage
  (e) measuring the watts being consumed by the circuit, the voltage drop in the circuit, and the current flowing through the circuit

8M-9 In a parallel resonant (antiresonant) circuit, which of the following conditions must be true for all values of the branch elements?
  (a) The susceptive elements are equal in magnitude and opposite in sign.
  (b) The branch circuit impedances are conjugates.
  (c) The voltage reactance drops are equal in magnitude.
  (d) The reactive components are equal in magnitude, but opposite in sign.
  (e) The branch circuit currents are conjugates.

8M-10 The revolutions per minute of an ac electric motor:
   (*a*) vary directly as the number of poles
   (*b*) vary inversely as the number of poles
   (*c*) are independent of the number of poles
   (*d*) are independent of the frequency
   (*e*) are directly proportional to the square of the frequency

9

ENGINEERING ECONOMICS

The principal emphasis has been on comparing the costs of alternative types of equipment purchase, payout periods, and costs of replacement of existing equipment. Included with these general subjects have been the complementary subjects of depreciation, interest, sinking funds, etc.

9-1 EXAMINATION COVERAGE

Economic analysis is covered in the morning session, and is also one of the required subjects in the afternoon session. The subdivisions of the major subject which may be covered, and which should be reviewed, include the following:

Time value of money

Annual cost

Percent return

Present worth

Future value

Capitalized cost

Break-even analysis

Valuation and depreciation

9-2 INTEREST

The simplest form of interest is termed, appropriately enough, "simple interest." This is the amount charged in dollars per year for the use of

$100 of capital; it is similar to yearly rental. Thus, if someone charged $3 interest for a loan of $60 for a period of 8 months, the simple interest rate would be $\frac{3}{60} \times \frac{12}{8} = 7.5$ percent. Simple interest is seldom used for periods of longer than 1 year except when the interest is paid to the owner of the capital instead of being added to the capital. If the money is deposited in an account of some type, the interest is ordinarily added to the account when it is due, and for the next period interest is calculated on the total amount, principal plus accrued interest. This is known as compound interest.

By this method the amount in the account at the end of the first interest period would be $S_1 = $ principal $+ i \times$ principal $= P(1 + i)$; at the end of the second interest period the sum would equal $S_2 = S_1 + iS_1 = S_1(1 + i) = P(1 + i)^2$. Similarly

$$S_3 = S_2 + iS_2 = S_2(1 + i) = P(1 + i)^3$$

For n periods, $S_n = P(1 + i)^n$, which is known as the "compound interest law." Restating this we have, "The total amount of money at the end of n periods resulting from the investment of a principal P at an interest rate of i per period can be determined from the equation $S_n = P(1 + i)^n$." Note that n is the number of investment periods and i is the interest rate per period. If interest should be compounded quarterly, the number of periods would be four times the number of years and the interest rate i would be one-quarter the yearly or nominal rate.

This leads to the distinction between nominal and effective rates of interest. Interest rates are usually quoted on an annual, or yearly, basis, and such an annually based rate is termed the "nominal" interest rate. Interest payments (or charges) are, however, frequently made semiannually, quarter-annually, or monthly. In such a case part of the yearly interest can also earn interest during part of the year; the equivalent total yearly interest is termed the "effective" interest rate. Thus an interest rate of 5 percent compounded semiannually would give a nominal interest rate of 5 percent but an effective interest rate equal to $(1 + 0.025)^2 - 1 = 0.0506$, or 5.06 percent. Similarly, 5 percent compounded quarterly would give an effective rate of interest of 5.09 percent, although the nominal rate would still be 5 percent.

- You may purchase government bonds at $750 each which mature in 10 years and have a face value of $1000 at the end of 10 years. Which of the following most nearly equals the average nominal interest rate earned by the purchase price of $750, assuming annual compounding? (*Note:* the interest rate may not reflect the rate of interest presently available, but that is not important to this example. The method of analysis will be the same regardless of the interest rate chosen.)

(a) 2.5%   (b) 2.7%   (c) 2.9%   (d) 3.0%   (e) 3.2%

Annual compounding would mean 10 interest periods, so $1000 = 750 \times (1 + i)^{10}$, giving $i = 2.92$ percent nominal interest rate. The correct answer is (c).

It might be noted here that the effective interest rate would be the same for this case, since the nominal and effective rates will be the same when the interest periods equal 1 year. They will be different only when the interest periods are shorter than a year.

Effective annual interest rate $= [1 + (i/m)]^m - 1$, where i is the nominal interest rate and m is the number of interest periods per year.

- A household finance company lends a young engineer $120 cash, to be repaid in monthly installments of $8.72 over an 18-month period. The engineer receives the $120 cash, begins payments at the end of the first month, and continues the monthly payments until 18 payments have been made to complete the contract. For this particular capital-recovery-with-return schedule, which of the following most nearly equals the nominal rate of interest earned by the company's capital?

(a) 22%   (b) 27%   (c) 32%   (d) 36%   (e) 41%

By paying $8.72 at the end of each month for 18 months, the young engineer is actually setting up a sinking fund, the present value of which is $120, since that is the amount this individual received in return for the agreement to pay $8.72 per month for 18 months. Theoretically the company can reinvest the $8.72 as it is received and receive interest on each payment from then on. Assuming 18 interest periods, each of 1 month's time, with a payment at the end of each period, the sum of money which would be available at the end of the 18 months would equal

$$S = 8.72(1 + i)^{17} + 8.72(1 + i)^{16} + \cdots + 8.72$$

since the first payment would earn interest for the 17 remaining periods, the second payment for the 16 remaining periods, etc. The last payment would be made at the end of the 18 periods and would earn no interest.

If we multiply both sides of the equation by $(1 + i)$ and then subtract the original equation from the new one, we obtain

$$S(1 + i) = 8.72(1 + i)^{18} + 8.72(1 + i)^{17} + \cdots + 8.72(1 + i)$$
$$(-)S = \qquad\qquad 8.72(1 + i)^{17} + \cdots + 8.72(1 + i) + 8.72$$
$$Si = 8.72(1 + i)^{18} - 8.72$$

which can be generalized to give

$$Si = R[(1 + i)^n - 1]$$

where

n = number of periods
i = interest rate per period
R = amount invested at the end of each period
S = total amount in the sinking fund at the end of n periods

By investing the $120 for the 18 periods of 1 month each at an interest rate of i per period, the young engineer would have had a total of $S = 120(1 + i)^{18}$ at the end of the 18 months. The more general relationship is $S = P(1 + i)^n$, where S is the amount at the end of n periods and P the amount initially invested. Substituting this equivalence for S in the previous equation and solving for R gives

$$R = P \frac{i(1 + i)^n}{(1 + i)^n - 1}$$

where R is the amount to be invested in the sinking fund at the end of each period and P is the present worth of the sinking fund, or the amount which could be invested in a lump sum now to give an amount equal to that in the sinking fund at the end of n periods.

The amount at the end of the n periods would be S, where $S = P(1 + i)^n$, or $S = R[(1 + i)^n - 1]/i$.

In the problem being considered, $R = \$8.72, P = \$120$, and $n = 18$. This problem cannot be solved directly for i; it would have to be solved by trial and error. This expression is listed in interest tables as the capital recovery factor: $R/P = 8.72/120 = 0.0727$, for the example being considered. From a capital-recovery-factor table with $n = 18$, the CRF = 0.0727 for $i = 3$ percent. The interest rate per period would then be 3 percent, and the effective annual rate of interest earned by the company's capital would equal $(1 + 0.03)^{12} - 1 = 1.426 - 1$, or 42.6 percent. Using the equation given previously, effective annual interest $= [1 + (i/m)]^m - 1$, we can calculate the nominal rate i from the effective rate:

$$0.426 = \left(1 + \frac{i}{12}\right)^{12} - 1$$

$i = 12 \times 0.030 = 0.36$, or 36.0 percent nominal interest

The correct answer is (d).

9-3 PRESENT WORTH

■ A young man with thoughts of the future is about to purchase a new automobile. A roadster just right for his present needs costs $1000. The addition of a rumble seat for his prospective in-laws will cost an additional $150 at some future date. A roadster equipped with a rumble seat costs $1100 now. If interest is at 8 percent, which of the following most nearly equals how soon he must get married to justify the purchase of the $1100 car?

(a) 2.6 years   (b) 3.7 years   (c) 4.5 years   (d) 5.0 years
(e) 5.3 years

Since the $1000 would be invested anyway, the question is, "When would $100 invested now be worth $150 if interest is at 8 percent?" The only question is whether to spend the $100 now or, theoretically, to invest it at 8 percent compounded annually. We have, then, that $150 = 100(1 + 0.08)^n$, which gives $n = 5.27$, or $100 invested today at 8 percent compounded annually would amount to $150 in 5.27 years. This means that the young man would have to get married before a lapse of 5.27 years to gain by the investment now in the rumble seat. The correct answer is (e).

9-4 ANNUITY

Another important concept in economics is the annuity which might be termed a "sinking fund in reverse." That is, instead of depositing a certain amount at the end of each period so as to have available a large lump sum at the end of that time, the annuity does just the opposite. A lump sum is deposited in an account at a particular rate of interest and then a certain amount may be withdrawn from that account at the end of each period for a prescribed number of periods. This is the way pension plans are established. The equation which is used to determine how much must be deposited to provide a particular amount at the end of each period for a specified number of periods with a given interest rate is the equation for the present value of a sinking fund.

$$R = P \frac{i(1 + i)^n}{(1 + i)^n - 1} \quad \text{or} \quad P = R \frac{(1 + i)^n - 1}{i(1 + i)^n}$$

where
  R = amount to be paid out at end of each period
  P = amount to be deposited at start of annuity

n = number of periods, or life of annuity
i = interest rate per period

- An engineer learns that for each additional year she works, her pension will increase by $30 per month. The pension will be available to her when she reaches age 65. She figures that she will have a life expectancy of 14 more years when she reaches age 65. Which of the following most nearly equals how much the added pension would be worth to her at age 40?

  (a) $578   (b) $747   (c) $832   (d) $946   (e) $1084

At age 65 she would have a life expectancy of approximately 14 years. So the issue becomes how much the engineer would have to invest today to buy a 14-year annuity of $30 per month at age 65.

The value of the annuity at age 65, assuming a yearly interest rate of 7 percent throughout the life of the annuity, would equal

$$P = R \frac{(1 + i)^n - 1}{i(1 + i)^n}$$

$$30 \times \frac{1.005833^{168} - 1}{0.005833 \times 1.005833^{168}} = \$3207$$

P = value of annuity at age 65

R = $30 per month

n = $12 \times 18 = 168$ periods of 1 month each

i = $\frac{7}{12}$ percent = 0.5833 percent per month

If it is assumed that the engineer could invest her money at a net 6 percent per year while she was still actively employed (6 percent after paying income tax on the interest she obtains), the present value at age 40 of $3207 at age 65 (25 years hence) would equal

$$\text{Value at age 40} = \frac{3207}{1.06^{25}} = \$747$$

The correct answer is (b).

9-5 COST FACTORS

Interest is an expense. It is an expense charged if we borrow, and it is an expense if we purchase a piece of equipment, in that we do not receive the interest we would otherwise get if we invested the money in an interest-producing enterprise. This loss of interest due to investment in a machine is a cost attributable to the machine and must be taken into account in any consideration of the advisability of purchasing a given machine. There are, of course, other cost factors involved, and any final decision as to whether to purchase must be based on a consideration of all the contributing factors. These can best be reviewed with the aid of a few examples.

- A company may furnish a car for use of its salesperson or pay the salesperson for the use of a car at the rate of 11 cents per mile. The following estimated data apply to company-furnished cars: a car costs $1800; it has a life of 4 years and a trade-in value of $700 at the end of that time. Monthly storage cost for the car is $3 and the cost of fuel, tires, and maintenance is $0.028 per mile. Which of the following most nearly equals the annual mileage that a salesperson must travel by car for the costs of the two methods to be equal if the interest rate is assumed to equal 8 percent?

  (a) 5150 mi   (b) 5870 mi   (c) 6410 mi   (d) 7870 mi
  (e) 9450 mi

The best way to handle such a problem is to itemize all the individual expenses of each alternative method and add them to determine the total costs of the two possibilities. One method is a flat 11 cents per mile. The other method includes costs of operation, storage, depreciation, and loss of interest due to the invested capital.

The type of depreciation has not been specified, so we shall use straight-line depreciation and average interest. This would give an annual depreciation cost of ($1800 − $700)/4 = $275. The cost of depreciation is assessed at the end of each year; the amount invested throughout the first year is $1800. Similarly, the amount invested throughout the second year is $1525, the third year $1250, and the fourth year $975. The average interest lost would then be ($144 + $122 + $100 + $78)/4 = $111 per year. This could also have been calculated by adding the first year's interest loss to the last year's and dividing by 2, or

$$i_{avg} = \frac{\$144 + \$78}{2} = \$111$$

per year. This can be generalized to give

$$i_{avg} = \frac{\text{first cost} + \text{salvage value} + \text{yearly depreciation}}{2} \times i$$

Tabulate all the costs and add them to obtain total cost:

Depreciation	$275
Loss of interest	111
Storage (12 × $3)	36
Operating costs	0.028 × miles
Total yearly costs	$422 + 0.028 × mileage

The problem asks when this would be equal to a unit cost of 11 cents per mile, so we have $0.11 \times m = 422 + 0.028 \times m$, giving a yearly mileage of 5146 mi. The correct answer is (a).

Using this problem, we can illustrate another concept, the cost of not operating something that is already owned. In this case, using the above figures and assuming that the salesperson owns a car, our salesperson is confronted with the need of deciding, on a purely economic basis, whether to drive to a town 400 mi distant or to take the train. The ticket agent says that the ticket for the trip will cost only $14, which is considerably less than the $44 it would cost to drive the car at a cost of 11 cents per mile. We must recognize, however, that our salesperson already owns a car. The question is not whether to buy a car but rather whether to use the car presently owned. The costs of depreciation, loss of interest, and storage continue regardless of whether the salesperson drives the car or not. We might, then, say that it costs the salesperson ($275 + $111 + $36)/5146 = $0.082 per mile to not drive the car, so the total cost of the train trip would be $14 + $400 × 0.082 = $46.80, which is more expensive than driving. Another, and perhaps more realistic, way of looking at this question is to compare the additional out-of-pocket cost of driving vs. the cost of the train ride. Depreciation, loss of interest, and storage are what might be termed "fixed costs." A fixed cost is one which does not vary over the operational life span of the equipment being considered. The yearly operating cost of the car varies, depending upon how many miles the car is driven. The fixed costs might be considered costs of owning a car; the operating cost is only the cost of driving a car. The actual cost of driving the car on the proposed trip would be $400 × 0.028 = $11.20, which again is $2.80 less than the cost of the train ride.

This is not a discussion of the pros and cons of car vs. train. These items are only incidental to the concept of fixed costs, which continue regardless of whether the equipment is operated or not; they help to illustrate the costs of nonoperation. It should be easy to see, with the aid of the illustration, why it is often more profitable—at least for a while—to operate a plant at a loss than to shut it down altogether.

▪ Machine A cost $4400 five years ago. The book value of the machine today is $2400. The highest offer for it today is $1000. In 5 more years the salvage value will be $40.00. The annual operational costs have been $1300 and are expected to continue in that amount for the next 5 years. Machine B costs $7200 and has a life of 10 years, at which time it will be worth $900. Estimated annual cost is $300. If, at this time, money is considered to be worth 8 percent, which of the following most nearly equals the difference in the average annual costs of the two machines?

(a) $175   (b) $215   (c) $250   (d) $297   (e) $324

9-6 BOOK VALUE

Note that the book value of the machine is $2400, while the best price it can be sold for is $1000. The concept of book value is confusing and causes many errors in investment judgment. Book value is a fictitious value and depends on the accuracy of the estimated depreciation. It has no real significance and should be ignored when considering the cost of replacement. Regardless of what the book value may be, the machine is worth only what it can be sold for, and that is $1000. It may help to consider the case in which the book value is less than the salvage value. In this case you would certainly not dispose of the machine for the book value, you would sell it for the best price you could get, and this is the figure you would use in estimating replacement cost, not the lower book value. The same is true when the book value is higher than the salvage value; you would still use the sale price and not the book value in the determination of replacement cost.

The yearly costs of the machines will be depreciation, loss of interest on the capital invested, and the operating costs. Tabulate these and compare them. Machine A now represents a capital investment of $1000 and at the end of 5 years will be worth only $40. The average yearly depreciation would then be ($1000 − $40)/5 = $192. The yearly operating cost would be $1300, and the average interest loss would equal ($1000 + $40 + $192)/2 × 0.08 = $49.28. The comparative costs would be:

Cost item	Average yearly cost over next 5 years, machine A	Average yearly cost over next 10 years, machine B
Depreciation (straight-line)	$ 192.00	$ 630.00
Operating cost	1300.00	300.00
Average interest	49.28	349.20
Total annual cost	$1531.28	$1279.20

Machine B would cost $252 per year less than machine A, so the correct answer is (c).

- An asset has a first cost of $13,000, an estimated life of 15 years, and a salvage of $1000. For depreciation, use the sinking-fund method. If interest is assumed to be at 5 percent compounded annually, which of the following most nearly equals the annual sinking-fund annuity or depreciation charge?

  (a) $545   (b) $556   (c) $573   (d) $583   (e) $592

Since the salvage value at the end of 15 years is $1000 and $13,000 will be required for replacement, there must be a total of $12,000 in the sinking fund at the end of the 15-year period. We need to determine, then, what amount we must deposit at the end of each year to amount to $12,000 at the end of 15 years with interest at 5 percent.

The sinking-fund formula previously derived is

$$R = \frac{Si}{(1 + i)^n - 1} = \frac{\$12,000 \times 0.05}{(1.05)^{15} - 1} = \$556$$

to be deposited at the end of each year for 15 years.

This could also have been determined with the aid of compound-interest tables. Under the heading "Sinking-Fund Factor" we find a value of 0.04634 for 5 percent compound interest and 15 periods. This is equal to $i/[(1 + i)^n - 1]$ and need only be multiplied by the value of S to give R: $R - \$12,000 \times 0.04634 = \$556$. The correct answer is (b).

- Utilizing the same information given in the example above, which of the following most nearly equals the balance which would be in the sinking fund, i.e., the amount which would have accumulated toward depreciation of the asset, at the end of 9 years?

  (a) $6490   (b) $6130   (c) $5970   (d) $5860   (e) $5790

At the end of 9 years there would be an amount equal to

$$S = R \frac{(1 + i)^9 - 1}{i} = \$556 \frac{1.05^9 - 1}{0.05} = \$6130$$

The correct answer is (b).

- Again, making use of the data in the previous examples, which of the following most nearly equals the net *book-value gain or loss* which would be realized if the asset were to be sold for $4000 at the end of 9 years?

(*a*) $1440   (*b*) $1800   (*c*) $2240   (*d*) $2460   (*e*) $2870

If the asset were sold at this time for $4000, this would give a total of $6130 + $4000 = $10,130 available to balance the replacement cost of $13,000. The book value at this time equals

$$\$13,000 - \$6130 = \$6870$$

or the first cost less the amount in the sinking fund. The actual value is only the market value of $4000, so there is a "book-value loss" of $2870. The correct answer is (*e*).

9-7 DEPRECIATION

Four types of depreciation curves are shown in Fig. 9-1.

The fixed-percentage-on-diminishing-balance method depreciates an asset by a fixed percentage of its value at the beginning of the year. The book value at the end of the first year is equal to cost × $(1 - D)$; at the end of the second year,

$$C(1 - D)(1 - D) = C(1 - D)^2$$

at the end of the third year, $C(1 - D)^2(1 - D) = C(1 - D)^3$; at the end of nth year, $C(1 - D)^n$.

For the case considered, $C = \$13,000$, and the book value at the end of the fifteenth year is $1000, so $13,000(1 - D)^{15} = \$1000$; $D = 0.1572$, and the depreciation allowance each year is 15.72 percent of the book value of the asset at the beginning of that year. The book value at the end of the first year would then be

$$\$13,000 \times 0.8428 = \$10,950$$

at the end of the second year, $13,000 × 0.8428^2 = \$9280$, or $10,950 × 0.8428; at the end of the fifth year,

$$\$13,000 \times 0.8428^5 = \$5530$$

at the end of the ninth year, $13,000 × 0.8428^9 = \$2800$; at the end of the fifteenth year, $13,000 × 0.8428^{15} = \$1000$.

Another method of depreciation authorized by U.S. tax law is the sum-of-the-year digits. With this method the sum of the digits correspond-

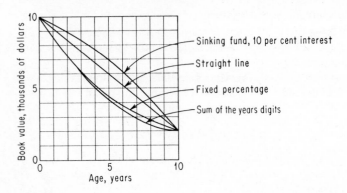

First cost $10,000
Salvage value at end of ten years $2,000
Straight-line depreciation ...... (10,000 − 2,000)/10 = $800 per year
Book value ....,.......... $10,000 − 800$n$
Fixed percentage .... 10,000$(1 − D)^{10}$ = 2,000
$$1 − D = 0.852$$
Book value .... 10,000 × 0.852^n
Sinking fund, interest at 10 percent

Amount in sinking fund .... $S = R\dfrac{(1 + i)^n − 1}{i}$

Book value .... $10,000 − 502 × \dfrac{1.10^n − 1}{0.10}$

There is no simple relationship that will give the book value at the end of each year for the sum-of-the-years-digits method of depreciation, so the data for this method are tabulated below:

Year	Sum-of-the-year digits book value at end of year
0	$10,000
1	8,545
2	7,236
3	6,072
4	5,054
5	4,181
6	3,454
7	2,872
8	2,436
9	2,145
10	2,000

Figure 9-1

ing to the estimated years of life are added together. For example, if the estimated life is 10 years, the sum-of-the-years digits $(10 + 9 + 8 + \cdots + 1)$ would equal 55. The depreciation the first year would equal $\frac{10}{55}$ times the depreciable cost. The second-year depreciation would equal $\frac{9}{55}$ times the depreciable cost, etc. (The tenth-year depreciation would equal $\frac{1}{55}$ times the depreciable cost.)

For the example used previously—first cost of $13,000, estimated life of 15 years, and salvage value of $1000—the first-year depreciation would equal $\frac{15}{120} \times \$12,000$ or $1500, and the book value would equal $11,500. The ninth-year depreciation would equal $\frac{7}{120} \times \$1200$, or $700, and the book value at the end of the year would equal $3100. The fifteenth-year depreciation would be $100 and the book value at the end of the year would equal $1000.

The only difference in these cases is the method of keeping the books. Since sinking funds are seldom if ever actually set up in practice and exist only as accounts in the books of the organization, these cases would be just the same insofar as the actual investment procedure and recommendations were concerned. Depreciation schedules are set up primarily for purposes of taxation and rate setting.

- An engineer is currently paying $125 per month house rent and can build a house for $12,000 which should have a life of 30 years. Taxes, insurance, and repairs will amount to $550 per year. This engineer has capital which is currently earning 6 percent on a mortgage which will be coming due very soon. Assume that the capital can be reinvested at 6 percent, and use the straight-line method for depreciation, allowing $3000 salvage. Which of the following most nearly equals the equivalent present worth of 30 years' rent at $125 per month, assuming that interest is 6 percent?

  (a) $15,870　　(b) $16,940　　(c) $18,930　　(d) $20,850
  (e) $21,670

The question can be restated to ask: "What is the present value of an annuity in which $125 is deposited at the end of each period with interest at $\frac{1}{2}$ percent per period, for 360 periods?"

The annuity, or sinking-fund, formula used previously is

$$R = P\,\frac{i(1 + i)^n}{(1 + i)^n - 1}$$

$$P = 125\,\frac{(1.005)^{360} - 1}{0.005(1.005)^{360}} = 125 \times 166.8 = \$20,850$$

The correct answer is (d).

Another way of stating this is that if $20,850 were invested in an annuity at 6 percent per year, that annuity would pay $125 at the end of each month for exactly 30 years. Thus $20,850 invested today at 6 percent would pay the rent of $125 per month for 30 years.

▪ Utilizing the data in the previous example, which of the following most nearly equals the equivalent present worth of 30 years' ownership of the house?

   (*a*) $13,600   (*b*) $14,200   (*c*) $14,700   (*d*) $15,200
   (*e*) $15,500

If the house were owned, $550 would have to be spent every year for taxes, insurance, and repairs. This would amount to an annuity with a present value of

$$\$550 \times 13.76 = \$7570$$

This means that of the $20,640 present worth of the rent annuity, $7570 would have to be used to set up a fund to return $550 each year for 30 years for expenses.

Another way of looking at this is to say that of the $1500 per year rental income $550 would have to be used for expenses, leaving a net of $950 to be deposited in the annuity at the end of each year. Either method will give the same value.

$$\$20,640 - \$7570 = \$13,070$$
$$\$950 \times 13.76 = \$13,072$$

In addition to the present worth of the net rental income annuity, there is also the estimated $3000 value of the house 30 years hence. The present worth of $3000 is

$$\frac{\$3000}{1.06^{30}} = \$522$$

The equivalent present worth of 30 years' owning the house is then

$$\$13,076 + \$522 = \$13,598$$

The correct answer is (*a*).

To pursue this example still further, we can determine which would be the most economical alternative, to rent or to build.

The present cost of owning the house for 30 years would equal the present cost of the house (of $12,000) plus the present cost of a sinking fund which would return $550 per year to cover the yearly expenses, less the present value of the $3000 future salvage value.

The present value of an annuity to provide $550 per year for 30 years would equal $550 × $(1.06^{30} - 1)/(0.06 \times 1.06^{30})$ = $550 × 13.766 = $7570.

The present value of $3000 thirty years hence = $3000/$1.06^{30}$ = $522, so the present cost of owning the house for 30 years, on the basis of yearly costs, equals $12,000 + $7570 − $522 = $19,048. This gives $20,640 − $19,048 = $1592 advantage in favor of owning the house, on the basis of present costs. This would appreciate to $1592 × 5.74 = $9140 at the end of the 30 years.

A comparison of yearly costs gives the following. The cost of borrowing $12,000 to be repaid in 30 yearly installments is $12,000/13.76 = $872 per year. From this would have to be subtracted the present yearly value of $3000 thirty years hence, which equals $3000/13.76 = $38 per year. Then the total cost of owning the house would equal $872 + $550 − $38 = $1384 per year. Comparison of this with the rental cost of $1500 per year gives a saving of $116 per year in favor of ownership: $116/year × $(1.06^{30} - 1)/0.06$ = 116 × 79.06 = $9170 future value, after 30 years. Another alternative is as follows: $1500 − $1522 = $78 per year; 78 × 79.06 = 6162, and $6162 future value plus $3000 equals $9162. (*Note:* These three values, $9140, $9170, and $9162, vary only because of rounding off.)

- It is desired to construct a bridge over a marshy area in a commercial recreation park. A wooden bridge would cost $8000 and would last an estimated 8 years. Maintenance costs would equal $700 per year. There would be no salvage value. The interest rate is 8 percent.

Utilizing these data, 10 individual problems are as follows:

- 1. Which of the following most nearly equals the yearly cost using the sinking-fund method?

  (*a*) $1875   (*b*) $2250   (*c*) $1680   (*d*) $2090   (*e*) $1920

- 2. Which of the following most nearly equals the yearly cost using straight-line depreciation and average interest?

  (*a*) $1800   (*b*) $1660   (*c*) $2060   (*d*) $2200   (*e*) $1980

- 3. Which of the following most nearly equals the first-year depreciation if the sum-of-the-years-digits method is used?

  (*a*) $1008   (*b*) $1576   (*c*) $1778   (*d*) $1876   (*e*) $1284

■ 4. Which of the following most nearly equals the amount which would have to be deposited in an account to pay for the $700-per-year maintenance costs over the 8-year life?

(a) $4022   (b) $3785   (c) $5600   (d) $4080   (e) $3628

■ 5. If the bridge is reinforced at the end of 5 years, the usable life could be extended an estimated 3 years (to 11 years total life). Which of the following most nearly equals the expense which could be justified to reinforce the bridge? (The same yearly maintenance costs would apply. Use sinking-fund cost.)

(a) $3026   (b) $2848   (c) $2972   (d) $2756   (e) $2932

■ 6. An alternative method of construction would be to make the bridge of prestressed concrete construction. The estimated life of such a structure would be 20 years. The first cost would be $15,000, and it would have a salvage value of $2000 at the end of its useful life. The yearly maintenance cost would be $200. Which of the following most nearly equals the total yearly cost using the sinking-fund method?

(a) $1684   (b) $2098   (c) $1875   (d) $1586   (e) $1714

■ 7. Which of the following most nearly equals the yearly cost of the prestressed concrete bridge using straight-line depreciation and average interest?

(a) $1722   (b) $1488   (c) $1684   (d) $1624   (e) $1556

■ 8. Which of the following most nearly equals the depreciation for the third year using the sum-of-the-years-digits method of depreciation?

(a) $1216   (b) $1114   (c) $1162   (d) $1098   (e) $1052

■ 9. Which of the following most nearly equals the depreciation for the third year if the fixed-percentage-on-diminishing-balance method is used?

(a) $1175   (b) $1150   (c) $1200   (d) $1135   (e) $1298

■ 10. Which of the following most nearly equals the rate of interest for which the yearly costs of the different structures in 1 and 6 would be the same?

(a) 6%   (b) 12%   (c) 10%   (d) 14%   (e) 16%

(1) The salvage value of the wooden bridge at the end of its useful life is zero, so there would have to be a total of $8000 plus the lost interest, or $8000(1.08)^8, at the end of 8 years. The present value of the sinking

fund is $8000 so the end-of-the-year deposits would equal

$$R = \frac{Si \times (1 + i)^n}{(1 + i)^n - 1} = \frac{\$8000 \times 0.08 \times 1.08^8}{1.08^8 - 1} = \$1392.12$$

To this must be added the maintenance cost of $700 per year, giving a total yearly cost of $2092.12. The correct answer is (d).

(2) Using straight-line depreciation, the yearly depreciation would equal $8000/8 = $1000 per year. The average yearly interest cost would equal

$$\frac{\$8000 + \$1000}{2} \times 0.08 = \$360 \text{ per year}$$

Total yearly costs would equal $1000 + $360 + $700 = $2060. The correct answer would be (c).

(3) The sum of the digits of an even number of years equals $(n + 1) \times n/2$, so the sum of the digits for 8 years would equal 9×4, or 36. The sum of the digits for an odd number of years is obtained with the same equation but is more easily understood if rearranged as $(n + 1)/2 \times n$. The sum-of-the-years digits for a life of 40 years, for example, equals 41×20, or 820. For 45 years the sum equals $(46/2) \times 45$, or 1035.

The sum-of-the-years digits for 8 years is 36. The first-year depreciation would then equal $\frac{8}{36} \times 8000 = \$1778$. The correct answer is (c).

(4) The amount required would be the amount necessary to set up a $700-per-year annuity with a life of 8 years (eight payments). This is the present value of a sinking fund. Thus

$$P = R \frac{(1 + i)^n - 1}{i(1 + i)^n} = \$700 \frac{(1.08)^8 - 1}{0.08(1.08)^8} = \$4022$$

The correct answer is (a).

(5) The capital-recovery-plus-a-return yearly payment for the original $8000 was found to equal $1392.12 in part 1 of the problem. This is the amount which would be paid each year for 3 additional years to pay for the original bridge plus the reinforcement.

At the end of 5 years the amount in the sinking fund would equal $1392.12 $\times$ $(1.08^5 - 1)/0.08$, or $8167.01. The original investment of $8000 would have appreciated to $8000 $\times$ 1.08^5, or $11,754.62, if it had been invested at 8 percent; so there would still remain $3587.61 to be paid off. That is, the book value of the bridge at the end of 5 years would equal $3587.61. The cost of reinforcement would be added to this to give a value of ($3587.61 + R) for the bridge at the end of 5 years, after it had been reinforced. This would appreciate to ($3587.61 + R) $\times$ 1.08^6

after 6 more years. This must equal the additional amount in the sinking fund resulting from deposits made at the end of the sixth through eleventh years. Thus

$$(\$3587.61 + R) \times 1.08^6 = \$1392.12 \times \frac{1.08^6 - 1}{0.08} = \$10,212.49$$

or

$$R = \frac{\$10,212.49}{1.08^6} - \$3587.61 = \$2847.99$$

which is the amount which could be justified for the reinforcement of the bridge. This assumes that the interest rate would remain constant at 8 percent for 11 years. The correct answer is (b).

(6) The depreciable cost equals $13,000. This is the capital which must be recovered. The $2000 would remain tied up throughout the 20-year life and would be recoverable at the end of that time. The present value of the sinking fund would then equal $13,000. The total cost would equal the capital-recovery-plus-a-return yearly payment for the $13,000 depreciable cost plus the interest lost on the $2000, which would be tied up for 20 years plus the yearly maintenance cost of $200:

$$R = P \frac{i(1 + i)^n}{(1 + i)^n - 1} = \$13,000 \frac{0.08 \times 1.08^{20}}{1.08^{20} - 1} = \$1324.08$$

The total yearly cost would equal

$$\$1324.08 + 0.08 \times \$2000 + \$200 = \$1684.08$$

The correct answer is (a).

(7) The depreciable cost is $13,000, so the straight-line depreciation would equal ($13,000/20), or $650 per year.

The average interest would equal

$$\frac{\$15,000 + \$2000 + \$650}{2} \times 0.08 = \$706.00 \text{ per year}$$

The total yearly cost would then equal $650.00 + $706.00 + $200.00 = $1556.00. The correct answer is (e).

(8) The sum-of-the-years digits would equal $(20 + 1) \times \frac{20}{2}$, or 210. The first-year depreciation would equal $\frac{20}{210} \times \$13,000$ or $1238. Second-year depreciation would equal $\frac{19}{210} \times \$13,000$, or $1176. And depreciation for the third year would equal $\frac{18}{210} \times \$13,000$, or $1114. The correct answer is (b).

(9) The first cost equals $15,000, and the salvage value at the end of

20 years would equal $2000:

$$\text{(First cost)} \times (1 - D)^n = \text{salvage value}$$

$$\$15,000 \times (1 - D)^{20} = \$2000$$

$$1 - D = 0.1333^{1/20} = 0.9042, \text{ giving } D = 0.0958, \text{ or } 9.58 \text{ percent}$$

The first-year depreciation would equal

$$0.0958 \times \$15,000 = \$1437.00$$

The amounts for depreciation for the next few years are:

Year	Book value	Depreciation
1	$15,000.00	$1437.00
2	13,563.00	1299.34
3	12,263.66	1174.86
4	11,088.80	1062.31
5	10,026.49	960.54

The correct answer is (a).

(10) The total cost for part 1 was found to equal

$$\$8000 \times \frac{i(1 + i)^8}{(1 + i)^8 - 1} + \$700$$

The total cost for part 6 was found to equal

$$\$13,000 \times \frac{i(1 + i)^{20}}{(1 + i)^{20} - 1} + \$2000 \times i + \$200$$

A table can be set up to compare the total costs of the wooden and concrete bridges for different interest rates.

Percent	Cost of wooden bridge per year	Cost of concrete bridge per year
8	$2092	$1684
9	2145	1804
12	2310	2180
14	2424	2442
13.75	2410	2410

The correct answer is (d).

9-8 CASH FLOW

Cash flow is the flow of money (cash) into and out of a business. It includes the money received from the sale of goods and services, and money paid out for business expenses. Accounts receivable represent assets, but they cannot be considered in the cash flow until they are collected. This is the reason for the existence of "factors," those who buy accounts receivable (at a discount) and then collect the money due.

As an example, assume that the Koe Company can save $5000 per year in materials-handling expenses by rearranging the storage area and materials-handling equipment in its warehouse. Rearrangement of the facilities will cost $37,000, and it is estimated that the savings will be realized for a total of 12 years. What would be the return on the investment?

Since no new equipment would be purchased, and no existing equipment would be repaired or modernized, there would be no depreciation charges. The $37,000 would constitute an operating expense, and the expense would all be incurred during the first year. The Koe Company pays a 50 percent tax on its income.

The cash flow the first year, assuming that a $5000 saving would be realized during the first year, would include the following:

Cash outlay for improvements	− $37,000
Savings in handling	+ 5,000
Net cash outlay before tax	− $32,000
Savings on income tax	+ $16,000
Cash flow for the first year after income tax	− $16,000

For the second year, there would be no additional expense and a savings of $5000 would be realized. Of this savings, 50 percent, or $2500, would have to be paid in taxes. The cash flow after income taxes for the second through the twelfth years would then equal + $2500 per year.

Thus the question becomes: If $16,000 is invested now for a return of $2500 per year for 11 years, what would be the rate of return? Or, if an annuity is purchased for $16,000 which will pay $2500 at the end of each year for 11 years, what is the rate of interest?

$$\$16,000 = \$2500 \frac{(1 + i)^{11} - 1}{(1 + i)^{11} \times i}$$

which gives an interest rate of 10.3 percent per year.

Sample Problems

9-1 To obtain a certain service, two plans are being considered. Plan A will require the purchasing of certain special equipment costing $12,000. The salvage value for this special equipment will be $500 regardless of the length of time it may have been used. Annual labor costs will average $3500. Plan B simply involves $5000 labor cost per year. All other incomes and expenses under the two plans will be the same. Allow 5 percent interest on capital, 5 percent compounded annually on amortization, or sinking fund for depreciation. Which of the following most nearly equals the number of years for the special equipment of plan A to pay for itself?

(a) 9.6 years   (b) 9.8 years   (c) 10.2 years   (d) 10.5 years
(e) 10.7 years

9-2 A young engineer's estimated annual earnings average $18,000, $25,000, and $37,000 per year in succeeding decades from the first job after graduation. Allow 9 percent interest compounded annually for both cost of money and return on capital. Which of the following most nearly equals the present worth (at graduation) in cash of the 30 years' earnings?

(a) $204,000   (b) $257,000   (c) $262,000   (d) $307,000
(e) $359,000

9-3 For the above-stated case, which of the following most nearly equals the equivalent uniform annual value of the 30 years' estimated income?

(a) $30,200   (b) $28,700   (c) $25,600   (d) $21,600   (e) $20,800

9-4 A piece of earth-moving equipment was purchased at a cash price of $25,000. The life of this equipment was estimated at 6 years with no salvage. However, at the end of 4 years the machine had become so inefficient, because of wearing of parts, that it was replaced. Depreciation was allowed on the company's books by the sinking-fund method with 4 percent interest. Which of the following most nearly equals the sunk cost at the time of replacement?

(a) $10,200   (b) $9000   (c) $8200   (d) $6400   (e) $4700

9-5 A certain type of automatic milling machine can be purchased for $6500. Expert estimates indicate a life of 10 years and a salvage value of $500 for this particular machine. Depreciation is calculated by means of the sinking-fund method with interest at 6 percent. Which of the following most nearly equals the *capital recovery plus a return* for this machine for the stated conditions?

(a) $845   (b) $852   (c) $858   (d) $864   (e) $873

9-6 An old light-capacity highway bridge may be strengthened at a cost of $9000, or it may be replaced by a new bridge of sufficient capacity at a cost of $40,000. The present net value of the old bridge is $13,000. It is estimated that the old bridge, when reinforced, will last for 20 years, with a maintenance cost of $500 per year and a salvage value of $10,000 at the end of 20 years. The estimated salvage value of the new bridge after 20 years of service is $15,000. The maintenance on the new bridge will be $100 per year. If interest is 6 percent, which of the following most nearly equals the difference in the yearly cost of the two alternatives? Use straight-line depreciation plus average interest.

(*a*) $865   (*b*) $880   (*c*) $895   (*d*) $925   (*e*) $960

9-7 A lathe costs $10,000 new. It has a life expectancy of 20 years and an estimated salvage value of $2000 at the end of 20 years. If interest is at 7 percent, which of the following most nearly equals the annual "capital-recovery-with-a-return" cost of the lathe on the basis of the preceding estimates? Use straight-line depreciation plus average interest in your calculations.

(*a*) $679   (*b*) $762   (*c*) $834   (*d*) $921   (*e*) $978

9-8 A company buys a machine for $12,000, which it agrees to pay for in five equal annual payments, beginning 1 year after the date of purchase, at an interest rate of 4 percent per annum. Immediately after the second payment, the terms of the agreement are changed to allow the balance to be paid off in a single payment at the end of the next year (end of the third year). Which of the following most nearly equals the final payment?

(*a*) $7970   (*b*) $7780   (*c*) $6450   (*d*) $5020   (*e*) $4860

9-9 Using the data of the above problem, which of the following most nearly equals the amount of the final payment if interest is 18 percent?

(*a*) $7780   (*b*) $7910   (*c*) $8670   (*d*) $9630   (*e*) $9845

9-10 A manufacturing plant has been purchasing the energy required for plant operation. It is considering building a power plant to supply a load estimated at 1000 kW (24 hours per day, 365 days per year). Purchased power will cost $0.023 per kilowatt-hour. The cost of the required plant is $1,000,000, and total operating expense, including interest on investment, is estimated at $95,000 per annum. Assuming 8 percent sinking-fund depreciation, 15-year life, and zero scrap value, which of the following most nearly equals the difference in the yearly costs of the two alternatives?

(*a*) $70,000   (*b*) $65,000   (*c*) $62,500   (*d*) $59,000   (*e*) $57,000

9-11 A manufacturer is planning to produce a new line of products which will require the purchase or rental of new machinery. A new machine will cost $17,000 and have an estimated value of $14,000 at the end of 5 years. Special tools for the new machine will cost $5000 and have an estimated value of $2500 at the end of 5 years. Maintenance costs for the machine and tools are estimated to be $200 per year. Which of the following most nearly equals the average annual cost of ownership during the next 5 years if interest is 6 percent? Assume that the annual cost equals straight-line depreciation plus average interest.

(*a*) $2410   (*b*) $2435   (*c*) $2460   (*d*) $2490   (*e*) $2510

9-12 Which of the following most nearly equals the amount that the owner of a building would be justified in paying for a sprinkler system that would save $500 per year in insurance premiums? The system will have to be renewed every 20 years and has a salvage value of 10 percent of its initial cost. Use sinking-fund analysis with interest at 5 percent.

(*a*) $6840   (*b*) $6690   (*c*) $6480   (*d*) $6320   (*e*) $6270

9-13 The owners of a concrete batching plant have in use a power steam shovel, which, if repaired at a cost of $2000, would last another 10 years. Maintenance costs, operating costs, taxes, and insurance would be $2500 per year. The present salvage value of the power steam shovel is $6000; if it is repaired and used for 10 years, the estimated salvage value at the end of that time is estimated to be $500. A new power shovel could be purchased at a cost of $30,000, and its estimated salvage value at the end of 10 years would be $20,000. The maintenance costs, operating costs, taxes, and insurance would be $1500 per year. Using straight-line depreciation and average interest, with interest at 6 percent, which of the following most nearly equals the difference of the two alternatives?
(a) $460 (b) $475 (c) $490 (d) $500 (e) $515

9-14 A debt of $10,000 with interest compounded annually at the rate of 4 percent is to be paid in a lump sum at the end of 5 years. To create a fund with which to pay the debt, the debtor decides to deposit equal sums at the end of each 6 months with a savings and loan association where the interest is compounded semiannually at an annual rate of 3 percent. Which of the following most nearly equals the amount of the semiannual deposit?
(a) $1126 (b) $1137 (c) $1145 (d) $1152 (e) $1161

9-15 An employer will furnish an automobile to certain employees or permit them to use their personal cars and will pay 20 cents per mile. It is assumed that a car would cost $12,350 and have a life of 5 years, with a trade-in value of $3000. There would be estimated costs of $75 per month for incidentals (including insurance); 3 cents per mile for oil, tires, and repairs; and 6 cents per mile for fuel. Which of the following most nearly equals the number of miles per year of travel required to justify the purchase of an automobile by an employee for use on the job? Use sinking-fund analysis with interest at 12 percent.
(a) 25,000 (b) 28,000 (c) 32,000 (d) 35,000 (e) 38,000

9-16 Utilizing the data in the above problem, which of the following most nearly equals the calculated amount of yearly travel required to justify the purchase of an automobile using a straight-line depreciation and average interest analysis?
(a) 25,000 (b) 28,000 (c) 32,000 (d) 35,000 (e) 38,000

Multiple-Choice Sample Quiz

For each question select the most nearly correct answer from the five given possibilities. For practice, try to complete the following 10 problems in 15 minutes or less.

9M-1 An individual wishes to deposit a certain quantity of money so that at the end of 5 years, at 4 percent interest, compounded semiannually, the account will have $500. The individual must deposit:
(a) $609.50 (b) $451.35

(c) $337.80   (e) none of these
(d) $410.15

9M-2 Money is invested at a nominal rate of interest of 5 percent per annum compounded semiannually. What is the effective rate per annum?
(a) 10 percent   (d) 5.06 percent
(b) 5.25 percent   (e) none of these
(c) 5 percent

9M-3 When a cost analysis of an engineering project is figured on a straight-line depreciation and an average interest basis, the average interest is:
(a) the interest for the first period plus the interest for the last period, divided by 2
(b) the interest figured on half the principal
(c) half the annual interest
(d) half the interest calculated at the end of the first period
(e) the sum of the interests for each period divided by 2

9M-4 The uniform annual end-of-year payment to repay a debt (the lender's investment) in n years, with an interest rate of i, is determined by multiplying the capital recovery factor by the:
(a) average investment
(b) initial investment, plus total interest
(c) average investment, plus interest
(d) initial investment, plus first year's interest
(e) initial investment

9M-5 The straight-line depreciation, plus average interest, is used to calculate the yearly cost of the investment in a machine because:
(a) It is a measure of the likelihood of bankruptcy.
(b) It is required by the Bureau of Internal Revenue in figuring income taxes.
(c) It is more accurate than the sinking-fund formula.
(d) It is easily understood and fairly accurate.
(e) It takes into account the "sunk cost" of the machine.

9M-6 The formula for determining average annual interest is

(a) $$(P - L)\frac{i}{2}\frac{n + 1}{n} + Li$$

(b) $$(P - L)\frac{1}{2}\frac{n + 1}{n}$$

(c) $$\frac{P + L}{2}i$$

(d) $$(P - L)\frac{1}{(1 + i)^n - 1} + Li$$

(e) $$(P - L)\frac{1}{(1 + i)^n - 1}$$

9M-7 The sum which must be paid yearly to pay a total sum at the end of a

given number of years and at a given rate of interest is the:
(a) sinking fund   (d) compound interest rate
(b) present worth   (e) discount rate
(c) capital recovery

9M-8   In the interest formula $A = P(1 + i/12)^{12n}$ the interest is compounded:
(a) daily   (d) semiannually
(b) monthly   (e) annually
(c) quarterly

9M-9   A sum of money invested at 4 percent compounded semiannually will double in amount in approximately:
(a) $15\frac{1}{2}$ years   (d) $18\frac{1}{2}$ years
(b) $21\frac{1}{2}$ years   (e) $19\frac{1}{2}$ years
(c) $17\frac{1}{2}$ years

9M-10 The present value of $5000.00 ten years hence with interest at 7.5 percent compounded annually most nearly equals:
(a) $2525.00   (d) $2500.00
(b) $2385.00   (e) $2400.00
(c) $2425.00

PROBLEM ANSWERS
AND SOLUTIONS

1-1 Answer (*e*). The problem can be sketched as shown in Fig. 1S-1.
$\alpha = \tan^{-1}(h/s)$ $d(\tan^{-1} x) = dx/(1 + x^2)$, where $x = h/s$
$d\alpha = (-hs^{-2})/[1 + (h/s)^2]\, ds = h/(s^2 + h^2)\, ds$ $d\alpha/dt = 2/(12.00 +$
4.00) $ds/dt = \frac{1}{8} \times 4$ mi/min $= 0.50$ rad/min $= 28.65°$/min   This
problem can also be solved by iteration. At $t = -0.001$ min, $s =$
3.4601 mi and $\alpha = 30.0287°$; at $t = +0.001$ min, $s = 3.4681$ mi and
$\alpha = 29.9714°$ $d\alpha/dt = 0.05731/0.002 = 28.65°$/min as before.

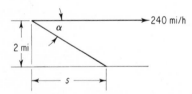

Figure 1S-1 $\alpha_0 = 30°$.

1-2 Answer (*e*). See Fig. 1P-2. The system is symmetrical, so $x^2 + y^2 = D^2$
$A = xy + y(x - y)$ $x = \sqrt{D^2 - y^2}$ Set $dA/dy = 0$
$D^4 - 5D^2y^2 + 5y^4 = 0$ $y = 0.851D$ or $0.526D$ (Solve for y^2,
using the general solution for a quadratic equation.) These round off to
$x = 0.85D$ and $y = 0.53D$.

1-3 Answer (*a*). See Fig. 1S-3. $BC = 1000 \cos 30°$ $BD = BC \sin 30°$

1-4 Answer (*b*). The general equation for a parabola is $y = ax^2 + bx + c$.
The coordinates of three points are given. Substitute in the general
equation and obtain three simultaneous equations. Solve for the three
unknowns: a, b, and c.

1-5 Answer (*a*). $A = bh$ $dA/dt = b(dh/dt) + h(db/dt)$

1-6 Answer (*c*). The general formula for a straight line is $y = mx + c$. The

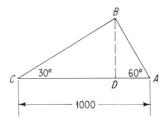

Figure 1S-3

coordinates of two points are given. Substitute for x and y, giving two equations which can be solved simultaneously for the two unknowns m and c.

1-7 Answer (c). $dA/dt = -6 \times 12 \times \frac{1}{8}$ in^2/s $\qquad dA/dt = 2\pi r\, dr/dt$
$dr/dt = -9/(2\pi \times 12) = -0.1193$ in/s, which rounds off to -0.12 in/s

1-8 Answer (e). See Fig. 1S-8. $400 \cos \alpha = 100 + 300 \sin \alpha$ $\qquad$ Square
and substitute $\cos^2 \alpha = 1 - \sin^2 \alpha$ $\qquad 25 \sin^2 \alpha + 6 \sin \alpha - 15 = 0$
$\alpha = 41.6°$ $\qquad$ Area $= 300 \times 100/\cos \alpha = 40{,}100$ ft^2

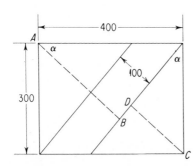

Figure 1S-8 $AB = 400 \cos \alpha$; $CD = 300 \sin \alpha$.

1-9 Answer (b). $\sec \theta = 1/\cos \theta$ $\qquad \cos^{-1}(1/1.40) = 44.4°$

1-10 Answer (a). See Fig. 1S-10. $\alpha = \sin^{-1} \frac{2}{6} = 19.47°$ $\qquad$ Area of sector of circle equals $(141.06/360) \times \pi r^2 = 44.31$ in^2. Area of triangle equals $2 \times 6 \cos \alpha = 11.31$ in^2. Area of liquid $= 44.31 - 11.31 = 33.00$ in^2 Volume of liquid $= 33 \times 60 = 1980$ in^3; $1980/231 = 8.57$ gal.

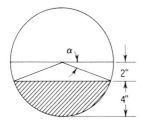

Figure 1S-10

1-11 Answer (c). $\sin x \, |_{\pi/2} = 0 - 1 = -1$

1-12 Answer (a). See Fig. 1S-12. $s^2 = x^2 + y^2$ $x = 12t$
$y = 32 - 15t$ $s = \sqrt{19.2t^2 - 2.60t + 2.78}$ $ds/dt =$
$(19.2t - 25.0)/\sqrt{t^2 - 2.60t + 2.78}$ When $t = 1$, $ds/dt =$
-5.34 mi/h

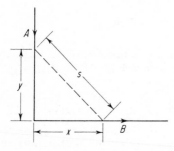

Figure 1S-12

1-13 Answer (a). $A = \int_{x=0}^{x=1} y \, dx = [wlx^2/4 - wx^3/6]_{x=0}^{x=1} = wl/4 - w/6$

1-14 Answer (d). $\sin^{-1}(0.25L - 0.75) - 3.989$, in radians   From a table
of integrals

1-15 Answer (a). Solve the two equations simultaneously
$y^2 - 9.231y - 6.769 = 0$. See Fig. 1S-15. Angle of curve $dy/dx =$
$-(x + 2)/(y - 3)$ Slope of curve $= \tan^{-1}(0.9938) = 44.82°$ or $\tan^{-1}$
$(2.450) = 67.803°$ Slope of line $= -\frac{2}{3} = \tan -33.69°$
$180 - (67.80 + 33.69) = 78.51°$ or $44.82 + 33.69 = 78.51°$

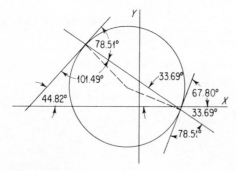

Figure 1S-15

1-16 Answer (d). See Fig. 1S-16. From the law of sines: $\sin C =$
$(AB/BC) \sin A$ $C = 18.24$ $B = 180 - (18.24 + 28.00) =$
$133.76°$ $AC = 1200 \times (\sin B)/(\sin A) = 1846$ ft

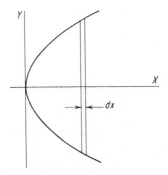

Figure 1S-16

1-17 Answer (b). The coordinates of the two points would be (1, 4) and (4, 8).
$y = mx + c$ Substitute the points in the equation for a straight line, and solve the two resulting equations simultaneously.

1-18 Answer (c). Differentiate the equation and solve for $dy/dx = (x^2/4a^2) - x/a$. The curve is parallel to the X axis when $dy/dx = 0$
$x = 4a^2/a = 4a$ $y = 16a/3 - 8a + \frac{5}{4} = -2.7a + 1.25$

1-19 Answer (c). See Fig. 1S-19. Volume $= \pi \int_0^6 y^2 \, dx = \int_0^6 8\pi x/dx = 4\pi x^2 \, |_0^6$

Y

X

dx

Figure 1S-19

1-20 Answer (c). See Fig. 1S-20. $L = (256/\sin \alpha) + (108/\cos \alpha)$
$dL/d\alpha = 256 \cos \alpha/\sin^2 \alpha - 108 \sin \alpha/\cos^2 \alpha = 0$ $108 \sin^3 \alpha = 256 \cos^3 \alpha$
$\alpha = \tan^{-1} \frac{4}{3} = 53.13°$

1-21 Answer (a). See Fig. 1S-21. Fence length $= L = 4A + 2B = 500$
$B = 250 - 2A$ $A^2 + B^2 = 12,500 = 62,500 - 1000A + 5A^2$
$A = 100$ $B = 50$

1-22 $\sec^2 \theta = 1/\cos^2 \theta = (\sin^2 \theta + \cos^2 \theta)/\cos^2 \theta = (\sin^2 \theta/\cos^2 \theta) + 1$, which equals $\tan^2 \theta + 1$ (*Note:* A problem of this type would not be given on a fundamentals examination since it is not a multiple-choice problem. It is included here for illustrative purposes.)

1-23 Answer (e).

$$\int_0^{\pi/2} \frac{\cos \theta \, d\theta}{1 + \sin^2 \theta} = \int_0^{\pi/2} \frac{d (\sin \theta)}{1 + \sin^2 \theta} = \tan^{-1} (\sin \theta) \Big|_{\theta=0}^{\theta=\pi/2} = \tan^{-1} 1 - \tan^{-1} 0 = \frac{\pi}{4}$$

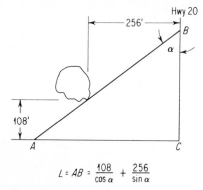

$$L = AB = \frac{108}{\cos \alpha} + \frac{256}{\sin \alpha}$$

Figure 1S-20

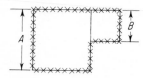

Figure 1S-21

1-24 Answer (a). $dL = \int_0^{x_1} \sqrt{(dx)^2 + (dy)^2}$ $dy = (dx/2)(e^{x/a} - e^{-x/a})$

$dx^2 + dy^2 = (dx^2/4)(e^{2x/a} + e^{-2x/a} + 2)$ $dL = \frac{1}{2}(e^{x/a} + e^{-x/a})\, dx$

1-25 Answer (e). See Fig. 1S-25. Angle $T = 180 - (105 + 30) = 45°$. Law of sines gives $TB = 70.7$ yd. Width of stream $= 70.7 \sin 75° = 68.3$ yd.

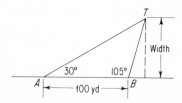

Figure 1S-25

1-26 Answer (c). At elevation 8 ft, diameter $= \frac{8}{20} \times 10 = 4$ ft $dh/dt = (dQ/dt)/A = 15/(4\pi) = 1.194$ ft/min, which rounds off to 1.2 ft/min

1-27 Answer (c). For $R = 0$, $e^{\tan x} = 1/\infty$, so $\tan x \to -\infty$, which is true for $x = -\pi/2$ or for $x = \pi/2$ if approach is from the right-hand side.

1-28 Answer (c). See Fig. 1S-28. Use law of cosines: $c^2 = 900 + 324 - 1080 \cos 45° = 460.32$ $(460.32)^{1/2} = 21.455$, which rounds off to 21.5 mi

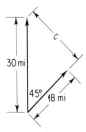

Figure 1S-28

1-29 Answer (c). $\log_4 (3x - 5) = 2$ $\quad 3x - 5 = 4^2 \quad x = 7$

1-30 Answer (b). See Fig. 1S-30. Use law of sines: $(\sin 35°20'/84 =$
$(\sin B)/48$ $\quad B = 19.30°$ $\quad C = 180 - (35.33 + 19.30) = 125.37°$
$(\sin 125.37°)/AB = (\sin 35.33°)/84 \quad AB = 118.4$ ft

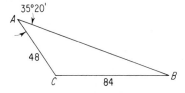

Figure 1S-30

1-31 Answer (b). See Fig. 1S-31.

$$A = \frac{2l + 2l + 2l \cos \alpha}{2} \, l \sin \alpha = 2l^2 \sin \alpha + l^2 \sin \alpha \cos \alpha$$

$dA/d\alpha = 2l^2 \cos \alpha + l^2 \cos^2 \alpha - l^2 \sin^2 \alpha = 0$ $\quad \sin^2 \alpha = 1 - \cos^2 \alpha$
$dA/d\alpha = 2 \cos^2 \alpha + 2 \cos \alpha - 1$ $\quad \cos \alpha = 0.366 \quad \alpha = 68.53°$
$180 - 68.53 = 111.47$ or $111.5°$

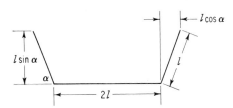

Figure 1S-31

1-32 Answer (c). $(1 - 0.35)^n = 0.50$ $\quad n = 1.609$ h $\quad$ Or $dA/dt = kA$
$\ln A = kt + c$ $\quad$ Or $A = A_0 \, e^{kt}$ $\quad A/A_0 = 0.65$ at $t = 1$

$0.65 = e^k$ $k = -0.4308$ For $A/A_0 = 0.5 = e^{-0.4308t}$
$t = 1.609$ h, which rounds off to 1.6 h

1-33 Answer (c). y approaches infinity as x approaches infinity; there is
no maximum ordinate. Minimum ordinate occurs when $dy/dx = 0$
$0 = 6x - 2$ $x = \frac{1}{3}$ and $y = -\frac{31}{3}$.

1-34 Answer (c). See Fig. 1S-34. Slope of curve = slope of tangent
$dy/dx = 3x^2/100 - \frac{5}{4}$ At $x = 5$, $y = -5$ and slope $= -0.5$
$-5 = -0.5 \times 5 + c$ $c = -2.5$ Equation of tangent is $y =$
$-0.5x - 2.5$ At $x = 0$, $y = -2.5$ $A = (2.5 + 5)/2 \times 5 =$
18.75, which rounds off to 18.8

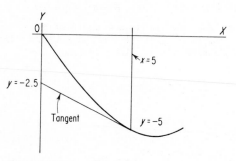

Figure 1S-34

1-35 Answer (b). $A = \pi R^2 \times 2 + 2\pi RL$ Volume $= 10,000 = \pi R^2 L$
$L = 10,000/(\pi R^2)$ $A = 2\pi r^2 + 2\pi R \times 10,000/(\pi R^2)$ $dA/dR =$
$4\pi R - 20,000/R^2 = 0$ $R = 11.675$ ft, which gives $D = 23.4$ ft and
$L = 23.4$ ft

1-36 Answer (d). See Fig. 1S-36. $OD = 190 \cos 20° = 178.54$
$AB = (190 + OD)/\cos 10° = 374.2$ ft; or, from the law of cosines,
$(AB)^2 = 190^2 + 190^2 - 2 \times 190^2 \cos 160°$

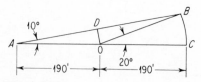

Figure 1S-36

1-37 Answer (c). Use the rule of l'Hospital.

$$\lim_{x \to 0} y = \frac{5x^3 - \cos x + e^x}{x^2 + 2x} = \lim_{x \to 0} \frac{15x^2 + \sin x + e^x}{2x + 2} = \frac{1}{2}$$

1-38 Answer (c).

$$\int_0^{\pi/2} \sin\theta\cos\theta\,d\theta = \int_0^{\pi/2} \sin\theta\,(d\sin\theta) = \frac{\sin^2\theta}{2}\bigg|_0^{\pi/2} = \frac{1}{2}$$

1M-1 Answer (d). $\log\frac{1}{2} = \log 1 - \log 2 = 0 - 0.301030 = 9.69897 - 10$

1M-2 Answer (b). $\sin A = \cos A\,(\sin A/\cos A) = \sin A$

1M-3 Answer (e). A negative number cannot have a real-number logarithm. There is no exponent which can make e^x negative.

1M-4 Answer (c). $\tan(A + B) = (\tan A + \tan B)/(1 - \tan A \tan B) = \frac{7}{12}/(1 - \frac{1}{12}) = \frac{7}{11}$ or $\tan(18.43° + 14.04°) = \frac{7}{11}$

1M-5 Answer (d). $8^{4/3} = 2^4 = 16$

1M-6 Answer (a). $A^{0.250} = 10$, so $0.250\log A = \log 10 = 1$ $\quad\quad \log A = 4$

1M-7 Answer (a). $\csc 960° = \csc(960° - 720°) = \csc 240° = 1/(\sin 240°)$
$\sin 240° = -\sin(240° - 180°) = -\sin 60° = -\sqrt{3}/2$
$\csc 960° = -2/\sqrt{3} = -2\sqrt{3}/3$

1M-8 Answer (c). $i^{27} = (i^2)^{13}i = (-1)^{13}i = -i$

1M-9 Answer (d). By definition, $B^x = N$.

1M-10 Answer (c). $(x - 6)(x + 4) = x^2 - 2x - 24$

2-1 Answer (d). See Fig. 2S-1. $\alpha = \tan^{-1}\frac{2}{1} = 63.4°$ $\quad\quad \mu = 0.50$ (from handbook) $\quad F_x = 447\cos\alpha = 200$ lb $\quad F_y = 447\sin\alpha = 400$ lb
Frictional resistance to motion $= 0.50 \times 200 = 100$ lb
Maximum weight $= 400 + 100 = 500$ lb $\quad\quad$ Minimum weight $=$
$400 - 100 = 300$ lb

Figure 2S-1

2-2 Answer (c). See Fig. 2S-2. $\alpha = \cos^{-1}\frac{20}{25} = 36.87°$ $\quad\quad L_1\sin\alpha =$
$L_2\sin\alpha + 5$ $\quad\quad L_1 + L_2 = 25$ $\quad\quad L_1 = 16.67$ ft $\quad\quad L_2 = 8.33$ ft
$x = L_1\cos\alpha = 13.33$ ft

2-3 Answer (a). $\bar{x} = (50 \times 5 - 12.57 \times 7)/(50 - 12.57) = 4.33$ in

2-4 Answer (c). Work per revolution $= 12F$ ft·lb $\quad\quad$ Work also $=$
$2300 \times 0.125/12 = 24$ ft·lb $\quad\quad F = 2.00$ lb $\quad\quad$ See Fig. 2S-4.

2-5 Answer (d). See Fig. 2S-5. $\Sigma M_0 = 0 = 6R - 2 \times 10,000 - 4 \times 15,000$
$R = 13,333$ lb $\quad\quad \Sigma F_y = 0 = 13,333 - U_4L_4$ $\quad\quad U_4L_4 = 13,333$ lb T

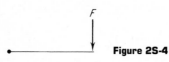

Figure 2S-2

Figure 2S-4

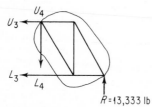

$R = 13,333$ lb   **Figure 2S-5**

2-6 Answer (a).   For the purposes of illustration, the forces in all three members will be calculated. See Fig. 2S-6.

			Components				
	Length	X	X/L	Y	Y/L	Z	Z/L
AB	6.00	6	1.00	0	0	0	0
AC	11.18	−6	0.537	8	0.716	5	0.447
AD	12.88	−6	0.466	11	0.854	−3	−0.233

$\Sigma F_x = 0 = AB - 0.537AC - 0.466AD$ $AB = 607.8$
$\Sigma F_y = 0 = 0.716AC + 0.854AD - 1000$ $AC = 424.7$
$\Sigma F_z = 0 = 0.447AC - 0.233AD$ $AD = 814.7$

Figure 2S-6

2-7   Answer (d).   See Fig. 2S-7. $T \cos 30° = (10 + T \sin 30°) \times 0.40$
$T = 6.00$ lb $P = 6.00 \cos 30° + (10 + 3 + 20) \times 0.15 = 10.146$ lb

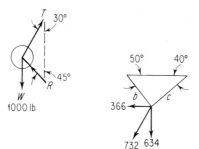

Figure 2S-7

2-8   Answer (c).   See Fig. 2S-8. $T/\sin 45° = R/\sin 30° = 1000/\sin 105°$
$T = 732$ lb, $R = 518$ lb $b \sin 50° + c \sin 40° = 634$
$b \cos 50° + 366 = c \cos 40°$ $c = 687.60$ lb $b = 250.49$ lb

Figure 2S-8

2-9   Answer (a).   See Fig. 2S-9.

$$\bar{y} = \frac{(3 \times 8 \times \frac{1}{2}) \times 2 + 4 \times 8 \times 5 - \pi/2 \times 9 \times (7 - 4/\pi)}{3 \times 8 \times \frac{1}{2} + 4 \times 8 - \pi/2 \times 9} = 3.45 \text{ in}$$

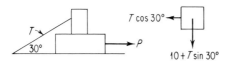

Figure 2S-9

2-10   Answer (b).

$$\bar{y} = \frac{(8 \times 8 \times 12) \times 6 - (\pi R^2 \times 6) \times 9}{8 \times 8 \times 12 - \pi R^2 \times 6} = 5$$

$4608 - 169.65R^2 = 3840 - 44.25R^2$ $R = 3.19$ in   Diameter =
$2 \times 3.19 = 6.38$

2-11 Answer (c). See Fig. 2S-11. $\Sigma M_a = 0 = 10 \times 3000 \times \cos 30° - 30 \times$
3000 $\sin 30° - 20 \times 1000 + 50F_y$ $F_y = 780.4$ lb $\Sigma F_x = 0 =$
$A_x - 3000 \cos 30°$ $A_x = 2598$ lb $A_y = 3000 \sin 30° + 1000 -$
$780.4 = 1719.6$ From $\Sigma F_y = 0 = 1719.6 - 1000 - CH \cos 45°$
$CH = 1017.7$ lb T

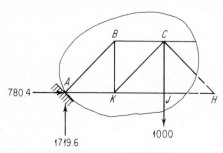

Figure 2S-11

2-12 Answer (d). $\Sigma M_0 = 0$ $C_x = 12 \times \frac{2000}{16} = 1500$ lb $\angle A = \tan^{-1}$
$\frac{2}{1} = 63.43°$ $\frac{1500}{2} = AC \sin A$ $AC = 838.5$ lb

2-13 Answer (e). Equivalent work $FS = WS \tan \alpha$ $F = 10F \tan \alpha$
$\alpha = 5.7°$

2-14 Answer (b). See Fig. 2S-14. $\Sigma F_y = 0 = 3000 - 2000 - L_1 U_2 \sin$
$26.57°$ $L_1 U_2 = 2235.7$ lb C

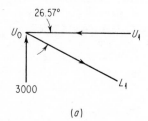

(a)

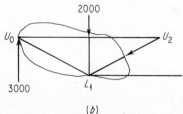

(b)   **Figure 2S-14**

2-15 Answer (d). $2T \sin \theta \ 4000$ $\sin \theta = \frac{2000}{2600}$ $\theta = 50.28°$

2-16 Answer (a). See Fig. 2S-16. $\alpha = \cos^{-1} \frac{50}{60}$ $\alpha = 33.56°$
$2T \sin \alpha = 5000$ $T = 4522.7$ lb   No torque can be withstood by
the pulley, so tension will be the same throughout the length of the line.

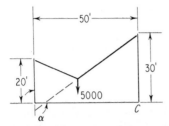

Figure 2S-16

2-17 Answer (a).

$$x = \frac{-(4\pi/2)(4x^2/3\pi) + 32 \times 4 + 4 \times \frac{1}{2} \times (8 + \frac{4}{3})}{-(4\pi/2) + 8 \times 4 + 4 \times \frac{1}{2}} = 5.80 \text{ in}$$

Semicircle $I = (0.1089 \times 2^4 + 2\pi \times 4.95^2 = 155.71$

Triangle $I = (4 \times 4^3/36) + 8 \times 3.53^2 = 106.80$

Rectangle $I = (4 \times 8^3/12) + 32 \times 1.80^2 = 274.35$

$I_{CG} = 106.80 + 274.35 - 155.71 = 225.44 \text{ in}^4$

2-18 Answer (c). $I_{x-x} = (3 \times 12^3/12) + 4[(4 \times 4^3/12) + 16 \times 4^2] = 1541.3 \text{ in}^4$

2-19 Answer (d). See Fig. 2S-19. $LM = 9.52/\sin 30° = 19.04$ kips $LP =$
$LM \cos 30° = 16.49$ kips $MP = 0$ (pin joint as M cannot take any
torque) $NP = 0$ from $\Sigma F_y = 0$ at point P $PQ = LP = 16.49$
kips   Force in $a = 16,490$ lb T

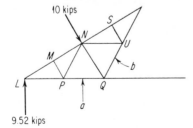

10 kips

9.52 kips

Figure 2S-19

2-20 Answer (c). See Fig. 2S-20. The force in the various members are shown
in the figure. The force in member A is 20,000 lb T

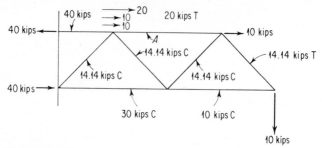

Figure 2S-20

2-21   Answer (a).   See Fig. 2S-21. For the purpose of illustration, calculate for the case of the tractor on level ground as well as on a slope. Analyze at the moment the front wheel lifts off the ground. Assume free body.
$\Sigma M_a = 0 = W \times 4 - 1.5X$ $X = 2.67W$ For the case of the 20° slope $3 \cos 20° - 4 \sin 20° = 1.45$ $\Sigma M_A = 0 = 1.45W - 1.50X$
$X = 0.967W$

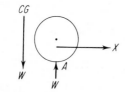

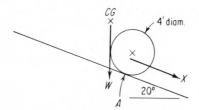

Figure 2S-21

2-22   Answer (a). $\Sigma M_A = 0 = 12 \times P/4 + 6 \times P/2 - 12 \times P/2 - 6P + 17B_y$ $B_y = 0.353P$ $\Sigma M_B = -6 \times P/2 - 12 \times P/4 + 6P + 12 \times P/2 - 17A_y$ $A_y = 0.354P$ At point A, $\Sigma F_y = 0 = 0.353P - 0.5P \times 0.707 + 0.707AE$, which gives $AE = 0$.

2-23   Answer (c).   See Fig. 2S-23. Take moments about the point of contact. The maximum moment will occur when P is at an angle of 90° to the radius. $\beta = \sin^{-1}(r - h)/r$ $\alpha = 90 - \sin^{-1}(1 - h/r)$ or

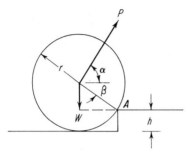

Figure 2S-22

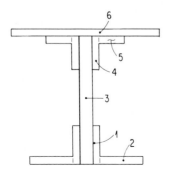

Figure 2S-23

$$\alpha = \sin^{-1}[(1/r)(2rh - h^2)^{1/2}] \qquad \alpha = 48.19° \qquad \Sigma M_A = 0 = -Pr +$$
$$Wr \cos 48.19° \qquad P = 66.7 \text{ lb}$$

2-24   Answer (b).   See Fig. 2S-24.

$$y = \frac{2(3 \times 0.5 \times 1.5) + 2 \times (3.5 \times 0.5 \times 0.25) + 1 \times 10 \times 5 + (2.5 \times 0.5 \times 8.75) +}{2 \times 3 \times 0.5 + 2 \times 3.5 \times 0.5 + 1 \times 10 + 2 \times 2.5 \times 0.5 +}$$
$$\frac{2(2 \times 0.5 \times 9.75) + 0.5 \times 12 \times 10.25}{2 \times 2 \times 0.5 + 0.5 \times 12} = \frac{158.25}{27} = 5.86 \text{ in}$$

Figure 2S-24

(1) $[(0.5 \times 3^3/12) + 1.5 \times 4.36^2] \times 2$ $I_{CG} = 59.28$
(2) $2 \times (1.75 \times 5.61^2)$ $I_{CG} = 110.15$
(3) $(1 \times 10.25) + 10 \times 0.86^2$ $I_{CG} = 90.73$
(4) $(0.5 \times 2.5^3/12 + 1.25 \times 2.89^2) \times 2$ $I_{CG} = 22.18$
(5) $(1 \times 3.89^2) \times 2$ $I_{CG} = 30.26$
(6) $0.5 \times 12 \times 4.39^2$ $I_{CG} = 115.63$

$$I_{CG} = 428.23 \text{ in}^4$$

2-25 Answer (a). See Fig. 2S-25. A 3-4-5 triangle. $\alpha = 36.87°$
$\beta = 53.13°$ From $\Sigma F_y = 0$ $T_1 \sin \alpha + T_2 \sin \beta = 200$
From $\Sigma F_x = 0$ $T_1 \cos \alpha = T_2 \cos \beta$ $T_1 = 120$ lb $T_2 = 160$ lb

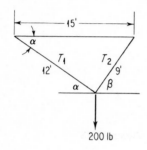

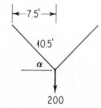

Figure 2S-25

2-26 Answer (a). See Fig. 2S-26. $\Sigma M_{L,8} = 0 = 12 \times 20$ kips $+ 44 \times$ 12 kips $- 16 \times 16$ kips $- 56 \times (L_0)_y$ $(L_0)_y = 9.14$ kips
For the section shown in the figure: $\Sigma F_y = 0 = 9.14$ kips $-$ 12 kips $+ U_2L_4 \sin 45°$ $U_2L_4 = 4.045$ kips C

2-27 Answer (b). See Fig. 2S-27. $T = \frac{50}{2} = 25$ lb $2T \sin \alpha = 30$
$\alpha = 36.87°$ $L_1 \cos \alpha + L_2 \cos \alpha = 8$ $L_1 + L_2 = 10$
$L_1 \sin \alpha = L_2 \sin \alpha + 2$ $L_1 = L_2 + 3.333$ ft $L_1 = 6.67$ ft
$L_2 = 3.33$ ft $L_1 \sin \alpha = 4.00$ ft

2-28 Answer (d). $\Sigma M_A = 0 = 50 \times 0.5 - B_y \times 1$ $B_y = 25$ lb
Frictional force $= [25 + (25 + 3)] \times \frac{1}{3} = 17.67$ lb $= P$

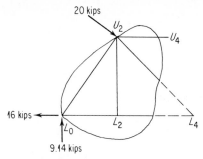

Figure 2S-26

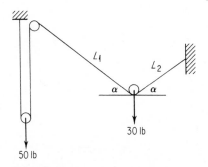

Figure 2S-27

2-29 Answer (*e*). See Fig. 2S-29. $I = 0.5 \times 11^3/12 + 4 \times (0.5 \times 3^3/12 + 0.5 \times 3 \times 4.5^2) + 4 \times (0.5 \times 2.5 \times 5.75^2) = 346.8$ in^4

2-30 Answer (*a*). See Fig. 2S-30. $\Sigma F_x = 0 = F - C \cos 45° - B \cos 45°$
$\Sigma F_y = 0 = -100 - 40 + C \sin 45° + B \sin 45°$ $F = 140$ lb

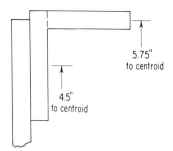

5.75"
to centroid

4.5"
to centroid

Figure 2S-29

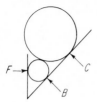

Figure 2S-30

2-31   Answer (d).   See Fig. 2S-31. $\Sigma F_x = 0 = F - R_1\mu - R_2\mu$
$F = (R_1 + R_2) \times 0.20$ $\Sigma M_0 = 0 = R_1 d + R_2 d - FX$
$FX = 4 \times (R_1 + R_2)$ $FX = 4 \times 5F$ $X = 20$ in

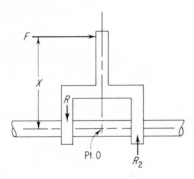

Figure 2S-31

2-32   Answer (e). $1/k = \frac{1}{5} + \frac{1}{10} + \frac{1}{15}$ $k = 2.73$ lb/in

2-33   Answer (c).   See Fig. 2S-33. Assume a specific weight of 1, and work
with volumes: $M = \int x \, dv = \int x\pi r^2 \, dx$ $r = 4 + (x/144) \times 2 =$
$4 + x/72$ $r^2 = 16 + x/9 + x^2/5184$ $M = \pi \int_0^{144} (16 + x/9 +$
$x^2/5184)x \, dx = 933{,}731$ in^3 $\times$ in $V = \pi \int_0^{144} (16 + x/9 + x^2/5184)$
$dx = 11{,}460.5$ in^3 $\bar{x} = 933{,}731/11{,}460.5 = 81.47$ in or 6.79
ft   Moment about small end $= 933{,}731$ in^3 $\times$ in $M = \frac{2}{3}Vl =$
$\frac{2}{3} \times 11{,}460.5l$ $l = 122$ in or 10.17 ft

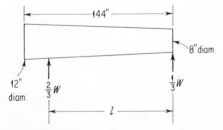

Figure 2S-33

2-34 Answer (b). Referring to Fig. 2P-34, denote the left end as point L and the right end as point R. $\Sigma M_L = 0 = -50 \times 9 - 25 \times 24 + 30 R_y$ $R_y = 35.00$ $F_2 = 35.00/\sin 60° = 40.41$ lb $R_x = 40.41 \cos 60° = 20.20 = H_2$ $L_y = 40.0$ lb $L_x = -20.20$ lb $\theta_1 = \tan^{-1} 40/20.2 = 63.2°$ $F_1 = 40^2 + 20.2^2 = 44.8$ lb

2-35 Answer (b). See Fig. 2S-35. $\Sigma M_{b-c} = 0 = T \times 10.5 - 16,000 \times 10.5$ $T = 16,000$ lb   Force along line $ed = \sqrt{13,856^2 + 24,000^2} = 27,710$ lb $F_{bd} = (27,710/2)/\sin 60° = 16,000$ lb C

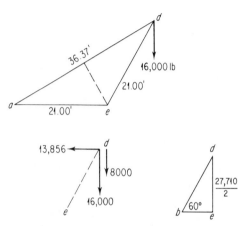

Figure 2S-35

2M-1 Answer (c). See Fig. 2M-1.

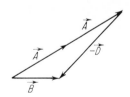

Figure 2M-1

2M-2 Answer (a). $M_B = 0 = 100 \times 0 + A_y \times 10$ $A_y = 0$

2M-3 Answer (c). By definition, 1 N $= 10^5$ dyn   dyn $= $ g $\times$ cm/s^2 N $= $ kg $\times$ m/s$^2 = 1000$ g $\times 100$ cm/s^2

2M-4 Answer (e). $F_x = 0$ $F_y = 0$ $M = 0$

2M-5 Answer (d). Resultant.

2M-6 Answer (c). $100 = T$ $2T \cos \theta = 50$ $\theta = 75.5°$

2M-7 Answer (d). $T = 50$ F_y = force up $- 2T - 50 = 0$ Force up = 150 lb

2M-8 Answer (c). See Fig. 2M-8. $T = 100/\cos 30° = 115$ lb

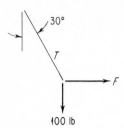

100 lb   **Figure 2M-8**

2M-9 Answer (d). See Fig. 2M-9.   Force parallel to face of plane
$F = 80 \times \sin 30° = 40$ lb

30°   **Figure 2M-9**

2M-10 Answer (a). See Fig. 2M-10. $2T \sin 60° = 100$ $T = 57.7$ or 58 lb

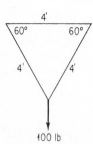

100 lb   **Figure 2M-10**

3-1 Answer (d). $F = l - (18 - l) - 1 = 2l - 19$ $F = ma$
$a = dv/dt = v\, dv/ds$ $2l - 19 = (18/g)v\, dv/dl$ $2l\, dl - 19\, dl = (18/g)v\, dv$ $l^2 - 19l = (9/g)v^2 = C$ $l = 10$ at $v = 0$, so
$C = -90$ $v^2 = (g/9)(l^2 - 19l + 90)$ $v = dl/dt = (\sqrt{g}/3)\sqrt{l^2 - 19l + 90}$ $dt = (3/\sqrt{g})\, dl/\sqrt{l^2 - 19l + 90}$
$t = (3/\sqrt{g}) \ln (2l - 19 + 2\sqrt{l^2 - 19l + 90}) = 1.865$ s

3-2 Answer (a). $M_1v_1 = (M_1 + M_2)v_r$ $v_1 = \sqrt{2gh} = 35.84$ ft/s
$(1000/g) \times 35.84 = (1800/g)v_r$ $v_r = 19.91$ ft/s $F = ma$
$F = 30{,}000 - 1800 = (1800/g)a$ $a = 504.0$ ft/s^2 $v^2 = v_0^2 +$
$2as$ $0 = 19.91^2 - 2 \times 504.0s$ $s = 0.393$ ft or 4.719 in

3-3 Answer (d). Work $= \int F\,ds = \int_{-1}^{1}(x^3 - x)\,dx = [(x^4/4) - (x^2/2)]_{x=-1}^{x=1} = 0$

3-4 Answer (b). $a = 12$ in $b = 6$ in $W = 20$ lb $m =$
0.61 slugs $d^2y/dt^2 + (k/m)y = 0$ $\omega^2 = k/m$ $F = ma$
See Fig. 3S-4. $d^2y/dt^2 = F/m$ $F = -ky = -(a/b)k_{sp}y'$
Force on weight $= -(a/b)^2k_{sp}y$ $k = (a/b)^2k_{sp}$ $\tau = 15$ s/10
cycles $= 1.5$ s $\tau = 2\pi\sqrt{(m/k_{sp})(b/a)^2}$ $k_{sp} = 2.72$ lb/ft

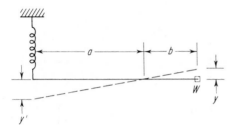

Figure 3S-4

3-5 Answer (e). $v = v_0 + at$ $a = F/m$ $m = 3220/32.2 = 100$
slugs $a = (8 - 1.2t)$ ft/s^2 $\int dv = \int a\,dt$ $\Delta v = \int_0^5 (8 -$
$1.2t)\,dt = [8t - 0.60t^2]_{t=0}^{t=5} = 25$ ft/s $v = v_0 + \Delta v = 35$ ft/s

3-6 Answer (c). $I = mk^2 = (64.4/32.2) \times 0.75^2 = 1.125$ slug·ft^2 ΔPE
$= \Delta KE$ $KE = \frac{1}{2}mv^2 + \frac{1}{2}I\omega^2 = \frac{1}{2}mv^2 + \frac{1}{2}I\,(v^2/r^2) = 1.5625v^2 = 10$
mg (ΔPE) $v = 20.3$ ft/s

3-7 Answer (a). $W = 60$ lb $m = 1.865$ slug $T = I\alpha = k\theta$
$k =$ spring constant $I = \int_0^3 r^2\,dm = \int_0^3 r^2\,(20/g)\,dr = 180/g$
$k = 1.91$ lb·ft/rad $= \frac{1}{30}$ lb·ft/deg   See Fig. 3S-7. Tq $= I\alpha =$

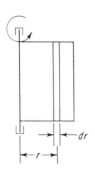

Figure 3S-7

$5.595\omega \ (d\omega/d\theta)$ $\mathrm{Tq} = -k\theta = -1.91\theta$ $5.595\omega \ d\omega/d\theta =$
$-1.91 \ \theta \ d\theta$ $2.929\omega^2 = -\theta^2 + C$ $\omega = 0$ at $\theta = \pi/2$
$C = 2.467$ $2.929\omega^2 = 2.467 - \theta^2$ At $\theta = \pi/4$, $\omega = 0.795$ rad/s

3-8 Answer (d). $S = \frac{1}{4}t^4 - 2t^3 + 4t^2$ $v = ds/dt = t^3 - 6t^2 + 4 =$
$t(t - 4)(t - 2)$ $v = 0$ at $t = 0$, $t = 4$, and $t = 2$ At $t = 3$, $v =$
-1 ft/s   Thus the body moves backward between $t = 2$ s and
$t = 4$ s.

3-9 Answer (b). $R = 1000$ ft, 40 mi/h$= 58.67$ ft/s. See Fig. 3S-9. $mv^2/R =$
$3.44m$ $\alpha = \tan^{-1}(3.44/g) = 6.10°$ Superelevation $=$
$60 \sin \alpha = 6.38$ in $= y$ $\Sigma M_A = 0 = 30W - 48(v^2/1000)(W/g)$
$v = 141.8$ ft/s or 96.7 mi/h

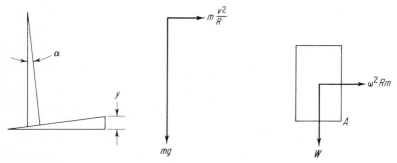

Figure 3S-9

3-10 Answer (e). $T = 200 \times 8 \times \frac{1}{4} = 400$ lb $\mathrm{Tq} = (400 - 300) \times 0.5$
$= 50$ ft·lb   See Fig. 3S-10.   Horsepower $= 2\pi N \ \mathrm{Tq}/33,000 =$
$(2\pi \times 1650 \times 50)/33,000 = 15.7$ hp

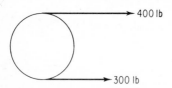

Figure 3S-10

3-11 Answer (a). $T_{\max} = (\pi/4) \times 1.5^2 \times 40,000 = 70,686$ lb $70,688 =$
$32.2m + 8.05m = 1.25$ mg   See Fig. 3S-11. Weight $= mg = 70,688/$
$1.25 = 56,549$ lb

3-12 Answer (c). $S_0 = \sqrt{150^2 - 120^2} = 90$ 4 mi/h $= 5.867$ ft/s
$L = \sqrt{120^2 + (90 + 5.867t)^2}$ See Fig. 3S-12. $dl/dt = (1056.06 +$
$68.82t)/(2 \times \sqrt{22,500 + 1056.06t + 34.42t^2}$ At $t = 0$, $dL/dt =$
$1056.06/300 = 3.52$ ft/s or 3.5 ft/s

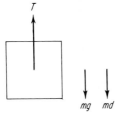

mg   md   **Figure 3S-11**

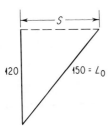

Figure 3S-12

3-13 Answer (d). $h = v_y^2/2g = 2700 \sin 45°/2g = 56{,}650$ ft
$v = v_0 + at$ $v_y = v_0 - gt$ $v_x = 2700 \cos 45° = 1908.9$ ft/s $=$
$v_{y,0}$ $t = v_{y,0}/g = 2700 \sin 45°/g = 59.38$ s   See Fig. 3S-13.
Total time of flight $= 2 \times 59.38 = 118.76$ s $S = v_x t =$
$2700 \cos 45° \times 59.38 = 226{,}700$ ft or $227{,}000$ ft

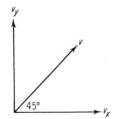

Figure 3S-13

3-14 Answer (e). $\sqrt{(240 - 28.28)^2 + 28.28^2} = 213.60$ mi/h ground speed
See Fig. 3S-14.

3-15 Answer (a). $(2900 \times 12)(\text{in/s})/10(\text{in/r}) = 3480$ r/s $= 208{,}000$ r/min

3-16 Answer (b). Use relative velocities $U^2 = U_0^2 + 2as$
60 mi/h $= 88$ ft/s   35 mi/h $= 51.33$ ft/s $0 = (v_1 - v_2)^2 +$
$2a \times 100$ $a = -36.67^2/200 = -6.72$ ft/s^2 $a = 6.72/g = 0.209g$
which rounds off to $0.2g$

3-17 Answer (d). $F = W - W \times v^2/256$ $F = ma = (W/g)\,dv/dt$

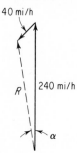

40 mi/h

R 240 mi/h

α

Figure 3S-14

$dv/dt = (g/256)(256 - v^2)$ $\qquad$ $dv/(v^2 - 256) = -(g/256)\,dt$
From table of integrals $\qquad$ $\int dx/(x^2 - a^2) = 1/2\alpha \times$
$\ln[(x - a)/(x + a)] + C$ $\qquad$ $\frac{1}{32} \times \ln[(v - 16)/(v + 16)] = -(g/256) +$
C $\quad$ or $\quad$ $(v - 16)/(v + 16) = Ce^{-4.021t}$ $\qquad$ $v = 176$ at $t = 0$
so $C = 0.8333$ $\qquad$ for $v = 20$ ft/s $\qquad$ $\frac{4}{36} = 0.8333e^{-4.021t}$
$t = 0.501$ s

3-18 Answer (c). Impulse = momentum $\qquad$ $Ft = Mv$ $\qquad$ $F \times \frac{1}{25} =$
$[1.62/(16g)] \times 249.3$ ft/s $\qquad$ $F = 19.62$ lb

3-19 Answer (e). $\omega Rm = mg$ $\qquad$ $\omega = \sqrt{g/(4500 \times 5280)} = 0.00116$ rad/s
or 16.00 r/day

3-20 Answer (b). $\Sigma F_x = 0 = P - (W/g)a - 0.333A_y = P - 49.74 - 0.333A_y$
$F_y = 0 = A_y + b_y - 200$ $\qquad$ $\Sigma M_A = 0 = 2P + 2W - 4B_y - 3(W/g)a =$
$2P + 250.78 - 4B_y$ $\qquad$ $A_y = 96.39$ lb $\qquad$ $B_y = 103.61$ lb
$P = 81.84$ lb

3-21 Answer (a). $\Sigma M_a = 0 = 30B - (30 - X - 8) \times 4 - (30 - X) \times 16$
$B = 18.933 - 0.667X$ $\qquad$ $M_B = 0 = 16X + 4(X + 8) - 30A$
$A = 1.067 + 0.667X$ $\qquad$ $M_X = BX = 18.933X = 0.667X^2$
$dM_X/dx = 0 = 18.933 - 1.333X$ $\qquad$ $X = 14.20$ ft
$A = 10.536T$ $\qquad$ $B = 9.464T$ $\qquad$ $M_{max} = 134.42$ ton·ft

3-22 Answer (e). Upward velocity positive. Height of balloon at
$t = 0 = 60v_0$ $\qquad$ $h = v_0 t - \frac{1}{2}gt^2$ $\qquad$ $60v_0 = 5v_0 - 12.5g$
$v_0 = 6.187$ ft/s $\qquad$ Stone drops $60v_0$ ft or 371.220 ft (height at
time of drop) $+ \Delta h$ $\qquad$ Δt to apogee $= v_0/g =$
0.192 s (stone has upward velocity at time of drop)
$v_0 t - \frac{1}{2}gt^2 = 0.595$ ft $\qquad$ Total height of stone = 371.815 ft or 371.8 ft

3-23 Answer (a). $F_1 = W \sin 30° = 0.500W$ $\qquad$ See Fig. 3S-23.
$F_2 = W(0.500 - 0.866\mu)$ $\qquad$ $S = v_0 t + \frac{1}{2}at^2$ $\qquad$ so $\frac{1}{2}a_1 t^2 = \frac{1}{2}a_2(2t^2)$
$a_1 = 4a_2$ $\qquad$ $a = F/m$ $\qquad$ $F_1 = 4F_2$ $\qquad$ $0.500 = 4(0.500 - 0.866\mu)$
$\mu = 1.500/3.464 = 0.433$

3-24 Answer (d). $F = \omega^2 Rm$ $\qquad$ $\omega = \frac{78}{60} \times 2\pi \times 8.168$ rad/s
Resisting force $= mg \times 0.30$ $\qquad$ $66.719Rm = 0.30mg$
$R = 0.1447$ ft = 1.736 in

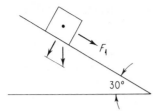

Figure 3S-23

3-25 Answer (e). $50 = (200/g)a + (50/g)a + 200 \times 0.15$ $a_1 =$
2.574 ft/s² $v = \sqrt{2as} = \sqrt{10.2944} = 3.208$ ft/s $a_2 = F/m =$
$30/(200/g) = 4.8255$ ft/s² $v^2 = v_0^2 + 2as$ $0 = 10.291 - 2 \times$
$4.8255s$ $s = 1.066$ Total distance $= 3.066$ ft   Distance from
edge of table $= 5 - 3.066 = 1.934$ ft

3-26 Answer (b). For aluminum, $w = 0.10$ lb/in³   Volume $=$
3.1416 ft³   Weight $= 543$ lb $I = \frac{1}{2}mr^2 = 8.437$ slug·ft²
Work $= 20 \times 15 = 300$ ft·lb   KE $= \frac{1}{2}I\omega^2 =$ work on drum
$2 \times 300/8.437 = \omega^2$ $\omega = 8.433$ rad/s $(8.433/2\pi) \times 60 =$
80.53 r/min

3-27 Answer (d). KE $= \frac{1}{2}mv^2 = \frac{1}{2} \times (50 \times 2000/g) \times 4.40^2 = 30,090$ ft·lb
PE $= \frac{1}{2}ks^2$ $30,090 = (\frac{1}{2} \times 40,000s^2 \times \frac{1}{12})$ ft·lb $s = 4.25$ in

3-28 Answer (e). $\omega^2rm = (v^2/r)m = mg \times 0.15$ $v = 69.5$ ft/s
60 mi/h $= 88$ ft/s $(v^2/r)m = 7.744m$ Slope $= \tan^{-1}(7.744/g) =$
13.5°

3-29 Answer (b). $m_1v_1 + m_2v_2 = (m_1 + m_2)v_r$ $v_r = (100 \times 20 + 150 \times$
$15)/(100 + 150) = 17.0$ ft/s   Initial KE $= \frac{1}{2} \times (100/g) \times 20^2 + \frac{1}{2} \times$
$(150/g) \times 15^2 = 1146.3$ ft·lb   Final KE $= \frac{1}{2} \times (250/g) \times 17.0^2 =$
1122.9 ft·lb   Loss in KE $= 23.4$ ft·lb

3-30 Answer (a). Net force acting $= 200 \times 0.866 - 0.5 \times 200 \times 0.25 =$
148.2 lb   Total work $= 148.2 \times (100 + s)$ lb·in $= \frac{1}{2}ks^2 = 50s^2$
$s^2 - 2.946s - 296.4 = 0$ $s = 18.76$ in

3-31 Answer (d). $I = k^2m = (7/12)^2 \times 100/g = 1.058$ slug·ft²
KE $= \frac{1}{2}I\omega^2 = \frac{1}{2} \times 1.058[(150/60) \times 2\pi]^2 = 130.53$ ft·lb   Work against
friction $= 1.5\pi \times 20 \times 0.25 \times$ revolutions $= 130.53$ ft·lb
Revolutions to come to rest $= 5.54$

3-32 Answer (b). $I = \frac{1}{2}mr^2 = \frac{1}{2} \times (12,880/g) \times 2^2 = 800.75$ slug·ft²
KE $= \frac{1}{2}mv^2 + \frac{1}{2}I\omega^2 = \frac{1}{2}mv^2 + \frac{1}{2}I(v^2/r^2) = 480,448$ ft·lb
PE $= Wh$ $h = 37.30$ ft   Horizontal travel $= 373.0$ ft
Distance up slope $= 374.86$ ft

3-33 Answer (b). Tq $= 33,000 \times$ hp$/(2\pi N) = 159.15$ ft·lb   Tq $= \int (2\pi r$
$dr)\mu pr = 2\pi\mu p \int r^2 \, dr = 2\pi \, \mu p r^3/3 \, |_{R_1}^{R_2}$ $159.15 = 2\pi \, \mu p/3 \times$
$[(5.12)^3 - (1.5/12)^3] = 0.0442P$ $P = 3598.7$ lb/ft²   or
24.99 lb/in²

3-34 Answer (c). $T = 70.7 + (70.7 \times 0.50) + (100/g)a = 137.13$ lb
$W - (W/g) \times 5 = 2T$ $0.8446W = 274.26$ $W = 324.7$ lb

3-35 Answer (d).   70 mi/h = 102.67 ft/s   60 hp = 33,000 ft·lb/s
Resisting force at 70 mi/h = 321.42 lb   Resistance at 50 mi/h =
163.99 lb   Force to push up slope = 157.43 lb $\sin^{-1}(157.43/4000) = 2.256°$ Slope = tan 2.256° = 0.0394 ft/ft

3-36 Answer (e).   3000 r/min = 50 r/s or 18,000°/s   18°/(18,000°/s) =
0.001°   3.5 ft/(0.001 s) = 3500 ft/s

3-37 Answer (a). $\Delta PE = 18.1875 \times \frac{8}{12} = 12.125$ ft·lb $\Delta PE = \Delta KE$
$\frac{1}{2} \times 18.1875/g \times v_R^2 = 12.125$ $v_R = 6.549$ ft/s
$m_1 v_1 = (m_1 + m_2)v_R$ $v_1 = (18.1875/0.1875) \times 6.549 = 635.3$ ft/s

3-38 Answer (d). $C_{\tan} = 4\pi \times \frac{60}{60} = 12.566$ ft/s = velocity of B in X
direction   Tangential velocity of B = 12.566/cos 15° = 13.01 ft/s

3-39 Answer (e). $I = \frac{1}{2}mR^2 - \Sigma r^2 \Delta m$ $W = (\pi/4) \times 30^2 \times 4 \times$
0.284 lb/in^3 = 803 lb $W_{\text{hole}} = (\pi/4) \times 3^2 \times 4 \times 0.284 = 8.03$ lb
$I = \frac{1}{2} \times (803/g) \times (15/12)^2 - 6 \times (8.03/g) \times (10/12)^2 = 18.64$ slug·ft^2
$KE = \frac{1}{2}I\omega^2 = \frac{1}{2} \times 18.46 \times [(300/60) \times 2\pi]^2 = 9109.6$ ft·lb

3-40 Answer (c).   PE = 16 × 16 = 256 ft·lb $KE = \frac{1}{2} \times (16/g)v^2 +$
$\frac{1}{2}I\omega^2 = 0.249v^2 + (1.875/2)(v^2/0.5^2)$ $256 = 3.9987v^2$
$v = 8.0$ ft/s $\frac{1}{2}I\omega^2 = (1.875/2)(8.0^2/0.5^2) = 240$ ft·lb

3-41 Answer (a). $S = \frac{1}{2}gt^2 = 64.34$ ft   10,000 × (64.34 + S) = 15,000S
$S = 128.68$ ft   Total drop = 64.34 + 128.68 = 193.02 ft

3-42 Answer (d). $F = (\pi/4) \times 12^2 \times 100 = 11,310$ lb   Force acting on
drop hammer = 2000 + 11,310 = ma $a = (13,310g)/2000 =$
214.1 ft/s^2 $v^2 = 2as$ $v = \sqrt{2 \times 214.1 \times 28/12} = 31.6$ ft/s

3-43 Answer (c).   [(1750 r/min)/60] × 2π = 183.26 rad/s
$I = (10 \text{ lb·ft}^2)/g = 0.3108$ slug·ft^2 $\alpha = \text{Tq}/I = 96.53$ rad/s^2
$KE = \frac{1}{2}I\omega^2 = \frac{1}{2} \times 0.3108 \times 183.26^2 = 5219$ lb·ft   Work per
revolution = 30 × 2π = 188.50 ft·lb, so 27.69 r $\theta = 26.79 \times 2\pi =$
173.97 rad = $\frac{1}{2}\alpha t^2$ $t = 1.90$ s

3-44 Answer (c).   74 mi/h = 110 ft/s   15 mi/h = 22 ft/s
85 mi/h = 124.67 ft/s $v^2 = v_0^2 + 2as$ $22^2 = 110^2 - 2 \times 1.8s$
$s = 3226.7$ ft $v = v_0 + at$ $t = 48.89$ s   5 mi at 22 ft/s
$t = 1200$ s   Increase speed 124.67 = 22 + a × 90
$a = 1.141$ ft/s^2 $v^2 = v_0^2 + 2as$ $s = 6598.9$ ft
$s_{\text{total}} = 36,225.6$ ft $t_{\text{total}} = 1338.9$ s   36,225.6/110 = 329.3 s
so lost 1009.6 s $\Delta v = 124.67 - 100 = 14.67$ ft/s
1009.6 × 110 ft/s = 111,056 ft   111,056/14.67 = 7570.3 s or 2.103 h

3-45 Answer (a). $KE_0 = \frac{1}{2} \times (30 \times 2000/g) \times 5.867^2 = 32,100$ ft·lb
After impact, KE = 0.75 × 32,100 = 24,075 ft·lb $KE = \frac{1}{2}m_1v_{r1}^2 +$
$\frac{1}{2}m_2v_{r2}^2$ $77.45 = 3v_{r1}^2 + 2v_{r2}^2$ $m_1v_0 = m_1v_{r1} + m_2v_{r2}$
$17.601 = 3v_{r1} + 2v_{r2}$ $v_{r2} = 8.800 - 1.5v_{r1}$ $v_{r1}^2 - 7.04v_{r1} +$
$10.324 = 0$ $v_{r2} = 5.676$ ft/s $v_{r1} = 2.083$ ft/s

3-46 Answer (e). $\Delta PE = \Delta KE = mgh = mg(2R - 2r)$ See Fig. 3P-46.
At top $\omega^2 \rho m = mg$ or $v_t^2 = g(R - r)$ KE at top $= \frac{1}{2}mv_t^2 = \frac{1}{2}mg(R - r)$ $R - r = 9.5$ ft   KE at bottom $= \frac{1}{2}mv_B^2 =$
$\frac{1}{2}mg(R - r) + 2\,mg(R - r)$ $v_B^2 = 5g(R - r) = 5g \times 9.5$
$v_B = 39.09$ ft/s

3-47 Answer (a).   KE at top $= \frac{1}{2}mv_t^2 + \frac{1}{2}I\omega^2 = \frac{1}{2}mv_t^2 + \frac{1}{2} \times \frac{2}{5}mv_t^2$
KE at top $= 0.7mv_t^2$ $\omega^2(R - r)m = mg = [v_t^2/(R - r)]m$
KE at bottom $= \frac{1}{2}mv_B^2 + \frac{1}{5}mv_B^2 = 0.7mv_t^2 + 2mg(R - r)$
$0.7mv_B^2 = 0.7\,mg(R - r) + 2mg(R - r)$ $v_B = 34.33$ ft/s

3-48 Answer (b). $\Delta PE = \Delta KE$ $\Delta PE = 25 \times 5.667 - 1.2 \times (20 \times 8^2) \times$
$\frac{1}{12} = 88.333$ ft·lb $\frac{1}{2}mv^2 = \frac{1}{2} \times (25/g)v^2 = 88.333$ $v = 15.08$ ft/s

3-49 Answer (b). $a = \alpha r = 24$ ft/s^2 vertical   See Fig. 3S-49. $\omega^2 r =$
$36 \times 3 = 108$ ft/s$^2 = $ radial acceleration   Acceleration of cylinder to
right $= 24$ ft/s^2 $\sqrt{24^2 + (108 + 0.24)^2} = 134.16$ ft/s^2

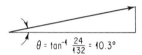

$\theta = \tan^{-1}\frac{24}{132} = 10.3°$ **Figure 3S-49**

3-50 Answer (c). $\Delta PE = 100 \times 36 + (100 + 25) \times S = \frac{1}{2} \times 100S^2$
$S^2 - 2.5S - 72 = 0$ $S = 9.83$ in

3-51 Answer (e).   Forces acting on cylinder $= F_1$ and F_2 Forces exerted
by cylinder $= 100$ lb down and $(100/g) \times 20 = 62.17$ lb to left
Resultant $= 117.5$ lb at $58.13°$ down to left $F_1 \cos 30° + F_2 \cos 60° =$
62.170 lb $F_1 = 3.86$ lb $F_2 \sin 60° = F_1 \sin 30° + 100$
$F_2 = 117.65$ lb

3-52 Answer (a). $I = \frac{1}{2}mr^2 = \frac{1}{2} \times (1000/g) \times 4^2 = 248.68$ slug·ft^2
$\omega = 120$ r/min $\times (2\pi/60) = 4\pi$ rad/s $\omega = \omega_0 + \alpha t$
$\alpha = -0.4\pi$ rad/s^2 $Tq = I\alpha = 4f$ $f = 78.125$ lb   Normal
force $F = 78.125/0.25 = 312.5$ lb $F = P \times (b + c)/c$
$3.125c = b + c$ $b = 2.125c$ $b/c = 2.125$

3-53 Answer (e). $T = 750 - (750/g)a = 500 + (500/g)a$
$a = 6.43$ ft/s^2 $T = 600$ lb $v^2 = v_0^2 + 2as = 2 \times 6.53 \times 7.5$
$v = 9.82$ ft/s

3-54 Answer (c). $h^2 + s^2 = 50^2$ $s = 3t$ $h^2 + 9t^2 = 2500$
$dh/dt = -(9t/h)$ For $s = 14$, $h = \sqrt{2500 - 196} = 48$ ft
$s = 3t$ $t = \frac{14}{3}$ min $dh/dt = -(3 \times 14)/48 = -0.875$ ft/min

3M-1 Answer (d). $(8/g) \times 4 = 0.995$

3M-2 Answer (b).   By definition.

3M-3 Answer (e). $\Sigma F_y = 0$ $T = 1000$ lb

3M-4   Answer (a). $\frac{1}{2} \times (2400/g) \times 100^2 = 373,018$ ft·lb

3M-5   Answer (c). $Tq = F_1r - F_2r = \frac{5}{12} \times (F_1 - F_2)$ $F_1 = 4F_2$
$250 = \frac{5}{12} \times 3F_2$ $F_2 = 200$ lb $F_1 = 800$ lb

3M-6   Answer (e). $d^2s/dt^2 = a$

3M-7   Answer (d). $F = \omega^2rm$

3M-8   Answer (a). $KE = \frac{1}{2}mv^2 =$ ft·lb

3M-9   Answer (a). $N = kg \times m/s^2$

3M-10   Answer (a). $v = \sqrt{2gh} = \sqrt{900.76} = 30.0$ ft/s

4-1   Answer (b). This is a composite beam (see Fig. 4S-1). $n = E_s/E_i =$
1.667 $\bar{y} = [4 \times (8 \times 1) + 8.5 \times (6 \times 1) + 9.125 \times$
$(10 \times 0.25)]/(8 + 6 + 2.5)$ $\bar{y} = 105.21/16.5 = 6.431$ in
$I = (1 \times 8^3/12) + 8 \times 2.413^2 + 6 \times 2.087^2 + 2.5 \times 2.712^2 =$
133.768 in^4 Check three possible critical points $M = S(I/c)$
Cast iron in compression: $M = 15,000 \times 133.768/6.413 = 312,880$ in·lb
Cast iron in tension: $M = 5000 \times 133.768/2.587 = 258,500$ in·lb
Steel in tension: $M = (20,000/1.667) \times 133.768/2.837 = 565.700$
(*Note:* Don't forget to apply the n factor to the allowable stress in the steel.
Cast iron in tension controls.)

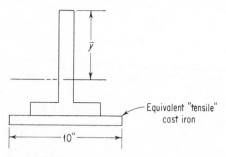

Figure 4S-1

4-2   Answer (a). $F = SA = E\delta A$ Elongations of the three wires will be
equal, so the strains will be equal. $1000 = 2E_c\delta0.20 + E_s\delta0.20 = 12.8 \times$
$10^6 \times \delta = 7.81 \times 10^{-5}$ in/in   Load held by copper wire $= E_cA_c\delta =$
531 lb or 265.5 lb each   Load held by steel wire $= E_sA_s\delta = 469$ lb

4-3   Answer (c). For all the load to be held by the steel wire, strain due
to load alone $= S/E_s = 1000/(0.20 \times 30 \times 10^6) = 1.667 \times 10^{-4}$ in/in.
Total strain in steel wire $= 1.667 \times 10^{-4} + \alpha_s \Delta T$ Strain in
copper wire due to thermal expansion alone $= \alpha_c \Delta T$ $1.667 \times 10^{-4} \times$
$6.5 \times 10^{-6} \times \Delta T = 9.3 \times 10^{-6} \times \Delta T$ $\Delta T = 59.5°F$
Temperature $= 119.5°F$

4-4   Answer (d). Total stress will equal bending stress plus direct tensile

stress. $Mc/I = 5000 \times 1.5 \times 0.5/\frac{1}{12} = 45,000$ lb/in^2 Maximum
tensile stress $= 45,000 + 5000/1.00 = 50,000$ lb/in^2

4-5 Answer (a).   Construct shear and moment diagrams as shown in Fig.
4S-5. Moment at center $= wl^2/8 + Pl/2 = (100 \times 64)/8 + (1000 \times 8)/2 = 4800$ lb·ft $= 57,600$ lb·in (*Note:* It is not necessary to construct
the shear and moment diagrams in this case, but it helps to give a more
complete picture.)

4-6 Answer (e).   Construct the shear and moment diagrams as shown in
Fig. 4S-6. The maximum moment occurs at a point of zero shear. The
maximum moment for this case equals 363,000 lb·in.

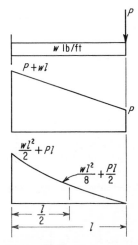

Figure 4S-5

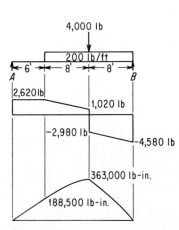

Figure 4S-6

4-7 Answer (b).   Treat the cable as a spring. $k = $ lb/in $\Delta L = L \times$
$S/E = L \times F/(AE) = 60,000 \times 1/(1 \times 12 \times 10^6)$ $\Delta L = 0.005$ in/lb
or $k = 200$ lb/in $= 2400$ lb/ft   Energy absorbed by spring $=$
$\frac{1}{2}mv^2 + Wh = 155.4 \times 29.33^2 + 10,000h$ Energy $= 133,710 +$
$10,000h = \frac{1}{2} \times 2400h^2$ $h^2 - 8.33h - 111.4 = 0$ $h = 15.51$ ft
Strain $= 15.51/5000 = 0.00310$ ft/ft or in/in   Stress $= E\delta =$
$12 \times 10^6 \times 0.00310 = 37,200$ lb/in^2

4-8 Answer (e).   Refer to Fig. 4S-8. $n = \frac{30}{2} = 15$ $3 \times 15 = 45$ in^2
Take moments about neutral axis $8\bar{y}(\bar{y}/2) = (12 - \bar{y}) \times 45$
$\bar{y}^2 + 11.25\bar{y} - 135 = 0$ $\bar{y} = 7.28$ in

4-9 Answer (e).   See Fig. 4S-9. $d^2y/dx^2 = M/EI$ $d^2y/dx^2 = Px/EI$
$dy/dx = (P/EI)x^2/2 + C_1$ $dy/dx = 0$ at $x = L$, so $C_1 = -L^2/2$
$y = (P/EI)(x^3/6 - L^2x/2 + C_2)$ $y = 0$ at $x = L$, so $C_2 = L^3/3$
$y = (P/EI)[x^3/6 - L^2x/2 + L^3/3]$ $I = 6 \times 8^3/12 = 256$ in^4
for $x = 72$ in $y = 100/(256 \times 1.8 \times 10^6)[62,208 - 746,496 + 995,328]$ $y = 0.0675$ in

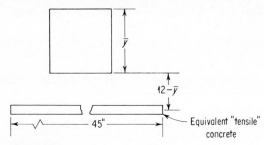

Figure 4S-8

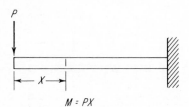

$M = PX$

Figure 4S-9

4-10 Answer (b). $I = r^2A$ Minimum $r = (790.2/49.09)^{1/2}$ Minimum $r = 4.012$ in $l/r = 600/4.012 = 149.55$ $P = 49.09 \times 18,000/[1 + (1/18,000)(149.55^2)] = 394,038$ lb

4-11 Answer (c). Construct shear and moment diagrams as shown in Fig. 4S-11. Maximum moment = 380,040 lb·in.

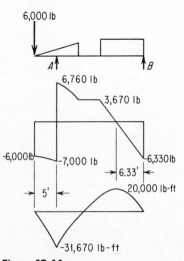

Figure 4S-11

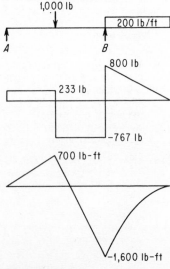

Figure 4S-13

4-12   Answer (d).   Work $= \int F \, dS = \int L \, dd$ $dd = -(2/10,000) \times$
$(L - 200) \, dL$ $\int L \, dd = -(1/5000) \times \int (L^2 - 200L) \, dL$

$[(L^3/3) - (200L^2/2)](-1/5000) \Big|_0^b = 450 - 225 = 225 \text{ in·lb}$

4-13   Answer (a).   From Fig. 4S-13, the maximum moment equals
$-1600 \text{ lb·ft} = -19,200 \text{ lb·in}$ $20,000 = 19,200/(I/c)$
$I/c = 0.96 \text{ in}^3$

4-14   Answer (e).   Construct the shear and moment diagrams as shown in Fig.
4S-14. The maximum moment $= -90,000 \text{ lb·ft} = 1,080,000 \text{ lb·in}$.

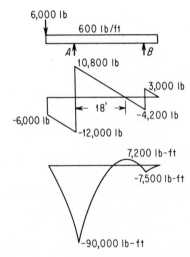

Figure 4S-14

4-15   Answer (c).   Construct the load, shear, and moment diagrams as shown
in Fig. 4S-15. The maximum moment equals 10,225 lb·ft.

4-16   Answer (b).   Construct the load, shear, and moment diagrams as shown
in Fig. 4S-16. The maximum moment will occur when the shear equals
zero. This condition will occur if the concentrated load of 10,000 lb is
applied at the point which is 16 ft from the left support.

4-17   Answer (a).   See Fig. 4S-17. $E = 30 \times 10^6 \text{ lb/in}^2$ $\Delta(\Delta L) =$
$(w \times dL/AE)L$ $\Delta L = \int (w/AE)L \, dL = w \times L^2/2AE$
ΔL due to 250 lb weight $= 250/(0.03E) \times 2000 = 0.556 \text{ ft}$ ΔL due
to weight of wire $wL = (0.03 \times 12 \times 2000) \text{ in}^3 \times 0.248 \text{ lb/in}^3 =$
204.5 lb $\Delta L = 204.5 \times 2000/(2 \times 0.03E) = 0.227 \text{ ft}$
Total $\Delta L = 0.783 \text{ ft}$ Depth of well $= 2000.783 \text{ ft}$

4-18   Answer (e). $S = (Tc/J)$ hp $= 2\pi NTq/33,000$ $J = \int \pi r^2 \, dA =$
$\int r^2 \times 2\pi r \, dr = \pi r^4/2 = \pi D^4/32$ Tq $= 40 \times 33,000/(2\pi \times 240) =$
$875.35 \text{ lb·ft} = 10,504 \text{ lb·in}$ $J/c = \pi r^3/2 = T/S = 1.167$
$r = 0.906 \text{ in}$, so the required shaft diameter equals 1.811 in. The
closest standard size is $1\frac{7}{8}$ in in diameter.

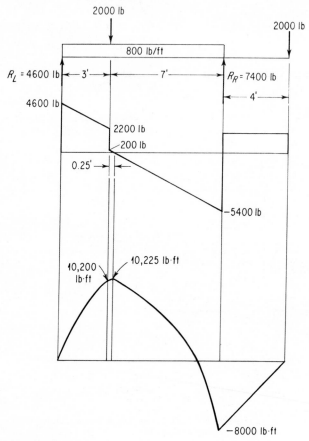

Figure 4S-15

4-19 Answer (d).

Component	Specific weight, lb/ft³	Percent void	Actual volume, ft³
1 cement	94	51.3	0.487
2 sand	111	35.4	1.292
3 agg.	108	32.0	2.040
			0.802

Water, 5 gal/sack of cement

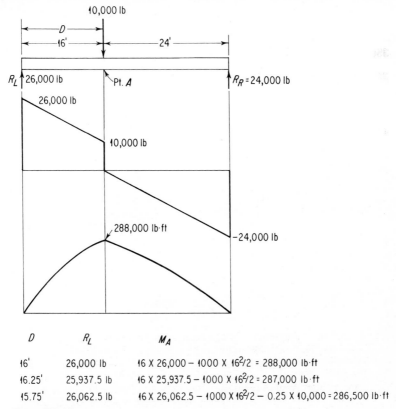

D	R_L	M_A
16'	26,000 lb	$16 \times 26,000 - 1000 \times 16^2/2 = 288,000$ lb·ft
16.25'	25,937.5 lb	$16 \times 25,937.5 - 1000 \times 16^2/2 = 287,000$ lb·ft
15.75'	26,062.5 lb	$16 \times 26,062.5 - 1000 \times 16^2/2 - 0.25 \times 10,000 = 286,500$ lb·ft

Figure 4S-16

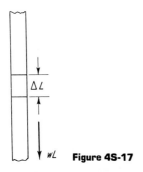

Figure 4S-17

Which gives a total volume of 4.622 ft³   27/4.622 = 5.842, so 5.842 batches are required to make 1 yd³ of concrete; 5.842 × 111 lb/ft³ × 2 ft³ = 1297 lb sand required per cubic yard of concrete.

4-20   Answer (d). Construct the load, shear, and moment diagrams as shown in Fig. 4S-20. The maximum bending moment equals 39,000 lb·ft = 468,000 lb·in.

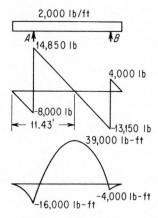

2,000 lb/ft

A 14,850 lb   B

4,000 lb

-8,000 lb

← 11.43' →

-13,150 lb

39,000 lb-ft

-4,000 lb-ft

-16,000 lb-ft

Figure 4S-20

4-21   Answer (b). The beam will be subject to double loading—a uniform downward load over the length of the beam and a concentrated vertical upward load at the center of the beam. Treat each load separately, and then add the results. For uniform loading, $y = 5WL^3/(384EI)$ = deflection at center point of beam. W = total uniform load on beam = wL $y = W \times 7.500 \times 10^{-6}$ For a concentrated load, $y = PL^3/(48\ EI)$ = $P \times 1.200 \times 10^{-5}$ But P will equal 30,000 times the deflection at the center of the beam. For a total deflection of 0.30 in, $P = 0.30 \times 30,000 = 9000$ lb. $0.30 = W \times 7.500 \times 10^{-6} - 0.108$ $W = 54,400$ lb or $w = 5440$ lb/ft   (This problem can also be solved directly; see Fig. 4S-21.) $EI\ d^2y/dx^2 = M = [(wL/2) - 4500]x - wx^2/2$

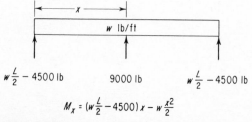

|← x →|

w lb/ft

$w\dfrac{L}{2} - 4500$ lb   9000 lb $w\dfrac{L}{2} - 4500$ lb

$$M_x = \left(w\frac{L}{2} - 4500\right)x - w\frac{x^2}{2}$$

Figure 4S-21

$EI\ dy/dx = \int M\ dx$ $EI\ dy/dx = wLx^2/4 - 4500x^2/2 - wx^3/6 + C_1$
$dy/dx = 0$ at $x = L/2$ $C_1 = 4500L^2/8 - wL^3/24$
$EIy = wLx^3/12 - 4500x^3/6 - wx^4/24 + 4500\ L^2x/8 + C_2$
$Y = 0$ at $X = 0$, so $C_2 = 0$ At $x = L/2$, $EIy = \frac{1}{24}(4500L^3) -$
$5wL^4/384$ For $L = 120$ in, $y = -9.00 \times 10^{-4}\ w + 0.108$
$y = -0.30$ $w = 453.33$ lb/in $= 5440$ lb/ft

4-22 Answer (c). Tq $=$ hp $\times 33,000/(2\pi N) = 787.82$ lb·ft or Tq $= 9453.8$
lb·in—a force of 9453.8 lb at a radius of 1.00 in $S = F/A =$
$9453.8/(3 \times 0.375) = 8403$ lb/in^2

4-23 Answer (a). See Fig. 4S-23. $\Sigma M_{RR} = 3000 \times 9 + 4000 \times 16 + 5000 \times$

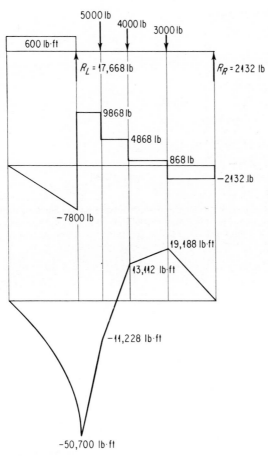

Figure 4S-23

$21 + 7800 \times 31.5 - 25R_L$ $R_L = 17{,}668$ lb $R_R = 2132$ lb
ΣM at 5000 lb load $= 4 \times 17{,}668 - 7800 \times 10.5 = 11{,}228$ lb·ft

4-24 Answer (b). Plot shear and moment diagrams (see Fig. 4S-24). There are two possible critical points, a bending moment of $-16{,}000$ lb·ft and a bending moment of 8000 lb·ft. Take moments about top edge to locate the neutral axis: $\bar{y} = [(4 \times 1) \times 0.5 + (5 \times 1) \times 3.5]/(4 + 5) = 2.167$ in
$I = (4 \times 1^3)/12 + 4 \times 1.667^2 + (1 \times 5^3)/12 + 5 \times 1.333^2$
$I = 30.75$ in^4 Maximum stress in top fibers $= 192{,}000 \times$
$2.167/30.75 = 13{,}530$ lb/in^2 Maximum stress in bottom fibers $=$
$96{,}000 \times 3.833/30.75 = 11{,}966$ lb/in^2

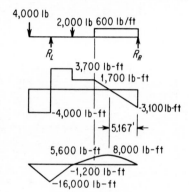

Figure 4S-24

4-25 Answer (c). Construct the load and shear diagrams as shown in Fig. 4S-25. The maximum bending moment will occur at the point of zero shear. $M_{\max} = -(3000 \times 5) \times 2.5 - (1000 \times 11) \times 5.5 + 26{,}000 \times 11 = -188{,}000$ lb·ft

4-26 Answer (e). The hole in the gear would expand $6.996 \times 70 \times 6.5 \times 10^{-6} = 0.00318$ in. Total ΔD required $= 0.004 + 0.005 = 0.009$ in, so shaft must contract by an amount 0.00582 in $= 7 \times 6.5 \times 10^{-6} \times \Delta T$
$\Delta T = 127.8°F$ Shaft must be cooled to $-57.8°F$.

4-27 Answer (a). Construct the shear and moment diagrams as shown in Fig. 4S-27. The maximum moment equals $108{,}000$ lb·in or 9050 lb·ft.

4-28 Answer (e). For a cantilever beam with a concentrated load at the free end, $y = PL^3/3EI$. For the rectangular beam, $I = 0.500 \times 0.600^3/12 = 0.00960$ in^4 and $y = P \times 6^3/(3 \times 30 \times 10^6 \times 0.00960) = 0.000250P$. For the circular cross-section beam, $I = \pi D^4/64 = 0.00307$ in^4 and $y = P \times 6^3/(3 \times 90 \times 10^6 \times 0.00307) = 0.000216P$. The rectangular beam would deflect 1.156 times as much as the circular beam.

4-29 Answer (c). See Fig. 4S-29. Weight held by each wire equals $SA = 0.01S$ $S = E\delta$ $\delta_c = \delta_s$ $300 = 0.01E_s\delta + 0.01E_c\delta =$

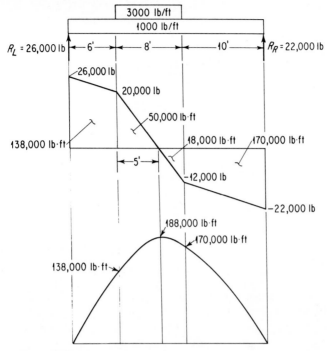

3000 lb/ft

1000 lb/ft

R_L = 26,000 lb ◄────6'────►◄────8'────►◄─────10'─────► R_R = 22,000 lb

26,000 lb

20,000 lb

50,000 lb·ft

138,000 lb·ft

◄─5'─►

18,000 lb·ft 170,000 lb·ft

−12,000 lb

−22,000 lb

188,000 lb·ft

170,000 lb·ft

138,000 lb·ft

Figure 4S-25

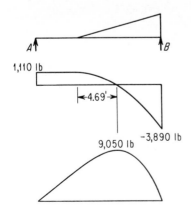

A↑ ↑B

1,110 lb

◄4.69'►

9,050 lb −3,890 lb

Figure 4S-27

$0.01 \times (30 + 15) \times \delta$ $\delta = 0.000667$ in/in $F_c = 100$ lb
$F_s = 200$ lb   Take moments about left end $100 \times 1 +$
$200 \times X = 300 \times 5$ $X = 7.00$ ft

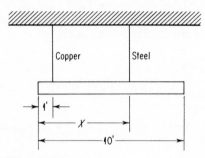

Figure 4S-29

4-30 Answer (a). The equation for calculating the bulk modulus of a solid is $K = E/(3 - 6v)$, where v = Poisson's modulus = 0.27 for steel and E = the modulus of elasticity, which equals 29×10^6 for steel $K = 29 \times 10^6/(3 - 6 \times 0.27) = 21 \times 10^6$ lb/in^2 The value given in the problem equals $1/(4.759 \times 10^{-8}) = 21.0 \times 10^6$ lb/in^2 $\Delta V = P/K$ $330 \times 10^{-6} = P/(21.0 \times 10^6)$ $P = 6930$ lb/in^2

4-31 Answer (b). The strains will be equal. 3500 lb $= 2 \times \delta \times 3 \times 20^7 + 2 \times 2 \times \delta \times 2 \times 10^7$ $\delta = 25.0 \times 10^{-6}$ in/in   Stress in $A = E\delta = 750$ lb/in^2

4-32 Answer (e). See Fig. 4S-32. 65 hp $= 2\pi \times 200 \times$ Tq/33,000
Tq $= 1707$ lb·ft $= 20,482$ lb·in $\theta = TL/JG$ $\Delta\theta = \Delta T \times L/JG$
For steel $G = 11.5 \times 10^6$ lb/in^2 ΔTq $= 4726$ lb·in $\Delta\theta = 4726 \times 60/(1.571 \times 11.5 \times 10^6)$ $\theta = 0.0157$ rad or $0.900°$

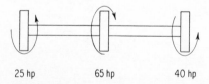

25 hp   65 hp   40 hp

Figure 4S-32

4-33 Answer (d). Each rivet will sustain a vertical shear of 3000 lb. Area of rivet = 0.442 in^2. Stress = 6787 lb/in^2. A torque equal to $8 \times 18,000 = 144,000$ lb·in will act on the section. The effect of the torque on each individual rivet can be calculated by remembering that stress is proportional to strain and that the strain in each rivet will be proportional to its distance from the neutral axis of the rivet section. This can be more easily expressed as: $I = \Sigma S^2 \, \Delta A = (4 \times 5^2 + 2 \times 3^2) \times 0.442 = 52.16 \text{ in}^4$. The stress in the outermost rivet due to the applied torque will equal $S = Mc/I = 144{,}000 \times 5/52.16 = 13{,}803 \text{ lb/in}^2$. These two stresses can be added vectorially to give the maximum rivet stress (see Fig. 4S-33). Total stress $= [(6787 + 8282)^2 + 11{,}042^2]^{1/2} = 18{,}681 \text{ lb/in}^2$.

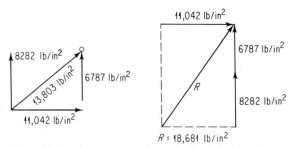

Figure 4S-33

4-34 Answer (c). $I = 4 \times 8^3/12 = 170.67 \text{ in}^4$ $\qquad Q = \bar{y}A = 2 \times 16 = 32$
$S_s = VQ/It = 2000 \times 32/(170.67 \times 4) = 93.7 \text{ lb/in}^2$

4-35 Answer (e). $\text{Tq} = 33{,}000 \times \text{hp}/(2\pi N)$ $\qquad$ For 2000 r/min, Tq $= 1000 \times 33{,}000/(2\pi \times 2000) = 2626 \text{ ft·lb} = 31{,}512 \text{ in·lb}$ $\qquad S = Tc/J = T/(\pi r^3/2) = 2508 \text{ lb/in}^2$ $\qquad$ For 10,000 r/min, Tq $= 1000 \times 33{,}000/(2\pi \times 10{,}000) = 525 \text{ ft·lb} = 6302 \text{ in·lb}$ $\qquad 2508 = 6302/(\pi r^3/2)$
$r^3 = 1.600$ $\qquad r = 1.17 \text{ in}$ $\qquad$ Diameter $= 2.34 \text{ in}$

4M-1 Answer (b). The wire will tend to twist as the load is applied, so it will be in torsion.

4M-2 Answer (c). If it fails along the grain, the allowable shear stress, parallel to the grain, will be exceeded.

4M-3 Answer (d). The equation is for the transverse moment of inertia of a rectangular section.

4M-4 Answer (a). By definition.

4M-5 Answer (a). Shown by the shear diagram.

4M-6 Answer (b). $M_2 - M_1 = \int_{x=1}^{x=2} V \, dx$

4M-7 Answer (a). Illustrated by the moment diagram for the loaded beam.

4M-8 Answer (b). $S_s = \frac{3}{2} \times V/(bh)$

4M-9 Answer (b). $0.625^2 \times 0.7854 \times 50{,}000 = 15{,}340 \text{ lb}$

4M-10 Answer (d). $(0.05/96) \times 30{,}000{,}000 \times 0.7854 = 12{,}272 \text{ lb}$

5-1 Answer (b).   See Fig. 5S-1. Depth of centroid of submerged surface $= \frac{1}{3}$
$\times 3 = 1$ ft   Pressure at centroid $= 1w = 62.4$ lb/ft^2 Force on
triangular area $= 62.4 \times \frac{1}{2} \times 5 \times 3 = 561.6$ lb   Or, $F = \int P \, dA$
$P = 62.4D$ $dA = W \, dD$ $W = 6 \times (3 - D)/3$ $F = 124.8$
$\int_0^3 (3D - D^2) \, dD = 124.8 \, (3D^2/2 - D^3/3) \, |_0^3 = 561.6$ lb

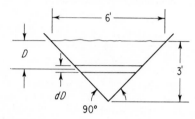

Figure 5S-1

5-2 Answer (c).   Refer to Fig. 5P-2. $P_A/w + v_A^2/2g + Z_A = P_B/w + V_B^2/2g + Z_B$ $P_A/w = 10 \times 144/62.4 + 1.5 = 24.58$ ft $P_B/w + 1$ ft $= 2.5$
$\times 0.49 \times 144/62.4 = 1.83$ ft $v_A = 2.5/(12.5/144) = 28.8$ ft/s
$v_A^2/2g = 12.89$ ft $24.58 + 12.89 + 0 = 1.83 + v_B^2/2g + 10$
$v_B = 1667.0^{1/2} = 40.8$ ft/s

5-3 Answer (e).   For a triangular weir, $Q = C_d(\frac{8}{15})(2g)^{1/2} \tan{(\theta/2)} \times H^{5/2}$
For the given case, $Q = 1.48 \times H^{5/2} = 1.48$ ft^3/s $1.48 \times 3600 \times$
$7.48 = 39,905$ gal/h   For a rectangular weir, $Q = C_d(\frac{2}{3})(2g)^{1/2} \times BH^{2/3}$,
where $B =$ width of weir in feet

5-4 Answer (c).   Glass displaces 50 cm^3 of water, then 50 cm^3 of gasoline
has a mass of $125 - 92 = 33$ g and the specific gravity of the gasoline $=$
$\frac{33}{50} = 0.66$.

5-5 Answer (a).   See Fig. 5S-5. Volume of timber submerged $= 12 \times (12 -$
$4.3)/144 \times 12 = 7.70$ ft^3 Weight of timber $= 7.70 \times 62.4 =$
480.48 lb   Take moments about point touching rock $\Sigma M_R =$
$480.48 \times 6 - 480.48 \times 6 + 8 \times \frac{1}{2} \times 12 \times 1 \times (4.3/12) \times 62.4 -$
$150 \times X$ $X = 7.16$ ft

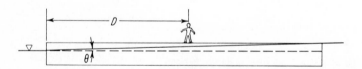

$\theta = 1.71°$ $\cos \theta = 0.9996$

Figure 5S-5 $\theta = 1.71°$, $\cos \theta = 0.9996$.

5-6 Answer (d).   8-in diameter gives $A = 50.27$ in$^2 = 0.349$ ft^2
$v = 8.59$ ft/s $v^2/2g = 1.148$ ft $P_1/w + v_1^2/2g + Z_1 = P_2/w +$

$v_2^2/2g + Z_2 + h_L$ (Ignore velocity head) $P_1/w = (Z_2 - Z_1) + h_L$
(P_2 = atmospheric pressure) $h_L = f(L/d)(v^2/2g)$ Re = $\rho Dv/\mu =$
$1.94 \times \frac{8}{12} \times 8.59 \times 47,800 = 5.3 \times 10^5$ $\epsilon/D = 0.00085/0.667 =$
0.0013 $f = 0.021$ $L/D = 3500/0.667 = 5250$ $h_L = 0.021$
$\times 5250 \times 1.148 = 126.6$ ft $P_1/w = 450 + 126.6 = 576.6$ ft
$p = 35,978$ lb/ft^2 = 250 lb/in^2

5-7 Answer (e). $Q = AC_d(2gh)^{1/2} = 0.0873$ ft$^2 \times 0.70 \times 3217^{1/2} = 3.47$ ft^3/s

5-8 Answer (c).   60 gal/min $\times \frac{1}{60} \times (1/7.48) = 0.1336$ ft^3/s
$A = 3.1416$ in$^2 = 0.0218$ ft^2 $v = 6.124$ ft/s $v^2/2g = 0.584$ ft
Re $= Dv/v = \frac{2}{12} \times 6.124 \times (1/1.2 \times 10^{-5}) = 8.5 \times 10^4$
$\epsilon/D = 0.000005/(2/12) = 0.00003$ $f = 0.0205$ (from Fig. 5-12)
$h_L = f(L/D)(v^2/2g) = 0.0205 \times (50 \times 6) \times 0.584 = 3.59$ Add
entrance loss of $0.5 \times 0.584 = 0.292$ ft $Z_1 - Z_2 = 4$ ft
$P_1/w + v_1^2/2g + Z_1 = P_2/w + v_2^2/2g + Z_2 + h_L +$ entrance loss
$P_1/w - P_2/w = 4 + 3.59 + 0.29 = 7.88$ ft or 3.41 lb/in^2

5-9 Answer (a). $A_{pipe} = 0.196$ ft^2 $A_{nozzle} = 0.0218$ ft^2 For pipe $v =$
5.10 ft/s $v^2/2g = 0.405$ ft   For nozzle $v = 45.84$ ft/s
$\epsilon/D = 0.00015/0.5 = 0.0003$ Re $= Dv/v = 0.5 \times 5.1/0.00001 =$
2.55×10^5 $f = 0.018$ $h_L = f(L/D)v^2/2g = 0.018 \times$
$(80/0.5) \times 0.405 = 1.17$ ft   Lose one-half velocity head at entrance to
pipe plus frictional head loss in line plus loss in nozzle. Pressures at surface
of reservoir and at exit of nozzle are equal. $h = 0.20 + 1.17 +$ head
at nozzle   For nozzle, $v = 0.90 \times (2gh)^{1/2}$ or $45.84 = 0.90 \times (2gh)^{1/2}$
$h = 40.32$ Head required $= 0.20 + 1.17 + 40.32 = 41.69$ ft

5-10 Answer (b).   Water potential energy = (2000 $\times$ 62.4) lb/min $\times$
100 ft/33,000 = 378.18 hp   Energy contained in discharge =
$(2000 \times 62.4) \times (12^2/2g)/33,000 = 8.46$ hp   Power to overcome friction:
hp $= 2\pi N$ Tq/33,000   hp $= 2\pi \times 120 \times 400/33,000 = 9.14$
Generator power $= 378.18 - 8.46 - 9.14 = 360.6$ hp

5-11 Answer (b).   Use Bernoulli's equation. $P = 3.5 \times 144/58.7 = 8.586$ ft
$v_{venturi}/v_{pipe} = (4/2)^2 = 4$ $P_1 - P_2/w = 8.586 = (16v_P^2 - v_P^2)/g$
$v_P = 6.069$ ft/s theoretical or $v_v = 24.275$ ft/s theoretical $Q =$
0.97×24.275 ft/s $\times 0.0218$ ft$^2 = 0.513$ ft^3/s $0.513 \times 60 \times$
(7.48 gal/ft^3) = 240.4 gal/min

5-12 Answer (d).   See Fig. 5S-12. Moment about hinge $= \int y\, dF$
$dF = 62.4 \times 4 \times (21 + y)\, dy$ $dF = (5241.6 + 249.6 \times y)\, dy$
$M = \int_0^6 y(5241.6 + 249.6)\, dy$ $M = 5241.6y^2/2 + 249.6y^3/3 \mid_0^6 =$
$94,348.8 + 17,971.2 = 112,320$ lb·ft $F = \frac{112,320}{6} = 18,720$ lb

5-13 Answer (a).   Total head above center of pump $= 39.2 \times 144/62.4 +$
$1.5 = 91.6$ ft   Add suction lift of 8 ft, giving total head to be supplied
by pump $= 100$ ft   Flow rate = 35 gal/min   Water hp =
(35×8.33)(lb/min) $\times$ 100 ft/33,000 = 0.883   Efficiency =
$0.883/1.25 = 71\%$

5-14 Answer (d).   See Fig. 5S-14. For a 1-ft width of the dam $W_1 = \frac{27}{2} \times$
$45 \times 150 = 91,125$ lb/ft $W_2 = 3 \times 45 \times 150 = 20,250$ lb

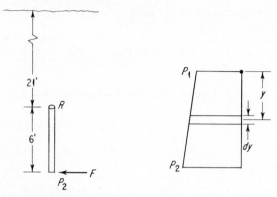

Figure 5S-12

Weight of dam per foot of width = 111,375 lb $F = \frac{42}{2} \times$
$62.4 \times 42 = 55,036.8$ lb   Area of base = 30 ft²/ft of width
Treat base of dam as a beam subjected to a bending moment; then
the bending stress due to the applied moment will be $S = Mc/I$
Take moments about center of base of dam (center of beam)
$M = 55,036.8 \times 14 - 91,125 \times 3 - 20,250 \times 14 = 213,640.2$ ft·lb
I of base = $bh^3/12 = 1 \times 30^3/12 = 2250$ ft⁴/ft of width
At downstream tip, $F = \frac{111,375}{30} + 213,640.2 \times \frac{15}{2250} = 5136.9$ lb/ft²
At upstream face, $F = \frac{111,375}{30} - 213,640.2 \times \frac{15}{2250} = 2282.2$ lb/ft²

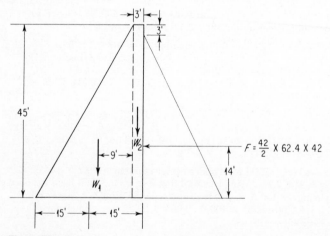

Figure 5S-14

5-15 Answer (b). $F = ma = (m/dt)\,dv$ $m/dt = 250 \times 8.33/(60g) =$ 1.079 slugs/s $v = (250 \times 231/60)/0.994 = 968.3$ in/s or 80.7 ft/s $F = 1.079 \times 80.7 = 87.1$ lb

5-16 Answer (e). See Fig. 5S-16. Apply Bernoulli's equation. $P_1/w + v_1^2/2g + Z_1 = P_2/w + v_2^2/2g + Z_2 + h_L$ $v_1 = 0$ $P_1 = 0$ $P_2/w = $ 18.48 ft $Z_2 = -28$ ft $0 = 18.48 - 28 + v_2^2/2g + h_L$ $P_3/w = 142.03$ ft $v_2 = v_3$ $h_L = $ frictional head loss in 500 ft $18.48 - 28 = 142.03 - 285 + \frac{8000}{500} h_L$ $h_L = 8.34$ ft $v_2^2/2g = 28 - 18.48 - 8.34 = 1.18$ $v_2 = 8.71$ ft/s $Q = 8.71 \times 0.7854 = 6.843$ ft³/s or 3071 gal/min

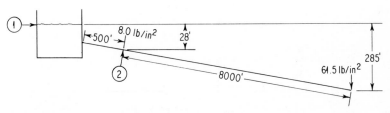

Figure 5S-16

5-17 Answer (d). $P_{\text{Inlet}} = -3 \times 13.6/12 + 0.05 = -2.9$ ft $P_{\text{outlet}} = $ $(17.5 \times 2.31) + 0.5 = 4.09$ ft   Total Δh across pump $= 40.9 +$ $2.9 + 1.0 = 44.8$ ft   Water power $= 240 \times 8.33 \times 44.8/33{,}000 =$ 2.71 hp   Power to pump $= 2\pi \times 3460 \times 7.89/33{,}000 = 5.20$ hp Pump efficiency $= 2.71/5.20 = 52.1\%$

5-18 Answer (b). Two forces act—pressure force and force due to change of momentum. $A = 1017.9$ in²   Pressure force $= 50{,}894$ lb Momentum force $= $ slugs/s $\times$ ft/s   Flow rate $= 7.1 \times 7.07 =$ 50.2 ft³/s   Mass flow rate $= 50.2 \times 62.4/32.17 = 97.37$ slugs/s $F = 97.37 \times 7.1 = 691.3$ lb $F_x = 50{,}894 + 691.3 - (50{,}894 +$ $691.3) \times \cos 60° = 25{,}793$ lb $F_y = 0 - (50{,}894 + 691.3) \times \sin 60°$ Total force acting on elbow $= [25{,}793^2 + (-44{,}674^2)]^{1/2} = 51{,}585$ lb

5-19 Answer (a). See Fig. 5S-19. Total force $= (10 \times 62.4 + 14 \times 62.4)/2 \times$

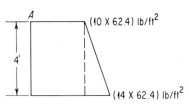

Figure 5S-19

$4 = 2995.2$ lb/ft or 11,981 lb total $\Sigma M_A = (10 \times 62.4 \times 4) \times$
$2 + (4 \times 62.4 \times \frac{1}{2}) \times \frac{8}{3} = 6323.2$ lb·ft/ft or total moment $=$
$25{,}292.8$ ft·lb $x = 25{,}292.8/11{,}981 = 2.11$ ft

5-20 Answer (b). See Fig. 5S-20. $v = 1.486S^{1/2}R^{2/3}/n$ $\alpha = \sin^{-1}\frac{2}{4} = 30°$
Wetted perimeter $= (180 + 60)/360 \times 8\pi = 16.76$ ft   Area of void $=$
$\frac{120}{360} \times 16\pi - 2 \times 4 \cos 30° = 16.76 - 6.93 = 9.83$ ft^2 Area
cross section $= 16\pi - 9.83 = 40.44$ ft^2 $R_H = 40.44/16.76 = 2.41$ ft
$v = 1.486 \times 0.001^{1/2} \times 2.41^{2/3}/0.013 = 6.50$ ft/s $Q = 6.50 \times 40.44 =$
263 ft^3/s

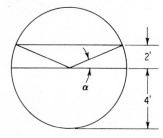

Figure 5S-20

5-21 Answer (d). $F = (m/t)v$ $v = 100 + 90 = 190$ ft/s $m/t =$
$200/g = 6.217$ slugs/s $F = 190 \times 6.217 = 1181$ lb

5-22 Answer (a). Use Manning's equation. $A = 4 \times 8 + \frac{1}{2} \times 1 \times 2 = 33$ ft^2
Wetted perimeter $= 4 + 4 + 3 + 3 + 2 \times 1.414 = 16.828$
$R_H = 33/16.828 = 1.961$ ft $v = 1.486 \times 0.01^{1/2} \times 1.961^{2/3}/0.013 =$
17.91 ft/s $Q = 17.91 \times 33 = 591$ ft^3/s

5-23 Answer (e). $A = 78.54$ ft^2 $v = 250/78.54 = 3.18$ ft/s
$F_x = (m/t)\,\Delta v = (250 \times 62.4/g) \times 3.18 \times \cos 45° = 1091$ lb

5-24 Answer (c). Displaced volumes $= 50/0.8 - 50/0.9 = 6.94$ cm^3
Distance between graduations $= 6.94$ cm

5-25 Answer (b). $P_1/w + v_1^2/2g = P_2/w + v_2^2/2g$ $v_2 = (24/3)^2 v_1 = 64 v_1$
$P_1 + 12$ in $H_2O = P_2 + 12$ in $Hg = P_2 + 163.2$ in H_2O
$P_1/w - P_2/w = 12.6$ ft 12.6 ft $= (v_2^2 - v_1^2)/2g$
$810.68 = (4096 - 1)v_1^2$ $v_1 = 0.445$ ft/s $v_2 = 28.476$ ft/s
$Q = 0.445 \times 3.142$ ft$^2 = 1.40$ ft^3/s theoretical   Actual $Q = 1.40 \times$
$0.98 = 1.37$ ft^3/s

5-26 Answer (e). Total head loss $= (50 \times 144/62.4) + 2 + 30 = 147.4$ ft
$h_L = f(L/D)(v^2/2g)$ $L = 200 + 30 + 370 = 600$ ft $v = (1.25/$
$12.7) \times 144 = 14.17$ ft/s $v^2/2g = 3.12$ ft $h_L = 0.024 \times (600 \times$
$12/4.02) \times 3.12 = 134.2$ ft   Head loss due to fittings $= 147.4 -$
$134.2 = 13.2$ ft

5-27 Answer (a). $m = I/Ay$ $y = 25$ ft $A = 4.909$ ft^2
$I = 2.5^4/64 = 1.917$ ft^4 $m = 1.917/(4.909 \times 25) = 0.0156$ ft $=$
0.187 in

5-28 Answer (*e*). 10-in standard pipe is 10.75 in OD   Total weight =
25 + 480 = 505 lb   505/(62.4 × 1.25) = 7.896 ft³ displaced
Outside area = 90.763 in² = 0.630 ft²   Pipe cap and weight displace
0.5 ft³, so pipe displaces 7.396 ft³   7.396/0.630 = 11.74 ft submerged,
so 3.26 ft show above surface

5-29 Answer (*c*). See Fig. 5S-29. a = area of orifice = 9 in² dV/dt =
$av = aC_d(2gh)^{1/2}$ = 0.306$h^{1/2}$ ft³/s $dV/dt = A\,dh/dt = (\pi/4)D^2\,dh/dt$
$D = 3 + (h/10) × 5 = 3 + h/2$ $dV/dt = (7.069 + 2.356h +$
$0.196h^2)\,dh/dt = 0.306h^{1/2}$ $dt = (23.101h^{1/2} + 7.699h^{1/2} + 0.641h^{3/2})$
dh $t = 46.202h^{1/2} + 5.133h^{3/2} + 0.256h^{5/2}\,|_0^{10}$ = 389.4 s = 6.49 min

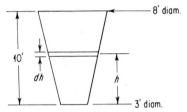

Figure 5S-29

5-30 Answer (*a*). $h_L = f(L/D)(v^2/2g)$ Re = $\rho\,Dv/\mu$ Viscosity of
gasoline at 60°F = 0.30 cP   Re = (42.43/g) × $\frac{10}{12}$ × 4/(0.30/47,800) =
7.0 × 10⁵ ϵ/D = 0.00015/0.833 = 0.00018 f = 0.015
h_L = 0.015 × (528,000/0.833) × 16/2g = 2364 ft   Total head =
2864 ft   2684 × 42.43/144 = 844 lb/in²

5-31 Answer (*b*). $v = 1.486S^{1/2}R^{2/3}/n$ A = [2 + (2 + 2 × 1.5/tan 60°)/2]
× 1.5 = 4.300 ft²   Wetted perimeter = 2 + 2 × 1.732 = 5.464
R = 4.30/5.464 = 0.787 ft   for Q = 200 ft³/s v = 200/4.300 =
46.512 ft/s   46.512 = 1.486$S^{1/2}$ × 0.787²/³/0.013 $S^{1/2}$ = 0.477
S = 0.227 or 227 ft/1000 ft

5-32 Answer (*d*). Head added to water = 10 + (22 × 144/62.4) + 1 =
61.77 ft   Water horsepower = 40 × 8.33 × 61.77/33,000 = 0.624
1105/746 = 1.48 hp into motor   Efficiency = 0.624/1.48 = 0.42
or 42%

5-33 Answer (*c*). 4 mi/h = 5.867 ft/s $F_2 = (\rho_2/\rho_1)(A_2/A_1)(V_2/V_1)^2$
F_2 = 2(63.8/62.4) × 10 × (5.867/5)² = 28.16 lb/ft

5-34 Answer (*e*). Poise = dyn·s/cm² = Pa·s $h_L = f(L/D)\,(v^2/2g)$
$Q = \frac{570}{5}$ = 114 cm³/s $v = Q/A$ = 114/0.283 cm² = 403.2 cm/min
or v = 6.72 cm/s   Assume laminar flow, then f = 64/Re =
64 $\mu/(\rho\,Dv)$ $h = (64\mu/\rho\,Dv)(L/D)(v^2/2g)$ Pressure = ρgh
$\Delta P\,(64\mu/Dv)(L/D)(v^2/2)$ $\mu = \Delta P × D^2/(32Lv)$ $\mu = 1.2 × 10^5 ×$
$0.6^2/(32 × 100 × 6.72)$ = 2.01 poises or 201 cP
Alternative solution: Q = 114 cm³/16.387 = 6.957 in³/min = 6.710 ×
10^{-5} ft³/s A = (0.6/2.54)² × $\pi/4$ = 0.0438 in² = 3.0434 × 10^{-4} ft²

$v = Q/A = 0.220$ ft/s $D = 0.6/(2.54 \times 12) = 0.0197$ ft
$L = 100/(2.54 \times 12) = 3.281$ ft $\Delta P = (1.2 \times 10^5/444,822)/(1/6.452) =$
1.741 lb/in^2 or $P = 250.64$ lb/ft^2 $\mu = 250.64 \times 0.0197^2/(32 \times$
$3.281 \times 0.220) = 0.004211$ lb·s/ft^2 $0.004211 \times 47,800 = 201$ cP

5-35 Answer (c). $h = 15$ ft = head loss plus velocity head $Q =$
$10/(60 \times 7.48) = 0.0223$ ft^3/s   Re $= \rho \, Dv/\mu = (1.025 \times 62.4/g)D \times$
$(0.0223/0.7854D^2)/(1.6/47,800) - 1686/D$ $v = 0.0223/(0.7854D^2) =$
$0.0284/D^2$ There are only a finite number of standard pipe sizes;
try $\frac{1}{2}$-, $\frac{3}{4}$-, and 1-in standard pipe sizes. For $\frac{1}{2}$-in standard pipe (schedule 40)
ID $= 0.622$ in $= 0.5818$ ft $\epsilon/D = 0.0029$ Re $= 3.3 \times 10^4$
$f = 0.03$ $h_L = 0.03 \times (55/0.0518) \times 1.740 = 55.4$ ft, which is
too much, so try $\frac{3}{4}$-in schedule 40 pipe   ID $= 0.824$ in $= 0.0687$ ft
Re $= 2.5 \times 10^4$ $\epsilon/D = 0.0022$ $f = 0.029$
$h_L = 0.029 \times (55/0.0687) \times 0.563 = 13.06$ ft   Add velocity head
of 0.563 to give 13.62 ft of head required

5-36 Answer (c).   See Fig. 5S-36. A floating block displaces its own mass
of liquid. Mass of block $= 1000 \times 0.50 = 500$ g   Mass of liquid
displaced $= 500$ g $= 100W + 100 \times 0.8 \times$ oil $W +$ oil $=$
$10 - 4 = 6$ $W = 5 - 0.8 \times$ oil   Oil $= 5$ cm thick

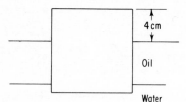

4 cm

Oil

Water   **Figure 5S-36**

5-37 Answer (a).   See Fig. 5S-37. $dV = \pi R^2 \, dh = \pi h \, dh$ $dV/dt = C_d a \times$
$(2gh)^{1/2} = 0.60 \times (0.7854/144) \times 8.021h^{1/2} = 0.0262h^{1/2}$ ft^3/s
$\pi h \, dh/dt = 0.0262h^{1/2}$ $h^{1/2} \, dh = 0.00836dt$ $\frac{2}{3}h^{3/2} = 0.00836t$
$\Big|_0^{10}$ $t = 2522$ s $= 42.0$ min

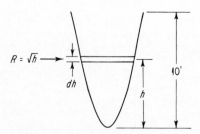

$R = \sqrt{h}$ 10'

dh h

Figure 5S-37

5-38 Answer (d). $\Delta h = h_L + 250$ $v = 3.0/0.7854 = 3.820$ ft/s
$v^2/2g = 0.227$ Re $= 1.94 \times 1 \times 3.820/(1/47,800) = 3.5 \times 10^5$
$\epsilon/D = 0.00085$ $f = 0.0205$ $h_L = 0.0205 \times \frac{4000}{1} \times 0.227 =$
18.6 ft   Total head required $= 268.6$ ft   Power $= (3 \times 62.4 \times$
$268.6)(\text{ft·lb/s})/550 = 91.46$ water hp $91.46/0.89 = 103$ hp required

5-39 Answer (a).   Velocity at center of line $v = (2gh)^{1/2}$ $h = (3.9/0.9)/$
$12 = 0.361$ ft $v = 4.820$ ft/s   Re $= \rho D v/\mu = (0.9 \times 62.4/g) \times$
$\frac{1}{12} \times 4.820/(35 \times 10^{-5}) = 2003$, so flow is in the laminar range and
$v_{avg} = \frac{1}{2}v_{max}$ $v_{avg} = 4.820/2 = 2.41$ ft/s

5-40 Answer (d).   For 2.3 lb/in^2, $h = 2.3 \times 144/(62.4/0.9) = 5.90$ ft
$v = 19.48$ ft/s   Re $= (0.9 \times 62.4/g) \times \frac{1}{12} \times 19.48/(35 \times 10^{-5}) =$
8096, so flow is turbulent and $v_{max}/v_{avg} = 1.43 f^{1/2} + 1$
$\epsilon/D = 0.000005/(1/12) = 0.00006$ $f = 0.033$ $v_{avg} = 19/84/$
$(1.43 \times 0.033^{1/2} + 1) = 15.75$ ft/s   Check Reynolds number to
determine change in friction factor   Re $= (15.75/19.48) \times 8096 =$
6426 $f = 0.035$ corrected $v_{avg} = 15.4$ ft/s

5-41 Answer (b).   See Fig. 5S-41. $Q = 60$ ft^3/s A to B: $v = 8.488$ ft/s;
$L/D = 1000$; $v^2/2g = 1.120$ ft; $h_L = 0.016 \times 1000 \times 1.120 = 17.9$ ft
B to C: top branch $h_L = 0.019 \times (2000/1.5) \times v_{18}^2/2g = 0.394 v_{18}^2$;
lower branch $h_L = 0.017 \times \frac{2400}{2} \times v_{24}^2/2g = 0.317 v_{24}^2$ The flow
splits, so $60 = v_{18} \times 1.767 + v_{24} \times 3.142$ $v_{18} = 33.956 - 1.778 v_{24}$
h_L through top branch $= h_L$ through bottom branch, so $0.628 v_{18} =$
$0.563 v_{24}$ $v_{18} = 0.563 v_{24}$ $v_{18} = 0.897 v_{24} = 33.956 - 1.778 v_{24}$
$v_{24} = 12.696$ $h_L = 51.1$ ft C to D: $v = 60/4.909 = 12.223$ ft/s;
$v^2/2g = 2.322$; $h_L = 0.017 \times (1500/2.5) \times 2.322 = 23.7$ ft   Total head
loss from A to $D = 92.7$ ft or $92.7 \times 0.433 = 40.2$ lb/in^2

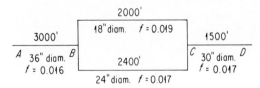

Figure 5S-41

5-42 Answer (e).   40 cm^3 $\times$ 0.25 g/cm^3 = 10 g L cm^3 $\times$ 11.3 g/cm^3 =
$11.3 \times L$ g $(40 + L)$ cm^3 = $(10 + 11.3 \times L)$ g $L = 30/10.3 =$
2.913 cm^3 Mass of lead $= 2.913 \times 11.3 = 32.19$ g

5-43 Answer (d).   See Fig. 5S-43. 4 ft^3 submerged $W = 4 \times 62.4 =$
249.6 lb   Displace 249.6 lb liquid $(1.0 \times 62.4 \times 0.8) + L \times$
$62.4 = 249.6$ $L = 3.2$ ft, so 4.2 ft immersed, 1.8 ft projects above
surface

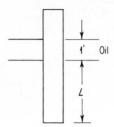

Figure 5S-43

5-44 Answer (*b*). $P_1/w = 10$ ft $v_{in} = 10$ ft/s $v_{throat} = (4/3)^2 \times v_{in} =$
$1.778v_{in}$ $v_T^2 = 3.160v_{in}^2$ $P_2/w = P_1 + v_1^2/2g - v_2^2/2g$
$P_2/w = 10 + 1.554 - 4.911 = 6.64$ ft

5-45 Answer (*d*).   Area of trapezoid $[(b - 0.29b) \times b/2 = 0.355b^2] = 0.355b^2$
Wetted perimeter $= b + 2 \times (b/2)/\sin 60° = 2.155b$
$R_H = 0.355b^2/2.155b = 0.165b$

5-46 Answer (*e*).   Volume of water displaced $= 523 - 447.5$ cm³   Volume
of copper $= 523/8.92 = 58.6$ cm³   Volume of void $= 75.5 - 58.5 =$
16.9 cm³

5-47 Answer (*b*). $P_1/w = 20 \times 144/62.4 = 46.15$ ft $v_1 = 3.0/0.349 =$
8.594 ft/s $v_1^2/2g = 1.148$ ft $v_2 = 3.0/0.196 = 15.279$ ft/s
$v_2^2/2g = 3.628$ ft $P_2/w = P_1/w + v_1^2/2g - v_2^2/2g = 46.15 +$
$1.148 - 3.628 = 43.67$ ft $P_2 = 43.67 \times 52.4/144 = 18.92$ lb/in²
$F_{M1} = m/t \times v_1 = (3 \times 62.4/g) \times 8.594 = 50.01$ lb $F_{M2} = m/t \times$
$v_2 = 5.819 \times 15.279 = 88.91$ lb   Force at inlet $= 20 \times 50.27 +$
$50.01 = 1055.3$ lb $= F_x$ Force at outlet $= 18.92 \times 28.27 + 88.91 =$
623.9 lb   Forces at outlet acting on bend $F_x = -311.9$ lb $F_y =$
-540.28 lb   Net force on elbow $=$ force at inlet $-$ force at outlet
$F_x = 1055.3 - 311.9 = 743.4$ lb $F_y = -540.3$ lb   Total force $=$
$(743.4^2 + 540.3^2)^{1/2} = 919.0$ lb

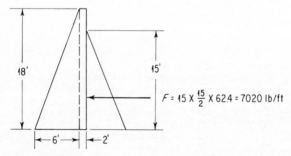

Figure 5S-48

5-48 Answer (a). See Fig. 5S-48. Take section of dam 1 ft wide. Force due to water pressure $= 15 \times \frac{15}{2} \times 62.4 = 7020$ lb/ft   Weight of dam $= (2 + 8)/2 \times 18 \times 141 = 12,690$ lb/ft I of base $= 1 \times 8^3/12 = 42.667$ ft^4 Uniform stress on base due to weight of dam $= 12,690/8 = 1586$ lb/ft^2 Take moments about center of base (center of beam). $M = 5 \times 7020 - 3 \times (2 \times 18 \times 141) = 19,872$ lb·ft   Bending stress $= Mc/I = 19,872 \times 4/42.667 = 1863$ lb/ft^2 Contact pressure at face of dam $= 1586 - 1863 = -277$ lb/ft^2 Contact pressure at toe of dam $= 1586 + 1863$ lb/ft$^2 = 3449$ lb/ft^2

5-49 Answer (c). See Fig. 5S-49. Calculate moments about hinge. $M_{upstream} = (2 \times 6 \times 62.4) \times 3 \times 4 + (3 \times 62.4 \times 6) \times 4 \times 4 = 26,956.8$ lb·ft
$M_{downstream} = (3 \times \frac{3}{2}) \times 62.4 \times (3 + 2) \times 4 = 5616$ lb·ft
$F = (26,956.9) - 5616)/6 = 3556.8$ lb

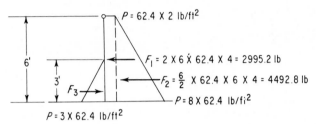

$$F_3 = \frac{3}{2} \times 62.4 \times 3 \times 4 = 1123.2 \text{ lb}$$

Figure 5S-49

5-50 Answer (b). See Fig. 5S-50. $P_A + 35 = P_1$ $P_1 - 299.2 = P_2$
$P_3 = P_2 + 12$ $P_B = P_3 - 244.2$ $P_A + 35 = P_3 - 12 + 299.2 = P_B + 244.2 - 12 + 299.2$ $P_A - P_B = 496.4$ in $= 41.37$ ft of water or 17.9 lb/in^2

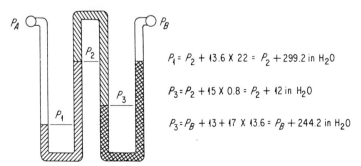

$P_1 = P_2 + 13.6 \times 22 = P_2 + 299.2$ in H_2O

$P_3 = P_2 + 15 \times 0.8 = P_2 + 12$ in H_2O

$P_3 = P_B + 13 + 17 \times 13.6 = P_B + 244.2$ in H_2O

Figure 5S-50

5-51 Answer (c). See Fig. 5S-51. Through 24-in branch: $h_L = 0.022 \times \frac{1500}{2} \times v_A^2/2g$; $h_L = 0.256v_A^2$ Through 12-in branch: $h_L = 0.022 \times \frac{3000}{1} \times v_B^2/2g$; $h_L = 1.026v_B^2$ Head losses from X to Y are equal $0.256v_A^2 = 1.026v_B^2$ $v_A = 2.006v_B$ $v_A \times 3.142 \text{ ft}^2 + v_B \times 0.785 \text{ ft}^2 = 60 \text{ ft}^3/s$ $7.087v_B = 60 \text{ ft}^3/s = 60 \text{ ft}^3/s$ $v_B = 8.466 \text{ ft/s}$ $Q_B = 6.65 \text{ ft}^3/s$

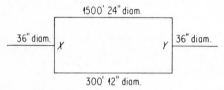

Figure 5S-51

5-52 Answer (d). See Fig. 5S-52. $A_C = 0.349 \text{ ft}^2$ $A_D = 0.348 \text{ ft}^2$ $A_E = 0.196 \text{ ft}^2$ To determine the flows and pressure drops through branches C and D - E, it is first necessary to estimate the flows and then recalculate using revised values. Branch C: $h_L = f_C (1000/0.667)$ $(v_C^2/2g) = 23.314f_C v_C^2$. Branch D - E: $v_E = (8/6)^2 v_D = 1.778v_D$; $h_L = f_D (500/0.667)(v_D^2/2g) + f_E (1000/0.500)(1.778v_D)^2/2g$. Assume $f_C = f_D = f_E$ for first approximation, then $h_L = 11.657f v_D^2 + 98.629f v_D^2 = 109.895v_D^2$. Head losses through branches C and D - E are equal, so $23,314f v_C^2 = 109.895f v_D^2$; $v_C = 2.171v_D$; $2 \text{ ft}^3/s = v_C \times 0.349 + v_D \times 0.349 = 1.107v_D$; $v_C = 3.923 \text{ ft/s}$; $v_D = 1.807 \text{ ft/s}$; $v_E = 3.213 \text{ ft/s}$. Using these values, recalculate Reynolds numbers and friction factors. Line C: $\text{Re} = 1.94 \times 0.667 \times 3.923/(1/47,800) = 2.4 \times 10^5$; $\epsilon/D = 0.0013$; $f_C = 0.022$ Line D: $\text{Re} = 1.94 \times 0.667 \times 1.807/(1/47,800) = 1.12 \times 10^5$; $\epsilon/D = 0.0013$; $f = 0.023$ Line E: $\text{Re} = 1.94 \times 0.500 \times 3.213/(1/47,800) = 1.5 \times 10^5$; $\epsilon/D = 0.00085/0.500 = 0.0017$; $f = 0.024$ Branch C: $h_L = 0.022 \times 1500 \times v_C^2/2g = 0.513v_C^2$ Branch D - E: $h_L = 0.023 \times 750 \times v_D^2/2g + 0.024 \times 2000 \times (1.778 \, v_D)^2/2g$; $h_L = 0.268v_D^2 + 2.385v_D^2 = 2.626v_D^2$; $0.513v_C^2 = 2.626v_D^2$; $v_C = 2.263v_D$; $2 = v_C \times 0.349 + v_D \times 0.349$; $v_D = 1.756 \text{ ft/s}$;

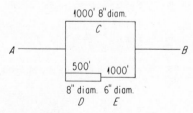

Figure 5S-52

$v_C = 3.974$ ft/s; $v_E = 3.122$ ft/s   Branch C: $h_L = 0.513 \times 3.974^2 =$ 8.10 ft   Branch D - E: $h_L = 0.023 \times 750 \times 1.756^2/2g + 0.024 \times 2000 \times 3.122^2/2g = 0.827 + 7.272 = 8.10$ ft, which checks   Section A: $v = 2/0.349 = 5.731$ ft/s; Re $= 1.94 \times 0.667 \times 5.731/(1/47,800) = 3.5 \times 10^5$; $\epsilon/D = 0.0013$; $f = 0.022$; $h_L = 0.022 \times 1500 \times 5.731^2/2g = 16.85$ ft   Section B: $v = 2/0.7854 = 2.546$ ft/s; Re $= 1.94 \times 2.546/(1/47,800) = 2.4 \times 10^5$; $\epsilon/D = 0.00085$; $f = 0.020$; $h_L = 0.020 \times 1500 \times 2.546^2/2g = 3.022$ ft   Total head loss $= 16.85 + 8.10 + 3.02 = 27.97$ ft or 12.11 lb/in^2

5M-1 Answer (c). $v = \mu/\rho$ (lb·s/ft^2)/(lb·s^2/ft^4) $=$ ft^2/s

5M-2 Answer (c).   Potential energy plus kinetic energy plus internal energy.

5M-3 Answer (c). $(100 - 25)/62.4 = 1.20$

5M-4 Answer (e). $h_L = f(L/D)(v^2/2g)$ (Velocity is a function of the diameter.)

5M-5 Answer (b).   Re $= \rho Dv/\mu$; if μ decreases, the Reynolds number increases.

5M-6 Answer (c).   When rocks are on barge, they displace their own weight of water. When they are dumped into the water, they displace only their own volume. Since the specific gravity of rock is greater than 1, the displacement of the water by the barge and rocks would reduce, and the level would fall.

5M-7 Answer (c).   By definition.

5M-8 Answer (e).   For laminar flow, $f = 64/$Re, and pipe roughness does not affect the value of the friction factor, unless it is an appreciable portion of the diameter of the pipe.

5M-9 Answer (a).   Cavitation results when the local pressure of the liquid drops below the vapor pressure at that temperature. Bubbles of vapor then form and collapse, exerting high, localized, impact forces.

5M-10 Answer (b).   If the air is removed, the buoyant force of the air will be lost.

6-1 Answer (b).   Write the balanced equation and determine the formula weights:

$$2Ag + 2HCl = H_2 \uparrow \ + 2AgCl$$

$$215.76 + 72.93 = 2.016 + 286.647$$

$7.20/286.647 = x/215.76$ $x = 5.419$ g Ag precipitate
$5.419/5.82 = 0.93$ or 93%

6-2 Answer (c).   Write the balanced equation and determine the formula weights:

$$Ca_3(PO_4)_2 + 3SiO_2 + 5C = 3CaSiO_3 + 2P + 5CO$$
$$_{310.2} {}_{61.69}$$

Formula weights show that 310.2 lb of phosphate rock is required to produce 61.96 lb of phosphorous. $2000/61.96 = x/310.2$
$x = 10,012$ lb or 5.01 tons

6-3 Answer (*b*).   200 A·h = 720,000 C = 7.461 faradays   Lead has a valence of 2, so 7.461 faradays would deposit $7.461 \times 207.2/2 = 1131$ g of Pb or $1131 \times (303.2/207.2) = 1655$ g of $PbSO_4$ or 3.65 lb

6-4 Answer (*d*).   Write the balanced equation and determine the formula weights:

$$NaCO_3 + Ca(OH)_2 \rightarrow 2NaOH + CaCO_3$$

$$106 + 74.1 \quad = 80 \quad + 100.1$$

$(11.023/106) \times 74.1 = 7.706$ lb of slaked lime $7.706 \times (56.1/74.1)$
$= 5.83$ lb lime (CaO)

6-5 Answer (*b*).   Heat given up = heat absorbed $Q_1 \times (140 - 80) \times 10 = Q_2 \times (80 - 65) \times 8.33$ $Q_1 = 8.62$ gal/min $Q_2 = 41.38$ gal/min $(8.62/50) \times 10 + (41.38/50) \times 8.33 = 8.618$ Btu/gal·F

6-6 Answer (*e*).   For each 100 ft^3 of gas:

CO	19.8 ft^3	$2CO + O_2 = 2CO_2$	9.9 ft^3 of O_2
H_2	15.1	$2H_2 + O_2 = 2H_2O$	7.55
CH_4	1.3	$CH_4 + 2O_2 = CO_2 + 2\,H_2O$	2.6
			20.05 ft^3

There would be 20.05 ft^3 of O_2 required for each 100 ft^3 of gas, but 1.3 ft^3 is contained in the gas, so an additional 18.75 ft of O_2 would be required. Air is 21 percent O_2 by volume. $18.75/0.21 = 89.3$ ft^3 of air.

6-7 Answer (*c*).   Combining the reaction equations gives $2ZnS + 4O_2 + 2H_2O \rightarrow 2ZnO + 2H_2SO_4$, so 1 mol of ZnS requires 2 mol of O_2. Atomic weights: Zn = 65.38, S = 32.07, and O = 16.00. Thus 95.45 lb of ZnS or 190.90 lb of ore requires 64 lb of O_2, so 1 ton of ore would require 670.5 lb of O_2. Volume = $WRT/P = 670.5 \times 48.25 \times 492/(14.7 \times 144) = 7519.4$ ft^3 of O_2 $7519.4/0.21 = 35,806$ ft^3 of air

6-8 Answer (*a*).   Write the balanced equation and determine the formula weights:

$$CaCO_3 + 2HCl \rightarrow CO_2 + H_2O + CaCl_2$$

$$100 + 72 \quad = 44 \quad + 18 \quad + 110$$

So $\frac{100}{44} = 2.272$ g of $CaCO_3$ would be required to produce 1 g of CO_2; 1 mol of CO_2 would weigh 44 g and occupy 22.4 liters at 0°C and 760 mm of pressure $V_2 = V_1(T_2/T_1)(P_1/P_2) = 10 \times \frac{273}{298} \times \frac{770}{760} = 9.28l$
$(9.28/22.4) \times 44 = 18.2$ g $18.2/44 = x/100$ $x = 41.4$ g of $CaCO_3$

6-9 Answer (d). Write the balanced equation and determine the formula weights:

$$FeS_2 + (H_2O) + (O_2) \rightarrow 2H_2SO_4 + (Fe)$$

(Atomic weights: Fe = 55.85, S = 32.07, O = 16.00, and H = 1.01)

$$119.99 \qquad\qquad 196.18$$

1 ton of pyrite will produce $(2000/119.99) \times 196.18 = 3270$ lb of 100 percent H_2SO_4 or 5450 lb of 60 percent H_2SO_4

6-10 Answer (e). Write the balanced equation and determine the formula weights:

$$3Ag + 4HNO_3 \rightarrow 3AgNO_3 + NO + 2H_2O$$

$$323.7 + 252 \quad = 509.7 \quad + 30 \; + 36$$

$323.7/30 = 10.79$ lb Ag to liberate 1 lb of NO $\qquad W = P \times$ Volume/ $RT = 14.7 \times 144 \times 100/(49.5 \times 520) = 8.22$ lb $\qquad 8.22 \times 10.79 = 88.73$ lb

6-11 Answer (e). From the given equations 2 mol of NaCl is required to produce 1 mol of soda ash, or 117 g of NaCl to produce 106 g of soda ash. $2200/106 \times 117 = 2428$ g of NaCl

6-12 Answer (a). From the formula weights: 100.1 lb $CaCO_3$ produces 44 lb of CO_2; 84.3 lb $MgCO_3$ produces 44 lb of CO_2. 1 ton of limestone contains 1200 lb of $CaCO_3$ and 800 lb of $MgCO_3$ $\qquad (1200/100.1) + (800/84.3) = 21.48$ or 21.48 mol of CO_2 produced per ton of limestone $21.48 \times 380 = 8161$ ft^3

6-13 Answer (b). 1 A·h = 3600 C = $\frac{3600}{96,500}$ = 0.373 faraday; 1 faraday = 1 Avogadro number of electrons = amount of electricity to deposit 1 gram-equivalent of hydrogen at the cathode. Simultaneously 1 gram-equivalent, or 8.00 g, of oxygen would be formed at the anode in the electrolysis of water.

Thus 0.0373 g of H^+ would be attracted to, and neutralized at, the cathode. Less electric energy is required to deposit hydrogen than to deposit sodium, so hydrogen would be deposited at the cathode and the sodium would remain in solution as Na^+ ions. The cathode reaction therefore would produce free hydrogen in the amount of 0.0373 g.

6-14 Answer (a). Write the balanced equation:

$$C_3H_8 + 5O_2 = 3CO_2 + 4H_2O$$

So 5 mol of oxygen is required to burn 1 mol of propane gas. ($5 \times 10,500)/0.21 = 250,000$ ft^3 air required.

6-15 Answer (c). Write the balanced equation and determine the formula weights:

$$Ca_3(PO_4)_2 + 3H_2SO_4 \rightarrow 2H_3PO_4 + 3CaSO_4$$

$$310 + 294 \quad = 196 \quad + 408$$

1 ton of rock contains 1700 lb of $Ca_3(PO_4)_2$ of which 95 percent or 1615 lb is converted to phosphoric acid, so $(1615/310) \times 196 = 1021$ lb phosphoric acid produced per ton of rock $\frac{1021}{2000} \times 100 = 51.05$ tons

6-16 Answer (*e*).

Gas	ft³/100 ft³			
CO_2	12	$H_2 + \frac{1}{2}O_2 = H_2O$	0.5×4	$=\ \ 2$ ft³ O_2
H_2	4	$CH_4 + 2O_2 = CO_2 + 2H_2O$	2.0×5	$= 10$
CO	23	$CO + \frac{1}{2}O_2 = CO_2$	0.5×23	$= 11.5$
N_2				23.5 ft³ O_2

23.5 ft³ of oxygen required for combustion of 100 ft³ of gas
$23.5/0.21 = 111.9$ ft³ of air

6-17 Answer (*d*).   Write the balanced equation and determine the formula weights:

$$Fe + S \rightarrow FeS$$

$$55.85 + 32.01 = 87.86$$

$$3\ lb + 2\ lb = 60\%\ FeS\ or\ 3\ lb\ FeS$$

3 lb of FeS requires: $3 \times (55.85/87.86) = 1.907$ lb Fe
$3 \times (32.01/87.86) = 1.093$ lb S
So the product contains 3 lb FeS + 1.093 lb Fe + 0.907 lb S, which gives 60 percent FeS, 21.9 percent Fe, and 18.1 percent S.

6-18 Answer (*a*).   Write the balanced equation.

$$4FeS_2 + 11O_2 + N_2 \rightarrow 2Fe_2O_3 + 8SO_2 + N_2$$

11 volumes of O_2 would carry along with it $3.76 \times 11 = 41.36$ volumes of N_2; 50 percent excess air would produce 8 volumes of SO_2 plus 5.5 volumes of O_2 plus 62.04 volumes of N_2 in the gases produced $8/75.54 = 0.106$ or 10.6 percent SO_2

6-19 Answer (*b*).   Sum the partial pressures: $300 + 100 + 50 + 189 + 12 = 651$ mm   Volume percent = pressure percent for gas mixture $300/651 = 0.461$ or 46.1%

6-20 Answer (*e*).   Write the balanced equation.

$$Ca_3(PO_4)_2 + 3SiO_2 + 5C = 3CaSiO_3 + 2P + 5CO$$

So 5 mol of C is required per mole of phosphate.

6-21 Answer (*c*).   Write the balanced equation and determine the formula weights:

$$Ca(HCO_3)_2 + Ca(OH)_2 \rightarrow 2CaCO_3 + 2H_2O$$

$$162 + 74 \quad = 200 \quad + 36$$

So 74 lb of lime would react with 162 lb of calcium carbonate, or $\frac{74}{162}$ = 0.457 lb/lb.

6-22 Answer (*c*).   Write the balanced equation and determine the formula weights:

$$2C_8H_{18} + 25O_2 \rightarrow 16CO_2 + 18H_2O$$

$$228 + 800 \;\; = 704 \;\;\; + 324$$

$\frac{800}{228}$ = 3.5 lb oxygen/lb fuel

6-23 Answer (*a*).   Write the balanced equation and determine the formula weights:

$$Na^+ + Ca^- \rightarrow Na + \tfrac{1}{2}Cl_2 \quad \text{Molecular wt of NaCl} = 58.5$$

$$= 23 \;\; + 35.5$$

For 50 percent efficiency 58.5 lb of NaCl would produce 17.75 lb Cl (2000/58.5) × 17.75 = 606.8 lb or 17.09 mol of Cl or 8.545 mol Cl$_2$
8.545 × 359 ft^3/lb·mol = 3067 ft^3

6-24 Answer (*e*).   Write the balanced equation:

$$C_{12}H_{22}O_{11} + 12O_2 \rightarrow 12CO_2 + 11H_2O$$

12 volumes of O$_2$ will carry along with it 12 × 3.76 = 45.12 volumes of N$_2$ 12/(12 + 11 + 45.12) = 0.176 or 17.6%

6-25 Answer (*b*).

C	$C + O_2 \rightarrow CO_2$	$\frac{32}{12}$ × 0.7161 = 1.909 lb O$_2$
H	$H_2 + \tfrac{1}{2}O_2 \rightarrow H_2O$	$\frac{16}{2}$ × 0.0526 = 0.421 lb O$_2$
S	$S + O_2 \rightarrow SO_2$	$\frac{32}{12}$ × 0.0074 = 0.0074 lb O$_2$
	(−.0979 lb in coal)	2.2395 lb O$_2$

Air is 23.3 percent O$_2$ by weight, so 2.2395 lb O$_2$ would carry with it (76.7/23.3) × 2.2395 = 7.3721 lb N$_2$

Flue gas by weight

CO$_2$	$\frac{44}{12}$ × 0.7161	=	2.6257 lb
H$_2$O	$\frac{18}{2}$ × 0.0526	=	0.4737 lb
N$_2$	7.3721 + 0.0123	=	7.3844 lb
SO$_2$	$\frac{64}{32}$ × 0.0074	=	0.0148 lb
			10,4986 lb

excluding ash, which would not be a gas.

6-26 Answer (d).   Write the balanced equation and determine the formula weights:

$$CuO + H_2 = Cu + H_2O$$
$$79.5 + 2 = 63.5 + 18$$

$(63.5/79.5) \times 10 = 7.99$ g

6M-1 Answer (b). $C + O_2 \rightarrow CO_2$

6M-2 Answer (a).   By definition.

6M-3 Answer (e).   By definition.

6M-4 Answer (a).   Avogadro's law.

6M-5 Answer (e).   Should be $Ba(NO_3)_2$.

6M-6 Answer (a).   By definition.

6M-7 Answer (c).   Avogadro's number by definition.

6M-8 Answer (a). $2.00 + 32.06 + 4 \times 16 = 98.06$

6M-9 Answer (a).   Law of le Chatelier; some of the free Cl_2 would combine with some of the excess H_2.

6M-10 Answer (c). $2Na + 2H_2O \rightarrow 2NaOH + H_2$

7-1 Answer (c).   Thermal efficiency = work/(heat added to working fluid) Heat into engine $h_1 = 1322.1$ Btu/lb (from steam table) Constant entropy process $s_1 = 1.6767$ Btu/(lb)(°F)   At discharge conditions, 4-in Hg: $s_{vap} = 1.9214$ Btu/(lb)(°F), $s_{liq} = 0.1738$ Btu/(lb)(°F); $h_{vap} = 1116.0$ Btu/lb, $h_{liq} = 93.34$ Btu/lb   Quality of discharge steam = $(1.6767 - 0.1738)/1.7476 = 0.860$ Heat in steam out of engine = $93.34 + 0.860 \times 1022.7 = 972.86$ Btu/lb   Heat extracted in engine = $1322.1 - 972.86 = 349.2$ Btu/lb   Heat added = $1322.1 - 93.34 = 1228.8$ Btu/lb   Efficiency = $349.2/1228.8 = 0.284$ or 28.4%

7-2 Answer (b).   Work = $P_1V_1 \ln (P_1/P_2)$; volume = WRT/P For 100 lb of mixture: $\frac{30}{44} = 0.682$ mol of CO_2; $\frac{70}{32} = 2.187$ mol of O_2, so 2.869 mol of gas $100/2.869 = 34.86$ lb/mol $R = 1544/34.86 = 44.30$ ft/°R $V_1 = 5 \times 44.30 \times 520/(15 \times 144) = 53.32$ ft^3 Work = $(15 \times 144) \times 53.32 \times \ln \frac{15}{150} = -265,190$ ft·lb (work done on gas)

7-3 Answer (e).   Steam at 20 lb/in² gauge saturated $h_{vap} = 960.1$ Btu/lb $T = 227.96$°F   Heat required = (200×8.40) lb $\times 0.85$ Btu/(lb)(°F) $\times (150 - 62)$°F = $125,664$ Btu   Heat from steam = $960.1 + (227.96 - 150) = 1038.1$ Btu/lb $125,664/1038.1 = 121.1$ lb steam

7-4 Answer (a). $PV^k = $ constant $k = c_p/c_v = 0.2025/0.1575 = 1.286$ $V_2 = V_1(P_1/P_2)^{1/k}$ $P_1 = 25 \times 0.49 = 12.25$ lb/in² abs $P_2 = 5 \times 14.7 = 73.5$ lb/in² abs $V_2 = 10 \times (12.25/73.5)^{0.778} = 2.48$ ft^3

7-5 Answer (d). 30 lb/in^2 abs and 400°F $\quad h = 1237.9$ Btu/lb $\quad V =$ 16.897 ft^3/lb $\quad$ water at 70°F $\quad V = 0.0161$ ft^3/lb $\quad h_1 = 38.04$ Btu/lb $\quad V_1 = 0.0161$ ft^3/lb $\quad h_2 = 1237.9$ Btu/lb $\quad V_2 =$ 16.897 ft^3/lb (400°F steam at 30 lb/in^2 abs) $\quad P\,\Delta V/J =$ $(30 \times 144) \times 16.881/778 = 93.73$ Btu/lb $\quad \Delta u = 1237.9 - 38.04 -$ $93.73 = 1106.1$ Btu/lb

7-6 Answer (d). Heat transfer through concrete and brick: $Q = \Delta t/[(L_1/k_1)$ $+ (L_2/k_2)] = 38/[(0.667/0.68) + (0.333/0.36)] = 19.94$ Btu/(h)(ft^2) $Q = h_C\,\Delta t = h_C \times 11 \quad h_C = 1.813$ Btu/(h)(°F) $\quad Q = H_B\,\Delta t =$ $h_B \times 7 \quad h_B = 2.849$ Btu/(h)(°F) $\quad 1/U = (1/h_C) + (L_1/k_1) +$ $(L_2/k_2) + (1/h_B) = (1/1.813) + (0.667/0.68) + (0.333/0.36) +$ $(1/2.849) = 2.808 \quad U = 0.356$ Btu/(h)(ft^2)(°F) or $U =$ 19.94 Btu/(h)(ft^2)/56°F $= 0.356$ Btu/(h)(ft^2)(°F)

7-7 Answer (a). h of water/steam: h of 115 lb/in^2 abs sat steam $=$ 1189.7 Btu/lb; h of 180°F saturated water $= 147.92$ Btu/lb; $\Delta h =$ 1041.8 Btu/lb; $Q = 1,041,800$ Btu; heat required $= 1,041,800/0.75 =$ 1,389,067 Btu/1000 lb steam; heat per barrel $= 42 \times 8.33 \times 1.008 \times$ $18,250 = 6,436,024$ Btu/bbl; cost of heat per 1000 lb of steam $=$ $(1,389,067/6,436,024) \times \$2.50 = \$0.540$

7-8 Answer (c). $\Delta U = 0.08 \times (760 - 560) + 0.002 \times (760^2 - 560^2) = 16 +$ $528 = 544$ Btu $\quad 544$ Btu $\times 778$ ft·lb/Btu $= 423,232$ ft·lb of work

7-9 Answer (e).

Gas	Vol %	MW	Gas, lb	Wt %
N$_2$	70	28	19.6	62.8
CO$_2$	20	44	8.8	28.2
CO	10	28	2.8	9.0
			31.2	100.0

7-10 Answer (b). Δmass $= 0.001 \times (1/32.17) = 3.108 \times 10^{-5}$ slugs $E = 3.108 \times 10^{-5}$ (lb·s^2/ft) $\times (186,000 \times 5280)^2$ (ft^2/s^2) $=$ 2.995×10^{13} ft·lb $\quad$ kW·h $= 3413$ Btu $\times 778$ ft·lb/Btu $=$ 2.655×10^6 ft·lb/kW·h $\quad 2.998 \times 10^{13}/(2.665 \times 10^6) =$ 1.125×10^7 kW·h

7-11 Answer (a). 60 lb/min steam flow $\quad 60 \times (1525 - 1300) = 13,500$ Btu/min or 10,503,000 ft·lb/min $\quad 60 \times (110^2 - 810^2)/2g =$ $-600,560$ ft·lb $\quad$ Net work $= 9,902,440$ ft·lb/min or 300 hp

7-12 Answer (e). Use Carrier's equation. $P_{swb} = 0.2951$ (from steam table) $P = 14.0$ lb/in^2 abs $\quad t_{wb} = 64$°F $\quad t = 85$°F

$$P_{wv} = 0.1951 - \frac{14.00 - 0.2951}{2830 - 1.44 \times 65}(85 - 64) = 0.1900$$

At 85°F, saturated pressure $= 0.5959$ lb/in^2 abs $\quad$ Relative humidity $= 0.1900/0.5959 = 0.319$ or 31.9% $\quad$ Note that from the steam table, 0.1900 = saturation pressure at 51.7°F, which is the dew-point temperature.

7-13 Answer (d). $P_2 = P_1(V_1/V_2)^k = 20 \times 18^{1.4} = 1144$ lb/in² abs $(P_2V_2 - P_1V_1)/(1 - k) = (1144 \times 0.0278 - 20 \times 0.500) \times 144/(1 - 1.4) = -7840$ ft·lb (work done on gas)

7-14 Answer (c). 1.25 in²/3.00 in = 0.4167 in average height, so average pressure = $0.4167 \times 420 = 175$ lb/in² gauge   Cylinder displacement = 10.18 in² × 3.00 in = 30.54 in³   Four-cycle, so $(2800/2) \times 8 = 11,200$ power strokes/min   11,200 × 30.54 × 175 × $\frac{1}{12}$ = 4,988,200 ft·lb/min   4,988,200/33,000 = 151.2 hp

7-15 Answer (b). $Q = UA\,\Delta t$ $U = 580/(2400 - 80) = 0.25$, so $1/U = 4.00$ $1/U = L_1/k_1 + L_2/k_2 = 0.8/0.75 + L_2/0.2$ $L_2 = 0.587$ ft or 7.04 in

7-16 Answer (a). $Q = UA\,\Delta T$ $1/U = L_1/k_1 = \frac{3}{12}$ Note that k is given in British thermal units per (hour × square feet × degrees Fahrenheit per *inch*) $Q = 4(32 - 2) = 120$ Btu/h   120/144 = 0.833 lb ice formed per hour per square foot of area   Ice has a specific weight of $0.9 \times 62.4 = 56.16$ lb/ft³   0.833/56.16 = 0.0148 ft³/h·ft² or 0.178 in/h

7-17 Answer (c). 0.90 cal/(cm²)(s)(°C)/cm $1/U = L/k = 15/0.09 = 16.67$ cal/(cm²)(s)(°C) $Q = UA\,\Delta t = 0.0600 \times 6.0 \times 100 = 36$ cal/s   Heat of fusion of ice = 79.7 cal/g $36 \times 60 \times 2/79.7 = 54.2$ g

7-18 Answer (e). $\Delta h = 1525 - 1300 = 225$ Btu/lb or 225 × 778 = 175,050 ft·lb/h $(v^2/2g) = (1100^2 - 810^2)/2g = 8608$ ft·lb/lb   Net energy removed from steam = 183,658 ft·lb/lb   60 × 183,658 = 11,019,480 ft·lb/min or 334 hp

7-19 Answer (e). This is an example of nonflow work, so the form of the general energy equation which applies is $U_1 + Q = U_2 + W$ $W = (P_2V_2 - P_1V_1)/(1 - n)$ $P_1/P_2 = (V_2/V_1)^n$ $300/40 = (30/6)^n$ $n = 1.25$ $W = (30 \times 40 - 300 \times 6) \times 144/(1 - 1.25) = -345,600$ ft·lb work done by gas $T = PV/R$ $R = 1544/28 = 55.1$ ft/°R $T_1 = (300 \times 144) \times 6/55.1 = 4704$°R $T_2 = (40 \times 144) \times 30/55.1 = 3136$°R $c_p = 0.248$ $c_v = 0.248/1.40 = 0.177$ $\Delta u = 0.177 \times (4704 - 3136) = 278$ Btu/lb $(345,600/778) - 278 = 166$ Btu added to gas

7-20 Answer (a). Heat added = $c_v\,\Delta T = 3 \times 4.6 \times (40 - 20) = 276$ g·cal   Gas expands adiabatically, 1 mol at 0°C and 1 atm = 22.4 liters $V_1 = 3 \times (293/273) \times 22.4 = 72.12$ liters   Constant volume, so $P_2/P_2 = T_1/T_2$ $P_2 = (313/293) \times 1.0 = 1.068$ atm $T_2 = 313$ K $P_3 = P_2(T_3/T_2)^{k/(k-1)} = 1.068 \times (293/313)^{1.4/0.4} = 0.848$ atm $V_3 = V_2(T_2/T_3)^{1/(k-1)} = V_2(313/293)^{2.5} = 85.06$ liters   1.0 atm = 1,013,000 dyn/cm²   Work = $(P_3V_3 - P_2V_2)/(1 - k) = (0.848 \times 85,060 - 1.0683 \times 72,120) \times 1,013,000/(1 - 1.4) = -1.245 \times 10^{10}$ dyn·cm   1 J = 10^7 dyn·cm, so work = 1245 J

7-21 Answer (a). System originally contains 5 qt of coolant, or 0.192 qt/qt. Require 9 qt of coolant, or 0.346 qt/qt $(26 - X) \times 0.192 + X = 26 \times 0.346 = 9$ $X = 4.95$ qt

7-22 Answer (c). $h_{in} = 1381.8$ Btu/lb   1350 hp $\times$ 33,000/778 = 57,262 Btu/min work   144,000/60 = 2400 Btu/min heat loss
Total heat extracted = 59,662 Btu/min or 397.7 Btu/lb of steam
$h_{out} = 984.1$ Btu/lb   0.8 lb/in^2 abs/0.49 = 1.63 in Hg
For 1.5 in Hg: $h_{vap} = 1101.7$ Btu/lb; $h_{evap} = 1042.0$ Btu/lb
1101.7 − 984.1 = 117.6 Btu/lb   117.6/1042.0 = 0.1129, or 11.29%
condensed, so quality = 88.71%

7-23 Answer (b). $dQ = Wc_p\, dT$ $\Delta s = \int dQ/T = \int (W \times c_p/T)\, dT$
$W = PV/RT$ $\Delta s = W \int (6.6 - 7.2 \times 10^{-4}T)/T\, dT$
$\int (6.6/T)\, dT - \int 7.2 \times 10^{-4}\, dT = [6.6 \ln T - 7.2 \times 10^{-4}\, T]$
$\Delta s = 8.891 - (1.44 - 0.374) = 7.825$ Btu/(mol)(°R)

7-24 Answer (d). Buoyancy = $35,000 \times (0.081 - 0.0056) = 2639$ lb
2639 − 425 = 2214 lb

7-25 Answer (b). $p_1 = 104.7$ lb/in^2 abs $T_1 = 535$°R $p_2 = 14.7 + (3.80/12) \times (62.4/144) = 14.837$ lb/in^2 abs $T_2 = 488$°R
$V_2 = V_1(p_1/p_2)(T_2/T_1) = 1 \times (104.7/14.837) \times (488/535) = 6.436$ ft^3/ft^3

7-26 Answer (e). Average height of diagram = 1.20/2.80 = 0.429 in
MEP = $0.429 \times 425 = 182.3$ lb/in^2 Volume/stroke = $(\pi/4) \times 3.5^2 \times 3 = 28.86$ in^3 Four-cycle, so $6 \times \frac{3000}{2} = 9000$ power strokes/min $9000 \times 28.86 \times 182.3 = 47,350,602$ in·lb/min or 119.6 hp

7-27 Answer (a). Btu/(lb)(°F) = dQ/dt $dQ/dt = 0.32 - 0.00016t = 0.296$ Btu/(lb)(°F)

7-28 Answer (c). 100 hp = 254,600 Btu/h output   Efficiency = (1000 − 600)/1000 = 0.40   254,600/0.40 = 636,500 Btu/h heat supplied   60% or 381,900 Btu/h lost

7-29 Answer (d). $v = C\sqrt{2gh}$ on ground, pressure = $(v_2^2/2g)w$
At altitude same pressure = $(v_2^2/2g)(w/2) = (v_1^2/2g)w$, so $v_2^2 = 2v_1^2$ or $v_2 = 282.8$ kn = true speed of plane

7-30 Answer (c). Constant temperature, so $P_2 = \frac{20}{8}P_1 = 2.5P_1$
$P_1 + 68$ cm = 1 atm = $80 \times 0.707 + P_2 = 56.56$ cm + $2.5P_1$
$P_1 = 7.627$ cm   Accurate barometer reading = $68 + P_1 = 75.627$ cm

7-31 Answer (b).

	1	2	3	4 and 1
P	p_1	$3P_1$	P_1	P_1
V	V_1	V_1	$3V_1$	V_1
T, K	300	900	900	300

1 to 2: $P_2 = (T_2/T_1)P_1 = 3P_1$; this is a nonflow process, so $\Delta Q = \Delta U = c_v \Delta T = 3 \times (900 - 300) = 1800$ cal   2 to 3: $V_3 = (P_2/P_1)V_2 = 3V_1$; work $= (P_2 V_2/J) \ln (V_3/V_2) = (RT_2/J) \ln 3 = 2 \times 900 \times 1.0986 = 1978$ cal; constant temperature, so c_v; $T = 0$ and U remains constant; thus heat is added to gas in the amount of 1978 cal 3 to 1: $c_p = R + c_v = 5$ cal/(mol)(K); heat removed $= 5 \times (900 - 300) = 3000$ cal; heat added $= 1800 + 1978 = 3778$ cal; heat exhausted $= 3000$ cal; net work done $= 778$ cal

7-32 Answer (a). $w = 1$ lb/ft³   At 350°F, $V_{liq} = 0.01799$ ft³/lb $V_{vap} = 3.342$ ft³/lb $\Delta V = 3.324$ ft³/lb   1 lb of H_2O would occupy 1.0 ft³ of volume; $X\%$ would be in the form of vapor, $(1 - X)\%$ would be in the form of liquid; then $X \times 3.342$ ft³/lb $+ (1 - X) \times 0.01799 = 1.0$ ft³; $X = 0.295$, so 29.5% of water would evaporate

7-33 Answer (e).   Work $= 50,000$ ft·lb or 64.27 Btu; 64.27 Btu work done plus 12 Btu heat lost $= 76.27$ Btu change in U

7-34 Answer (e).   29 in Hg $\times 0.490$ lb/in²/in Hg $= 14.21$ lb/in² abs Pressure $= 145 + 14.21 = 159.21$ lb/in² abs   Internal energy $= h - PV/J$ enthalpy at 159 lb/in² abs $= 1129.8$ Btu/lb; enthalpy at 160 lb/in² abs $= 1130.2$ Btu/lb; enthalpy at 159.21 lb/in² abs $= 1129.9$ Btu/lb

7-35 Answer (b). $1400 \times (85 - T) = 200 \times (T - 54)$ $T = 81.126$°F The relative humidity can be found from the psychrometric chart or from the steam table.
FROM PSYCHROMETRIC CHART:
   85°F 50% RH $V = 13.75$ ft³/lb
                      90 grains moisture/lb dry air
   54°F 40% RH $V = 12.90$ ft³/lb
                      30 grains moisture/lb dry air
$1400/13.75 \times 90 + 200/12.90 \times 30 = 9164 + 465 = 9629$ grains moisture   At 81.1°F, $V = 13.62$ ft³/lb $1600/13.62 = 117.5$ lb dry air or $1400/13.75 + 200/12.90 = 117.3$ lb dry air, take average $9629/117.4 = 82.0$ grains moisture/lb dry air for 81°F, which gives RH $= 52\%$
FROM STEAM TABLE:
   Saturated steam at 85°F $V = 543.5$ ft³/lb
For 50% RH $(1400/543.5) \times 0.50 = 1.288$ lb moisture, or 9016 grains at 54°F; $V = 1481.0$ ft³/lb. For 40% RH $(200/1481.0) \times 0.40 = 0.0540$ lb moisture, or 378 grains. Total moisture $= 1.342$ lb in 1600 ft³ of air at 81°F; $1600/1.342 = 1192.2$ ft³/lb. Saturated steam at 81°F $= 613.9$ ft³/lb; RH $= 613.9/1192.2 = 51.5\%$.

7-36 Answer (d).   This problem can be solved by using the psychrometric chart or the steam tables.
FROM PSYCHROMETRIC CHART: Read directly by following line from 81°F and 50% RH; wet-bulb temperature $= 67.5$°F
FROM STEAM TABLE: Air cools as water evaporates until the air becomes saturated at the temperature to which it has cooled. In the range of 81°F

to a slightly lower temperature, the heat of vaporization of water is about 1095 Btu/lb. $W = PV/RT = 14.696 \times 144 \times 1600/(53.3 \times 541) = 117.4$ lb air; c_p for air = 0.24, so $117.4 \times 0.24 = 28.18$ Btu to cool air 1.0°F. The problem can be solved by iteration. For a first try, cool to 70°F, or cool 11.1°F. $11.1 \times 28.18/1095 = 0.2857$ lb water into vapor, giving $1.342 + 0.2857 = 1.6277$ lb of water vapor. Similarly, by iteration, using 1600 ft^3 of air:

T, °F	Water vapor, lb	V, vapor, ft^3/lb	V, sat steam, ft^3/lb
81.1	1.342	1192.2	613.9
70	1.628	983.0	867.9
65	1.756	911.0	1021.4
67	1.705	938.5	956.6
68	1.679	952.9	925.9

Thus the wet bulb temperature lies between 67 and 68°F, by interpolation, $t = 67.4$°F. Similarly, the dew point is that temperature at which the moisture contained in the original volume of air at 81°F would start to condense, or the temperature at which the specific volume of saturated steam would equal 1192 ft^3/lb which is just under 60°F.

7-37 Answer (c). hp·h = 2546 Btu $2546/(0.56 \times 19,800) = 0.230$ or 23.0%

7-38 Answer (b). Nine-cylinder, four-cycle engine running at 1950 r/min $9 \times 1950/2 = 8775$ power strokes/min   Volume swept out = $(\pi/4) \times 4.1825^2 \times 6 \times 8775 = 957,703^3/\text{min}$ 225 hp $\times 33,000 = 7,425,000$ ft·lb/min output $(7,425,000 \times 12 \text{ in·lb/min})/(957,703 \text{ in}^3/\text{min}) = 93.04$ lb/in^2

7-39 Answer (e). $P_2 = P_1(V_1/V_2)^k = 13 \times 12^{1.4} = 421.5$ lb/in^2 abs $T_2 = T_1(V_1/V_2)^{k-1} = 580 \times 12^{0.4} = 1567$°R or 1107°F   Or $V_1 = RT_1/P_1 = 53.3 \times 580/(113 \times 144) = 1651$ ft^3/lb $V_2 = 16.51/12 = 1.376$ ft^3/lb $T_2 = (421.5 \times 144) \times 1.376/53.3 = 1567$°R

7-40 Answer (a). $p_1 = 14.4$ lb/in^2 abs; $p_2 = 39.7$ lb/in^2 abs   Atmospheric pressure = 14.4 lb/in^2 abs $V_2 = V_1(P_1/P_2)^{1/1.35} = 20,000 \times (14.4/39.4)^{0.741} = 9486$ ft^3/min $T_2 = T_1(V_1/V_2)^{1.35-1} = 530 \times (2.108)^{0.35} = 688$°R   At constant pressure $V_2 = V_1(T_2/T_1) = 9486 \times (530/686) = 7306$ ft^3/min

7-41 Answer (c). 95 hp $\times \frac{2546}{60} = 4031$ Btu/min $\Delta T = 131 - 65 = 66$ Btu/lb of H$_2$O $\frac{4031}{66} = 61.08$ lb/min $61.08/8.33 = 7.33$ gal/min

7-42 Answer (b). $du/dt = 0.08 + 0.004t$ $\int du = \int (0.08 + 0.004t)\, dt = 0.08 \times (300 - 100) + 0.004 \times (300^2 - 100^2)/2$ $\Delta u = 16 + 160 = +176$ Btu/lb or 176 Btu work done by the gas

7-43 Answer (b).   Assume 100 lb of gas.

	Weight, lb	MW	Moles	Vol %	c_p	$c_p \times$ Wt %
N_2	80	28	2.857	83.91	0.248	0.1984
CO_2	12	44	0.273	8.02	0.202	0.0242
O_2	7	32	0.219	6.43	0.219	0.0153
H_2O	1	18	0.056	1.64	0.430	0.0043
			3.405	100.00		0.2422

Partial pressure of water vapor $= 0.0164 \times 14.696 = 0.241$ lb/in^2 abs
Cool at constant pressure until water starts to condense
0.241 lb/in^2 abs $=$ saturated pressure of steam at 58.25°F
c_p of steam determined from steam table: $h_{100} - h_{60} =$
$1105.2 - 1088 = 17.2$ $17.2/40 = 0.430$ Btu/(lb)(°F)
$(100 - 58.25) \times 0.2422 = 10.11$ Btu

7-44 Answer (d). $PV^n =$ constant $P_1 = 140$ lb/in^2 abs $V_1 = 5.0$ ft^3
$P_2 = 50$ lb/in^2 abs $V_2 = 10$ ft^3 $\frac{140}{50} = (10.0/5.0)^n$
$n = 1.485$ $T_2 = T_1(V_2/V_1)^{1-k} = 580 \times (10.0/5.0)^{-0.485} = 414.4$°R
$T_2 = -45.6$°F $\varDelta T = 165.6$°F $c_v = c_p - (96.2/778) =$
0.426 Btu/(lb)(°F) $W = PV/RT = 140 \times 144 \times 5.0/(96.2 \times 580) =$
1.807 lb $\varDelta U = 1.807 \times 165.6 \times 0.426 = 127.4$ Btu reduction in
internal energy   Work $= (P_2V_2 - P_1V_1)/(1 - n) = (50 \times 144 \times 10 - 140 \times 144 \times 5)/(1 - 1.485) = 59,381$ ft·lb work done by gas. This is a
nonflow process, so $\frac{59,381}{778} - 127.4 = -51.1$ Btu reduction in internal energy
which gas gives up in addition to doing work. *An alternative method
is:* $Q = wc_v \dfrac{k - n}{1 - n}(T_2 - T_1)$ $k = c_p/c_v = 0.55/0.426 = 1.291$
$Q = 1.807 \times 0.426 \times (1.291 - 1.485)/(1 - 1.485) \times (414.4 - 580) = -51.0$ Btu

7-45 Answer (a). $\varDelta S = Wc_n \ln(T_2/T_1)$ $c_n = c_v(k - n)/(1 - n) =$
$0.426 \times (0.194/0.485) = 0.1704$ Btu/(lb)(°F) $\varDelta S = 1.807 \times 0.1704 \times \ln(414.4/580) = -0.1035$ Btu/°F

7-46 Answer (c).   hp $= 2\pi \times 150 \times 1500/33,000 = 42.84$ hp $42.84 \times 2545 \times \frac{15}{60} = 27,257$ Btu work in 15 min   Heat supplied $= 0.70 \times 8.33 \times 19,100 = 111,372$ Btu   Efficiency $= 27,257/111,372 = 0.245$
or 24.5%

7-47 Answer (e).   5 lb of CH_4 plus 10 lb of C_2H_6 Methane pressure $=$
$WRT/V = 5 \times \frac{1544}{16} \times \frac{520}{19} = 13,205$ lb/ft$^2 = 91.70$ lb/in^2 abs
For ethane $P = 10 \times \frac{154}{30} \times \frac{520}{19} = 14,085$ lb/ft^2 or 97.82 lb/in^2 abs
Total pressure $= 189.5$ lb/in^2 abs or 174.8 lb/in^2 gauge

7-48 Answer (b).   Enthalpy for saturated steam at 1 atm and 280°F $=$
1183.3 Btu/lb   At 150 lb/in^2 abs $h_{sat} = 1194.1$ Btu/lb; $h_{evap} =$
863.6 Btu/lb $(1194.1 - 1183.3)/863.6 = 0.0125$ liquid, so
98.75% quality

7-49 Answer (d). $1/U = 1/h_1 + L_1/k_1 + L_2/k_2 + 1/h_2$ $\qquad$ $1/U = 1.402 +$ $(14/12)/5.00 + 1/1.40 = 1.196$ $\qquad$ $U = 0.8356$ Btu/(h)(ft^2)($°$F)

7-50 Answer (c). From psychrometric chart RH = 30% $\qquad$ $V = 13.8$ ft^3/lb
Saturated pressure at 87$°$F = 0.6351 lb/in^2 abs; for saturated vapor $V =$
511.7 ft^3/lb, so $w = 0.30 \times (1/511.7) = 5.863 \times 10^{-4}$ lb/ft^3 or 4.104
grains/ft^3 $\qquad$ At 87$°$F, $V = RT/P = 53.3 \times 547/(14.7 \times 144) =$
13.77 ft^3/lb; 13.77 $\times$ 4.104 = 56.5 grains H$_2$O/lb dry air. Or, using
Carrier's equation:

$$P_{\text{wv}} = 0.3056 - \frac{14.7 - 0.3056}{2830 - 1.44 \times 65} \times (87 - 65) = 0.1899 \text{ lb/in}^2 \text{ abs}$$

Saturated pressure at 87$°$F = 0.6351 lb/in^2 abs $\qquad$ RH =
0.1899/0.6351 = 0.299 or 29.9% $\qquad$ 0.299 $\times$ (1/511.7) =
0.0005843 lb/ft^3 or 4.090 grains/ft^3 $\qquad$ 4.090 $\times$ 13.77 = 56.3 grains
moisture/lb dry air

7-51 Answer (a).

$$Q = \frac{2\pi r_m l(t_1 - t_2)}{(r_o - r_i)/r_m} = \frac{2\pi l(t_1 - t_2)k_m}{\ln(r_o/r_i)}$$

$l = 144/(\pi \times 1.315) = 34.86$ in = 2.905 ft $\qquad$ $38,000 = 2\pi \times 2.905 \times$
$(212 - t_0) \times 19.1/0.226 = 1542 \times (212 - t_0)$ $\qquad$ $t_0 = 212 - 24.6 =$
187.41$°$F

7-52 Answer (e). 460 mi/h = 675 ft/s $\qquad$ 40 lb/s air enters at 675 ft/s;
40.9 lb/s gas exits at 2100 ft/s $\qquad$ Treat jet engine as a pump which
supplies (adds) head to a fluid: power = (lb/s) $\times$ (ft of head) = (lb/s) $\times$
$(v^2/2g)$ ft·lb/s; air entering $(40 \times 675^2/2g) = 238,261$ ft·lb/s; gas leaving
$(40.9 \times 2100^2/2g) = 2,803,372$ ft·lb/s; energy supplied in jet engine =
2,326,850 ft·lb/s or 4230 hp; efficiency = (2,326,850/778)/(0.9 $\times$ 20,000)
= 0.166 or 16.6%.

7-53 Answer (b). Saturated water at 200 lb/in^2 abs and 381.79$°$F has an
enthalpy $h_{\text{liq}} = 355.36$ Btu/lb. Cool at constant pressure to 300$°$F. ($c_p =$
1.00) $\qquad$ $\Delta h = 355.36 - 81.79 = 273.57$ Btu/lb $\qquad$ At 50 lb/in^2 abs
$h_{\text{vap}} = 1174.1$ Btu/lb, $h_{\text{liq}} = 250.09$ Btu/lb, $h_{\text{evap}} = 924.0$ Btu/lb
(273.57 $-$ 250.09)/924.0 = 0.0254, or 2.54% steam

7-54 Answer (c). $PV = WRT$ $\qquad$ At point 1: $W_1 = (800 \times 144V)/(53.3 \times$
540) = 4.003V; remove 700 ft^3; $\Delta W = (14.7 \times 144 \times 700)/(53.3 \times 520)$
= 53.46 lb $\qquad$ At point 2: $W_2 = (300 \times 144V)/(53.3 \times 540) = 1.501V$;
4.003V $-$ 53.46 = 1.501V; volume = 21.37 ft^3

7-55 Answer (d). Critical pressure ratio for superheated steam ($k = 1.3$) =
$0.545P_1 = 54.5$ lb/in^2 abs = pressure at throat. Flow through ideal
nozzle is isentropic. $h_1 = 1227.6$ Btu/lb; $s_1 = 1.6518$ Btu/(lb)($°$F)
At discharge (20 lb/in^2 abs): $h_{\text{liq}} = 196.16$ Btu/lb, $h_{\text{evap}} = 960.1$
Btu/lb, $h_{\text{vap}} = 1156.3$ Btu/lb; $s_{\text{liq}} = 0.3356$ Btu/(lb)($°$F), $s_{\text{evap}} =$
1.3962 Btu/(lb)($°$F), $s_{\text{vap}} = 1.7319$ Btu/(lb)($°$F) $\qquad$ $s_{\text{disch}} = 1.6518 =$
0.3356 + 1.3962Q $\qquad$ $Q = 0.9427$ $\qquad$ So $h_{\text{disch}} = 196.16 + 0.9427 \times$
960.1 = 1101.2 Btu/lb $\qquad$ $\Delta h = 1227.6 - 1101.2 = 126.4$ Btu/lb
$v = (126.4 \times 778 \times 2g)^{1/2} = 2515$ ft/s

7-56   Answer (e). $aC + bH + d \times O_2 + 81N_2 = 12CO_2 + 7O_2 + 81N_2 +$ $c \times H_2O$ 81 volumes of N_2 carries with it 21.53 volumes of O_2, of which 19 volumes are accounted for, so $c = 2 \times 2.53 = 5.06$, and the combustion equation becomes $12C + 10.12H + 21.53O_2 + 81N_2 = 12CO_2 + 7O_2 + 81N_2 + 5.06H_2O$; $10.12H + 12C = 10.12 + 12 \times 12 = 154.12$; $10.12/154.12 = 0.0657$, so hydrogen consitutes 6.57 percent of the fuel by weight

7-57   Answer (b). $W_1 = PV/RT = (15 \times 144 \times 4)/(53.3T) = 162.10/T$ $W_2 = (60 \times 144 \times 4)/(53.3T) = 684.4/T$ $\Delta W = 486.3/T =$ $(15 \times 144V)/(53.3T) = 40.5 \times V/T$ $V = 486.3/40.5 = 12.0 \text{ ft}^3$

7-58   Answer (c).   Refer to Fig. 7P-58.

1	2	3
$P_1 = 14.7 \text{ lb/in}^2 \text{ abs}$	$P_2 = 38.79 \text{ lb/in}^2 \text{ abs}$	$P_3 = 14.7 \text{ lb/in}^2 \text{ abs}$
$V_1 = 1 \text{ ft}^3$	$V_2 = 1 \text{ ft}^3$	$V_3 = 2 \text{ ft}^3$
$T_1 = 520°R$	$T_2 = 1372°R$	$T_3 = 1040°R$

$W = PV/RT = 14.7 \times 144 \times 1.0/(53.3 \times 520) = 0.0764 \text{ lb}$ $T_3 = PV/WR = 1040°R$ or $T_3 = T_1(P_3/P_1) = 1040°R$ $P_2 = P_3(V_3/V_2)^k = 2.639 \text{ atm} = 38.79 \text{ lb/in}^2 \text{ abs}$ $T_2 = P_2V_2/WR = 1372°R$ Heat added, 1 to 2: $c_v \, \Delta T = 0.1715 \times 852 = 146.12 \text{ Btu/lb}$, or 11.163 Btu Work done, 2 to 3: $W = (P_3V_3 - P_2V_2)/(1 - k) = 3381 \text{ ft·lb}$ or 4.346 Btu work done by gas   Work done, 3 to 1: $W = P(V_3 - V_1) = -2116.8 \text{ ft·lb} = -2.72 \text{ Btu}$ work done on gas; heat lost $= c_p \, \Delta T = 0.24 \times 520 = 124.8 \text{ Btu/lb}$, or 9.535 Btu; heat added $= 11.163 \text{ Btu} - 9.535 \text{ Btu}$ heat lost $= 1.628 \text{ Btu}$ net heat added; 4.346 Btu work done by gas $- 2.721 \text{ Btu}$ work done on gas $= 1.625 \text{ Btu}$ net work done Efficiency $= 1.625/11.163 = 0.146$, or 14.6%

7-59   Answer (a).   Refer to Fig. 7P-59. 3 lb/in² gauge $= 17.7 \text{ lb/in}^2 \text{ abs}$ For saturated steam $T = 221.5°F$ $h_{\text{liq}} = 189.6$, $h_{\text{vap}} = 1153.8 \text{ Btu/lb}$, $h_{\text{evap}} = 964.2 \text{ Btu/lb}$ For water at 60°F, $h = 189.6 - (221.5 - 60) = 28.1 \text{ Btu/lb}$ $\text{heat}_{\text{in}} = \text{heat}_{\text{out}} = 500 \times 28.1 + \text{Steam} \times (1153.8 - 189.6) = (500 + \text{steam}) \times 189.6$ steam $= 83.75 \text{ lb/min}$

7-60   Answer (d).   First sketch the process on a temperature-entropy diagram (Fig. 7S-60). Saturated liquid in the condenser at point a expands adiabatically (constant enthalpy) through an expansion valve (orifice) and flashes into a mixture of cold gas and cold liquid at point b. This is not a reversible process and thus is not isentropic, and the entropy will increase. From point b to point c the cold liquid evaporates and absorbs heat at a constant temperature and constant pressure. The cold gas is then compressed isentropically to point d, where it is in the superheated region. The hot, compressed gas then passes into a condenser, where it is cooled at a constant pressure and condenses into a liquid as shown at point a. The properties of ammonia at the different state points can be obtained from a table.

Point	Temperature, °F	Pressure, lb/in² abs	Entropy, Btu/(lb)(°F)	Enthalpy, Btu/lb
a Saturated liquid	66.02	120		116.0
b Mixture	−27.29	15		116.0
c Saturated vapor	−27.29	15	1.3938	602.4
d Superheated vapor		120	1.3938	733.4

Coefficient of performance = (heat absorbed in evap)/(work of compression) = 486.4/131.0 = 3.71   Work of compression = $h_d - h_c = -131.0$ Btu/lb (negative since work done on gas) Refrigerant effect = heat absorbed in evaporator = $h_c - h_b$ = 486.4 Btu/lb   [30 tons refrigerant × 200 Btu/(min)(ton)]/486 = 12.4 lb/min of refrigerant   hp/ton = 4.71/c.o.p. = 4.71/3.71 = 1.27 hp/ton of refrigerant   Heat absorbed in condenser = $h_d - h_c = 617.4$ Btu/lb $(617.4 × 12.4)/15.0 =$ 510 lb/min of cooling water

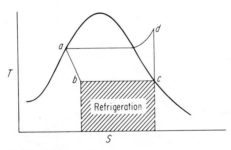

Figure 7S-60

7-61   Answer (b).   Insulated tank, so no heat lost or gained. Volume of balloon = $\frac{4}{3}\pi R^3 = 33.51$ ft³ $W = PV/RT = (16 × 144 × 33.51)/(520 × 1544/4) = 0.385$ lb helium in balloon $T_2 = T_1(P_2/P_1)^{(k-1)/k}$, where k for helium = 1.66 $T_2 = 520 × (215/3015)^{0.398} = 181.8°R$ $W_2 = W_1 - 0.385$ $(3015 × 144V)/(386 × 520) - 0.385 = (215 × 144V)/(386 × 181.8) = 2.163V - 0.385 = 0.4412V$ Volume = 0.224 ft³

7M-1   Answer (c). $(5000 × 80 + 1000 × 105)/6000 = 84.2°F$

7M-2   Answer (a).   Standard practice.

7M-3   Answer (b). $W = h_1 - h_2 = 34$ Btu/lb (change in kinetic energy would be negligible)

7M-4   Answer (c). PV = constant.

7M-5   Answer (d).   No work done, no heat transferred.

8-7 Answer (d). The voltage peaks first, so the current lags and the circuit is inductive. $Z = V/I = 14.67\ \Omega$ Power $= I^2R$ $R = 600/7.5^2 = 10.67\ \Omega$ $X_c = (Z^2 - R^2)^{1/2} = 10.07\ \Omega$ inductive reactance

8-8 Answer (c). See Fig. 8S-8. $X_L = 2\pi fL = 100\ \Omega$ $Z_1 = 10 + j100$ $X_C = 1/(2\pi fC) = -j100\ \Omega$ $Y_1 = 1/(10 + j100) = 9.90099 \times 10^{-5} \times (10 - j100) = 0.0009901 - j0.009901$ $Y_1 = 1/(0 - j100) = 10^{-4} \times (0 + j100) = 0 + j0.010$ $Y = Y_1 + Y_2 = 0.0009901 + j0.0000990$ $Z = 1/Y = 1000.00 - j99.99$ Total $Z = 1200.00 - j99.99$ $I = V/Z = 0.1034 + j0.00862$ or $I = 0.1038\ A$ Drop across resistor $R_1 = 20.76\ V$ Drop across capacitor $= 125 - 20.76 = 104.24\ V$

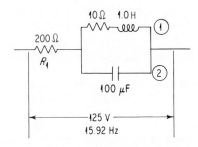

$10\ \Omega$ $1.0\ H$

$200\ \Omega$

R_1

① ②

$100\ \mu F$

$125\ V$

$15.92\ Hz$

Figure 8S-8

8-9 Answer (a). $25\ kW/230\ V = 108.70\ A$ in phase or $153.74\ A$ total at $45°$ lagging For 90 percent power factor out-of-phase component $= 0.484 \times 108.70 = 52.65\ A$ Correction $= 108.7 - 52.65 = 56.05\ A = I_C$ See Fig. 8S-9. $X_C = V/I_C = 4.103\ \Omega$ $C = 1/(2\pi fX_C) = 6.465 \times 10^{-4}$ or $646.5\ \mu F$

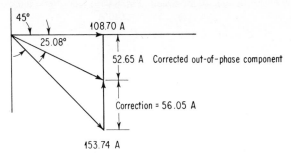

$45°$

$108.70\ A$

$25.08°$

$52.65\ A$ Corrected out-of-phase component

Correction $= 56.05\ A$

$153.74\ A$

Figure 8S-9

8-10 Answer (d). Assume Y connection (see Fig. 8S-10). Take one branch— $33.333\ kVA$ at $120\ V$, $I = 277.77\ A$ total, $222.22\ A$ in phase. Phase angle $= 36.87°$, correct to $25.84°$, or out-of-phase component to equal

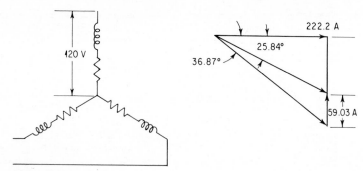

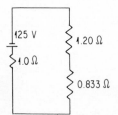

Figure 8S-10

$222.22 \times \tan 25.84° = 107.63$ A. Correction of $166.66 - 107.63 = 59.03$ A out of phase. Volt-amperes $= 3 \times 120 \times 59.03 = 21,252$ or 21.25 kVA reactance. For a delta connection $100,000/(3 \times 208) = 160.26$ A, branch current; 128.21 A in phase, 96.15 A out of phase, correct to 62.09 A out of phase or 34.06 A per phase; $34.06 \times 3 \times 208 = 21.25$ kVA reactance.

8-11 Answer (b). $1000 \times 746/(2200 \times 0.92 \times 3) = 122.86$ A in phase per branch. $122.86 \times \tan 29.54° = 69.63$ A out of phase To correct to 100 percent power factor, $3 \times 2200 \times 69.63/1000 = 459.5$ kVA reactance

8-12 Answer (c). Number 18 copper wire, from handbook, has a resistance of 6.38 Ω/1000 ft, so $R_{\text{wire}} = 0.638$ Ω dc power $= V^2/R$, so $R_{\text{lamp}} = \frac{100}{60} = 1.667$ Ω Total $R = 2.305$ Ω Power to lamp $= I^2R = 45.2$ W $I = 12/2.305 = 5.206$ A

8-13 Answer (e). Sketch equivalent circuit (Fig. 8S-13). Total $R = 3.033$ Ω $I = 41.21$ A $V_{R1} = 41.21 \times 1.20 = 49.45$ A

Figure 8S-13

8-14 Answer (a). See Fig. 8S-14. $Z = (37.7^2 + 12^2)^{1/2} = 39.56$ Ω $I = V/Z = 2.78$ A Power $= I^2R = 92.76$ W

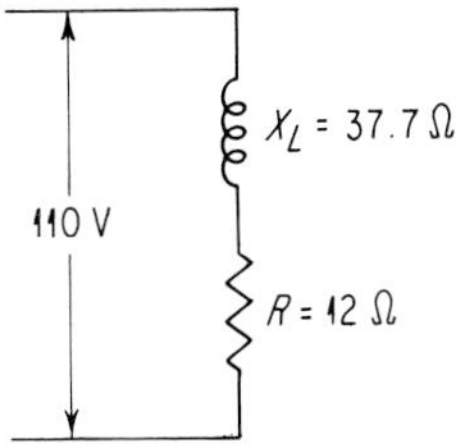

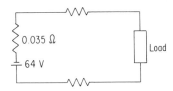

110 V

$X_L = 37.7\ \Omega$

$R = 12\ \Omega$

Figure 8S-14

8-15 Answer (b). Load current $= 1500/13.2 = 113.636$ A at 80 percent power factor leading $\quad I = 90.909 + j68.182 \quad Z_{\text{load}} = V/I = 13{,}2000/$ $(90.909 + j68.182) = 1.0222 \times (90.909 - j68.182) \quad Z_{\text{load}} = 92.928$ $- j69.696 \quad Z_{\text{line}} = 24 \times (0.27 + j0.62) = 6.48 + j14.88$ Series circuit, so total $Z = 99.408 - j54.816 \quad V = IZ = (90.909 +$ $j68.182) \times (99.408 - j54.816) = 12{,}775 + j1795$ or powerhouse voltage $= 12{,}900$ V

8-16 Answer (d). $X_L = 2\pi f L = 377 \times 0.010 = 3.77\ \Omega \quad X_C =$ $1/(2\pi f C) = 1/(377 \times 10 \times 10^{-6}) = 265.3\ \Omega \quad Z = 10 - j261.53$ $I = V/Z = 120/(10 - j261.53) = 0.00175 \times (10 + j261.53)$ $I = 0.0175 + j0.458$ or $I = 0.459$ A, $Z = [R^2 + (X_C - X_L)^2]^{1/2} =$ $261.69\ \Omega \quad I = 120/261.69 = 0.459$ A

8-17 Answer (e). Energy $= \frac{1}{2}CV^2 = \frac{1}{2} \times 2 \times 10^{-6} \times 10^6 = 1$ J Charge on plates $= CV = 2 \times 10^{-6} \times 1000 = 0.002$ C

8-18 Answer (c). To load A: $1200/(120 \times 0.8) = 12.5$ A, or $I =$ $(10 - j7.5)$ A $\quad$ To load C: $0.5 \times 746/(0.6 \times 0.8) =$ 777 VA $\quad \frac{777}{240} = 3.24$ A, or $(1.94 - j2.59)$ A $\quad$ Current in line $L_1 =$ $(1.94 - j2.59) + (10 - j7.5) = 11.95 - j10.09$ or 15.63 A

8-19 Answer (a). See Fig. 8S-19. $I = 15$ A $\quad R = 0.035 + \frac{400}{1000} \times 1.835 =$ $0.769\ \Omega \quad IR = 15 \times 0.769 = 11.535$ V $\quad$ Voltage across load $=$ $64 - 11.535 = 52.465$ V

0.035 Ω

64 V

Load

Figure 8S-19

8-20 Answer (b). See Fig. 8S-20. For series connection: $2V - 2R \times 0.2 =$ $5 \times 0.2 \quad V = 0.5 + 0.2R \quad$ For parallel connection: $V -$ $0.08R = 0.16 \times 5 \quad V = 0.80 - 0.08R \quad$ Simultaneous solution gives $R = 2.5\ \Omega$ and $V = 1.00$ V

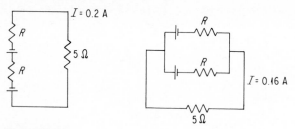

Figure 8S-20

8-21 Answer (d). $I = 7.395 - j4.191$ $X_C = 1/(2\pi f C) = 1/(2\pi \times 50 \times 150 \times 10^{-6}) = 17.684\ \Omega$ $I_C = 0 + j6.786$ Final current = $7.395 + j2.595$ $\phi = \tan^{-1}(2.595/7.395) = 19.337°$ Power factor = $\cos \phi = 0.944$ leading

8-22 Answer (c). ϕ = flux per pole $N = (\text{r/min})/60 = 20$ r/s
$V/\text{conductor} = P\phi N \times 10^{-8} = 2 \times (12 \times 40,000) \times 20 \times 10^{-8}$
$V = 0.192$ V per conductor, or turn

8-23 Answer (e). $Z = 4 + j3$ $I = V/Z$ $I = 200/(4 + j3) = 8 \times (4 - j3) = 32 - j24$, or $I = 40$ A

8-24 Answer (a). See Fig. 8S-24. Field loss = $3 \times 550 = 1650$ W
Armature loss = $I^2R = 76^2 \times 0.35 = 2021.6$ W   Total current = 79 A   Total losses = $79 \times 550 - 50 \times 746 = 6150$ W
Stray power losses = $6150 - (1650 + 2021.6) = 2478.4$ W

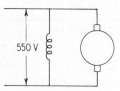

Figure 8S-24

8-25 Answer (b). Total $R = 15,000 + 12,000 = 27,000\ \Omega$ $I = 225/27,000 = 0.00833$ A   For full-scale reading of first voltmeter: $I = 150/15,000 = 0.0100$ A, so reading would equal $(0.0083/0.0100) \times 150 = 125$ V

8-26 Answer (d). $L_1/L_2 = R_x/R_s$ $R_x = R_s \times L_1/L_2 = 50.5 \times 15/25 = 30.3\ \Omega$

8-27 Answer (c). 500 hp $\times 746/0.93 = 401,075$ W $401,075/0.90 = 445,639$ VA   Line current to motor = $445,639/(2300 \times 1.732) = 111.9$ A $111.9 \times 2.300 = 257.3$ kVA per transformer

8-28 Answer (e). $i = 4 + 2t^2$ C = A·s $\int_{t=5}^{t=10} i\ dt = 4t + \frac{2}{3}t^3 \big|_{t=5}^{t=10} = (40 + 667) - (20 + 83) = 604$ C

8-29 Answer (a). See Fig. 8S-29. Use Kirchhoff's laws. $9 - 2i_1 - (i_1 + i_2) \times$
$3 = 0$ $6.5 - i_2 - (i_1 + i_2) \times 3 = 0$ $i_1 = 1.5$ A; $i_2 = 0.5$ A
Current through resistor $= 2.0$ A

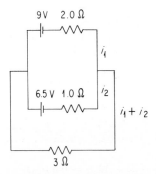

9V   2.0 Ω

i_1

6.5 V   1.0 Ω i_2

$i_1 + i_2$

3 Ω   **Figure 8S-29**

8-30 Answer (c). Add the three loads vectorially. 250 hp $\times$ 0.746/0.86 $=$
216.86 kW $\phi = \cos^{-1} 0.90 = 25.84°$ $\tan \phi = 0.4843$
0.4843 $\times$ 216.86 $= 105.03$ kVA out-of-phase component for motor load
$21.25 - j13.17$
$50 \quad\ - j0$
$\underline{216.86 - j105.03}$
      $- j118.20 = 118.20$ kVA reactance required

8-31 Answer (b). $I = 2500/2.300 = 1086.96$ A or $I = 869.57 - j652.18$
For rated use, $Z = V/I = 0.001947 \times (869.57 + j652.18)$
$Z = 1.693 + j1.270$ or $Z = 2.116$ Ω, resistance $= 1.693$ Ω
Voltage, line to neutral for wye $= 2300/1.732 = 1327.94$ V
Current $= V/Z = 1327.94/2.116 = 627.57$ A   Power draw $= I^2R =$
666.78 kW per furnace, or 2000 kW total   As an item of interest, if the
furnaces had been connected in a delta, the voltage across each would have
equaled 2300 V. In that case $I = 2300/2.116 = 1086.96$ A and the total
power draw would equal 3 $\times$ 1086.96^2 $\times$ 1.693 $= 6000$ kW.

8-32 Answer (d). See Fig. 8S-32. The voltage drops from A to C, A to D, and

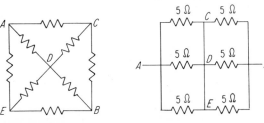

Figure 8S-32

A to E would all be the same. Also, the voltage drops between B and C, B and D, and B and E would all be the same. So points C, D, and E can be connected together without changing the characteristics of the circuit. This gives the equivalent circuit shown in the figure: $1/R_{\text{equiv.}} = \frac{1}{10} + \frac{1}{10} + \frac{1}{10}$ $R_{\text{equiv.}} = 1/0.300 = 3.333 \ \Omega$

8-33 Answer (e). Use Kirchhoff's laws. *Branch 1*: $1.5 - 0.1I_1 - 0.5I_1 - 1.0I_3 = 0$; $1.5 = 0.60I_1 + I_3$ *Branch 2*: $1.5 - 0.1I_2 + 1.5 - 0.1I_2 + 1.0I_3 - 0.5I_2 = 0$; $3 = 0.7I_2 - I_3$ $I_3 = I_1 + I_2$ $I_1 = 3.227$ A $I_2 = 3.663$ A $I_3 = 0.436$ A

8-34 Answer (b). $I = 20.00 - j15.00$ $X_C = 1/(2\pi fC) = 1/(377 \times 300 \times 10^{-6}) = 8.842 \ \Omega$ $I_C = 125/(0 - j8.842) = 0 + j14.137$ The resulting current $= 20.00 - j0.862$, or current $= 20.02$ A

8-35 Answer (d). $X_L = 2\pi fL = 2\pi \times 25 \times 0.1 = 15.71 \ \Omega$ $I = V/Z = 100/(20 + j15.71) = 0.155 \times (20 - j15.71) = 3.09 - j2.43$ or $I = 3.93$ A   Power $= I^2R = 309.0$ W

8-36 Answer (a). $X_C = 1/(2\pi fC) = 1/(377 \times 25 \times 10^{-6}) = 106.10 \ \Omega$ $X_L = 2\pi fL = 377 \times 0.15 = 56.55 \ \Omega$ $Z = 50 - j49.55$ $I = 120/(50 - j49.55) = 0.0242 \times (50 + j49.55) = 1.212 + j1.201$, or $I = 1.706$ A

8-37 Answer (c). Torque $= (K\phi V_2 R_2)/(R_2^2 + X_2^2)$, so the starting torque is proportional to the square of the stator voltage since ϕ is proportional to stator voltage. $V = (220^2/1.25)^{1/2} = 196.6$ V

8-38 Answer (d). $I = 6 - j8$, or $I = 10$ A $Z = V/I = 115/(6 - j8) = 1.15 \times (6 + j8) = 6.9 + j9.2$ Required $Z = \frac{230}{10} = 23.0 \ \Omega = (R^2 + X_L^2)^{1/2}$ $R = (529 - 84.64)^{1/2} = 21.08 \ \Omega$ total. Added $R = 21.08 - 6.9 = 14.18 \ \Omega$ Power to load $= I^2R$, same power as before

8-39 Answer (b). $I = 25$ A $V = 223$ V   Motor output per phase $= 25 \times 223 \times 0.85 \times 0.90 = 4265$ W   Horsepower $= 3 \times \frac{4265}{746} = 17.15$ hp

8-40 Answer (e). This is a parallel circuit. $Z_1 = 20 + j12$ $Z_2 = 10 + j10$ $I_1 = 120/(20 + j12) = 0.2206 \times (20 - j12) = 4.412 - j2.647$ $I_2 = 120/(10 + j10) = 0.6000 \times (10 - j10) = 6.00 - j6.00$ Total current $= 10.412 - j8.647$, or $I = 13.53$ A

8-41 Answer (c). $R = \frac{60}{2} = 30 \ \Omega$ $Z = \frac{50}{1} = 50 \ \Omega$ $Z^2 = R^2 + X_L^2$ $2500 = 900 + X_L^2$ $X_L = 40 \ \Omega$ $40 = 2\pi fL$ $L = 0.106$ H

8-42 Answer (a). See Fig. 8S-42. $1/R = 1/0.20 + 1/0.30 + 1/0.60$ $R = 0.10 \ \Omega$ Total equivalent resistance $= 0.12 \ \Omega$ $I = 6/0.2 = 50$ A   Terminal voltage $= 6 - 50 \times 0.02 = 5.0$ V $I_{0.02 \ \Omega} = 5/0.2 = 25$ A

8-43 Answer (d). 200 hp/$(0.85 \times 0.91) \times \frac{746}{1000} = 192.89$ kW $440I \times \sqrt{3}/1000 = 192.89$ $I = 253.1$ A

8-44 Answer (b). See Fig. 8S-44. $V = IR$ $V = 2.00 + 1.95 + 1.84 = 5.79$ $R = 0.05 + 0.075 + 0.064 + 0.110 = 0.299$ $5.79 =$

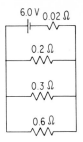

Figure 8S-42

$I \times 0.299$ $I = 19.36$ A
$19.36 \times 0.110 = 2.13$ V

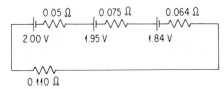

Figure 8S-44

Voltage across resistance =

8-45 Answer (e). Total resistance $= 2000 + 500 + 50 = 2550\ \Omega$ $I = \frac{20}{2550} = 0.007843$ A   Voltage drop across meter $= 50 \times 0.007843 = 0.3922$ V   Meter reading $= (0.3922/0.588) \times 15 = 10.00$

8-46 Answer (c). Add the loads vectorially:
$$
\begin{array}{r}
500 + j0 \\
212.50 - j131.70 \\
727.50 + j182.33 \\
\hline
\end{array}
$$
$1440.00 + j50.63 = 1440$ kW

8-47 Answer (b). $T_2/T_1 = I_2\phi_2/I_1\phi_1$ Field is constant, so torque is proportional to armature current $N = (V_L - I_aR_a)/K_g\phi$, where K_g = generator constant   Field is constant, so $N_2/N_1 = [V_L - (I_aR_a)_2]/[V_L - (I_aR_a)_1]$ $N_2 = 1200 \times (230 - 0.15I_a)_2/(230 - 0.15I_a)_1$ $(I_a)_2 = \frac{1}{2}(I_a)_1$ $N_2 = 1200 \times (230 - 0.075I_a)/(230 - 0.15I_a)$, where $I_a = (I_a)_1$ Assume full-load armature current $= 58$ A $N_2 = 1200 \times (225.65/221.30) = 1223.6$ r/min

8-48 Answer (d). See Fig. 8S-48. 5 kVA/230 $= 21.74$ A rated $\frac{2530}{2300} \times 21.74 = 23.91$ A $23.91 \times 2300 = 55,000$ or 55 kVA

8-49 Answer (e). This is a series circuit.
$$
\begin{array}{rl}
V_1 = & 20.84 + j118.18 \\
V_2 = & 82.08 - j225.53 \\
V_3 = & 129.90 + j75.00 \\
\hline
\end{array}
$$
$V = 232.82 - j32.35$ or $V = 235.06$ V

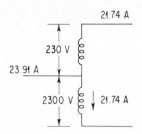

21.74 A

230 V

23.91 A

2300 V   ↓ 21.74 A

Figure 8S-48

8-50 Answer (c). $Z = 4.5 + j15$ $I = 240/(4.5 + j15) = 0.979 \times$
$(4.5 - j15) = 4.40 - j14.68$ or $I = 15.32$ A   Power $= I^2R =$
1.06 kW

8-51 Answer (a). Effective ac voltage is rms volts (refer to Fig. 8P-51):
rms volts $= [\int_0^\pi e^2 \, d\theta$ divided by $\pi]^{1/2} = [(1/\pi)V_m^2 \sin^2 \theta \, d\theta]^{1/2} =$
$[(V_m^2/\pi)(\theta/2 - \sin 2\theta/4)_0^\pi]^{1/2} = [(V_m^2/\pi)(\pi/2)]^{1/2}$ Effective voltage $=$
0.707 $V_m = 141.4$ V

8-52 Answer (e). Use no. 14 copper wire. According to a handbook, the
resistances of copper wire of different sizes at 77°F (25°C) are:

Gauge no.	8	10	12	14	16	18
R, Ω/1000 ft	0.641	1.02	1.62	2.58	4.09	6.38

$I_{\text{lamp}} = 250$ W/120 V $= 2.083$ A $=$ circuit current   Line $R =$
2.58 Ω for 1000 ft of wire   Voltage drop $= 2.083 \times 2.58 = 5.37$ V
Source voltage $= 125.37$ V

8-53 Answer (a). $1/R_1 = \frac{1}{10} + \frac{1}{30}$ $R_1 = 7.5 \, \Omega$ $1/R_2 = \frac{1}{50} + \frac{1}{100}$
$R_2 = 33.33 \, \Omega$ Total equivalent resistance $= 7.50 + 33.33 =$
40.83 Ω Current $= 120/40.83 = 2.94$ A   Drop across $R_2 =$
$2.94 \times 33.33 = 97.99$ V $I_{50} = 97.99/50 = 1.96$ A

8-54 Answer (c).

Load $= 750 - j562.50$
Capacitor $= 0 + j300$

Resulting load $= 750 - j262.5$ $\phi = \tan^{-1}(262.5/750) =$
19.29°   Power factor $= \cos 19.29° = 0.943$

8-55 Answer (a). $1/R_{A-B} = \frac{1}{200} + \frac{1}{600} + \frac{1}{300}$ $R_{A-B} = 100 \, \Omega$
Total equivalent resistance $= 1600 \, \Omega$ $I = \frac{110}{1600} = 0.06875$ A
Drop from A to $B = 0.06875 \times 100 = 6.875$ V   Current through
200-Ω resistor $= 6.875/200 = 0.344$ A

8M-1 Answer (c). $I^2R = 13^2 \times 37 = 6253$

8M-2 Answer (a). By definition.

8M-3 Answer (a). Voltage is proportional to the lines of flux cut per second, or the number of conductors which pass through the field per second.

8M-4 Answer (b). By definition. Current is a maximum when the impedance is a minimum.

8M-5 Answer (d). By definition.

8M-6 Answer (b). Power equals I^2R.

8M-7 Answer (b). $X_L = 2\pi \times 100 \times 0.150 = 94.25\ \Omega = X_C$ for resonance. Capacitance $= 1/(2\pi \times 100 \times 94.25) = 16.88\ \mu F$.

8M-8 Answer (e). Phase angle $= \cos^{-1}$ (watts/volt-amperes).

8M-9 Answer (d). By definition.

8M-10 Answer (b). Speed $=$ frequency/(pairs of poles).

9-1 Answer (c). Plan A: $11,500 = yearly payment $\times (1.05^n - 1)/(1.05^n \times 0.05)$ Yearly cost = yearly payment $+ 3500 + 500 \times 0.05 = 5000$ Yearly payment $= \$1475$ Then $\frac{11,500}{1475}(1.05^n \times 0.05) = 1.05^n - 1$ $0.3898 \times 1.05^n = 1.05^n - 1$ $1.05^n = 1.6389$ $n = 10.16$ years This problem can also be solved by assuming different times (years) and calculating the resulting yearly payments.

9-2 Answer (a). Present value of $18,000 per year for 10 years $= \$18,000 \times (1.10^{10} - 1)/(1.10^{10} \times 0.10) = \$18,000 \times 6.1446 = \$110,602$ Value of second 10 years' earnings at end of tenth year would equal $25,000 \times 6.1446 = \$153,615$ Present value of $153,615, 10 years hence $= \$153,615/1.10^{10} = \$59,225$ Value of third 10 years' earnings at end of year 20 $= 6.1446 \times \$37,000 = \$227,350$ Present value of $227,350, 20 years hence $= \$227,350/1.10^{20} = \$33,794$ Total present value of next 30 years' earnings $= \$203,621$

9-3 Answer (d). The equivalent uniform annual value of the 30 years' earnings of the previous problem would equal the yearly return from an annuity which could be purchased with the present value of those earnings. $203,621 = yearly income $\times (1.10^{30} - 1)/(1.10^{30} \times 0.10)$ Equivalent yearly income $= \$21,600$

9-4 Answer (b). $25,000 = yearly cost $\times (1.04^6 - 1)/(1.04^6 \times 0.04)$ Yearly cost $= \$4769$ At end of 4 years amount in sinking fund $= \$4769 \times (1.04^4 - 1)/0.04 = \$20,251$ In 4 years $25,000 would have appreciated to $25,000 \times 1.04^4 = \$29,246$ Sunk cost $= \$29,246 - \$20,251 = \$8995$

9-5 Answer (a). Cost to be recovered $= \$6000$ Yearly cost $= \$6000 \times (1.06^{10} \times 0.06)/(1.06^{10} - 1) + \$500 \times 0.06$ Yearly cost $= \$815 + \$30 = \$845$

9-6 Answer (e). Compare costs.
OLD BRIDGE:
    Present value $= \$13,000 + \$9000 = \$22,000$
    Depreciation $= (\$22,000 - \$10,000)/20 = \$600$/year

Average interest = ($22,000 + $10,000 + $600)/2 × 0.06 = $978/year
Total yearly cost = $600 + $978 + $500 = $2078
NEW BRIDGE:
Depreciation = ($40,000 − $15,000)/20 = $1250/year
Average interest = ($40,000 + $15,000 + $1250)/2 × 0.06 =
$1687.50/year
Total yearly cost = $1250 + $1687 + $100 = $3037
Cost advantage in favor of repairing old bridge of $959/year

9-7 Answer (c). Depreciation = $10,000 − $2000)/20 = $400/year
Average interest = ($10,000 + $2000 + $400)/2 × 0.07 = $434/year
Annual capital recovery plus a return = $834

9-8 Answer (b). Yearly payment = $12,000 × ($1.04^5$ × 0.04)/(1.04^5 − 1) =
$2696   At end of second year amount in fund = $2696 + 1.04 ×
$2696 = $5500   During third year this would appreciate to $5500 ×
1.04 = $5720   The $12,000 would have appreciated to $12,000 ×
1.04^3 = $13,498   The final payment would equal $13,498 − $5720 =
$7778.

9-9 Answer (e). If interest had been 18 percent: Yearly payment =
$12,000 × ($1.18^5$ × 0.18)/1.18^5 − 1) = $3837   Amount in fund at
end of second year = 2.18 × $3837 = $8365   At end of third year
this would have appreciated to 1.18 × $8365 = $9871   The original
$12,000 would appreciate to $12,000 × 1.18^3 = $19,716
Final payment = $19,716 − $9871 = $9845

9-10 Answer (σ). Power plant, depreciation only. Yearly depreciation cost =
$1,000,000 × 0.08/($1.08^{15}$ − 1) = $36,830   Total yearly cost =
$95,000 + $36,830 = $131,830   Cost of power = 1000 × 24 ×
365 × $0.023 = $201,480   Difference in cost in favor of building
power plant = $69,650/year

9-11 Answer (d). First cost = present value = $17,000 + $5000 = $22,000
Salvage value = $14,000 + $2500 = $16,500
Depreciation = ($22,000 − $16,500)/5 = $1100/year
Average interest = ($22,000 + $16,500 + $1100)/2 × 0.06 = $1188
Total cost = $1100 + $1188 + $200 = $2488/year

9-12 Answer (c). Value of sprinkler system = V. Yearly cost = 0.05 × 0.10V
+ 0.90V(1.05^{20} × 0.05)/(1.05^{20} − 1) $500 = 0.05 × 0.10V + 0.90V
× 0.0802 = 0.07722V   Justifiable value = $6475

9-13 Answer (d). Present value of repaired shovel = $2000 + $6000 = $8000
Depreciation = ($8000 − $500)/10 = $750/year
Average interest = ($8000 + $500 + $750)/2 × 0.06 = $277.50/year
Total cost of repaired shovel = $750 + $277 + $2500 = $3527/year
New shovel depreciation = ($30,000 − $20,000)/10 = $1000/year;
average interest = ($30,000 + $20,000 + $1000)/2 × 0.06 =
$1530/year   Total cost of new shovel = $1000 + $1530 + $1500 =
$4030/year   Differential in favor of repairing old shovel = $503/year

9-14 Answer (*b*). Debt would appreciate to $10,000 × 1.04^5 = $12,167
Establish sinking fund with 10 periods with 1.5 percent per period,
with a future value of $12,167 $12,167 = semiannual deposit ×
(1.015^{10} − 1)/0.015   Semiannual deposit = $1137

9-15 Answer (*d*). $12,350 − $3000 = $9350   Yearly deposit = $9350 ×
(1.12^5 × 0.12)/(1.12^5 − 1) = $2594/year   Total yearly cost =
$2594 + $900 + ($3000 × 0.12) + 0.09 × miles = 0.20 × miles
Miles = 3854/0.11 = 35,036

9-16 Answer (*d*). Depreciation = ($12,350 − $3000)/5 = $1870/year
Average interest = ($12,350 + $3000 + $1870)/2 × 0.12 = $1033.20/
year   Total cost = $1870 + $1033 + $900 + 0.09 × miles = 0.20
× miles   Miles = 3803/0.11 = 34,572

9M-1 Answer (*d*). 500/1.02^{10} = 410.17

9M-2 Answer (*d*). 1.025^2 = 1.0506

9M-3 Answer (*a*). By definition.

9M-4 Answer (*e*). By definition.

9M-5 Answer (*d*). Self-explanatory.

9M-6 Answer (*c*). By definition.

9M-7 Answer (*c*). By definition.

9M-8 Answer (*b*). $i/12$ = interest rate per month

9M-9 Answer (*c*). 1.02^n = 2   ln 2/ln 1.02 = 35.00   35.00/2 = 17$\frac{1}{2}$

9M-10 Answer (*c*). 5000/1.075^{10} = 2425.97

7M-6 Answer (d). By definition.

7M-7 Answer (e). $\Delta Q = 1$ cal/g $\times$ 10 g $= 10$ cal

7M-8 Answer (d). $P/T = $ constant

7M-9 Answer (a). Velocity increases, enthalpy decreases.

7M-10 Answer (b). $(2000 - 800)/2000 = 0.60$

8-1 Answer (b). Flux is proportional to current. The resistances are the same, but not the impedances. For 60 Hz, $525 = K_1\beta^{1.6} \times 60 + K_2\beta^2 \times 3600$; for 25 Hz, $140 = K_1\beta^{1.6} \times 25 + K_2\beta^2 \times 625$, which give $K_1\beta^{1.6} = 3.35$ Hysteresis loss at 60 Hz $= 60 \times K_1\beta^{1.6} = 210$ W

8-2 Answer (d). See Fig. 8S-2. 5 hp $\times 746/0.80 = 4662.5$ W $4662.5/(220 \times 0.707) = 29.98$ A total $4662.5/220 = 21.19$ A in phase $\phi = \cos^{-1} 0.707 = 45°$ Reactance $= X_c = V/I_c = 10.38$ Ω

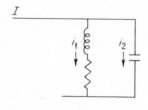
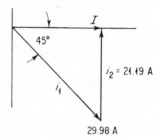

Figure 8S-2

8-3 Answer (c). $\tan \phi = 3 \times (W_1 - W_2)/(W_1 + W_2) = 1.732 \times \frac{2400}{1600} = 2.598$ $\phi = 68.9°$ Or $W_1 + W_2 = \sqrt{3} \times V_{\text{line}}I_{\text{line}} \cos \phi = 1.732 \times 220 \times 12 \cos \phi$ $\cos \phi = 1600/4572.6 = 0.35$ $\phi = 69.5°$

8-4 Answer (a). $\frac{1}{15} + \frac{1}{25} = 1/9.375$ $\frac{1}{60} + \frac{1}{200} = 1/46.154$
Equivalent circuit resistance $= 9.375 + 46.154 = 55.259$ Ω
$I = 120/55.529 = 2.161$ A, so V across 25-Ω resistor $= 20.260$ V
$I = 20.260/25 = 0.81$ A

8-5 Answer (c). $V = IR$ For shunt $220 = I_s \times 55$ $I_s = 4$ A
Power loss in shunt $= I_s^2 R = 16 \times 55 = 880$ W   Current through armature $= 50$ A   Armature loss $= 50^2 \times 0.08 = 200$ W
Total losses $= 180 + 880 + 200 = 1260$ W   Efficiency $= (54 \times 220 - 1260)/(54 \times 220) = 0.894$ or 89.4%

8-6 Answer (e). $C_1 = 6$ μF $C_2 = 12$ μF $Q_1 = C_1V_{a-b}$
$Q_2 = C_2V_{b-c}$ $Q_1 = Q_2$ $V_{a-b} = 80$ V $V_{b-c} = 40$ V
If touch + to + and $-$ to $-$ the voltages become equal, although the total charge remains the same. $Q_1 = 60 \times 80 = 480$ μC $Q_2 = 12 \times 40 = 480$ μC, so total $Q = 960$ μC   After joining terminals
$C_1V_1 + C_2V_2 = 960$; voltages are equal, so $V = 53.33$ V
Charge on 6 μF capacitor $= 6 \times 53.3 = 320$ μC

INDEX

ABOUT THE AUTHOR

Lloyd M. Polentz, P.E., holds an M.S. degree in engineering and has 45 years' experience in the field, the last ten of which have been as a consultant. For 30 years he has taught college-level engineering courses evenings, and he is presently an instructor in the University of California's Extension Division. He has written a correspondence course in engineering fundamentals for the University of California as well as authored many articles and book chapters on various engineering subjects.